T0276216

LONDON MATHEMATICAL SOCIETY LECTURE NOTE SERIES

Managing Editor: Professor N.J. Hitchin, Mathematical Institute,
University of Oxford, 24–29 St Giles, Oxford OX1 3LB, United Kingdom

The titles below are available from booksellers, or, in case of difficulty, from Cambridge University Press.

London Mathematical Society Lecture Note Series. 284

Foundations of Computational Mathematics

Edited by

Ronald A. DeVore
University of South Carolina

Arieh Iserles
University of Cambridge

Endre Süli
University of Oxford

CAMBRIDGE
UNIVERSITY PRESS

PUBLISHED BY THE PRESS SYNDICATE OF THE UNIVERSITY OF CAMBRIDGE
The Pitt Building, Trumpington Street, Cambridge, United Kingdom

CAMBRIDGE UNIVERSITY PRESS
The Edinburgh Building, Cambridge CB2 2RU, UK
40 West 20th Street, New York, NY 10011–4211, USA
10 Stamford Road, Oakleigh, VIC 3166, Australia
Ruiz de Alarcón 13, 28014 Madrid, Spain
Dock House, The Waterfront, Cape Town 8001, South Africa

http://www.cambridge.org

First published 2001

A catalogue record for this book is available from the British Library

Library of Congress Cataloguing-in-Publication Data
Foundations of computational mathematics / edited by Ronald A. DeVore, Arieh Iserles,
Endre Süli.
    p.   cm. – (London Mathematical Society lecture note series; 284)
    Includes bibliographical references and index.
    ISBN 0 521 00349 0 (pb.)
    1. Numerical analysis–Congresses. I. DeVore, Ronald A. II. Iserles, A. III. Süli,
Endre, 1956–   IV. Series.

QA297 .F6347 2001
519.4–dc21   2001025226

ISBN 0 521 00349 0 paperback

Transferred to digital printing 2001

Contents

Preface

The Society for the Foundations of Computational Mathematics supports fundamental research in a wide spectrum of computational mathematics and its application areas. As part of its endeavor to promote research in computational mathematics, it regularly organises conferences and workshops which bring together leading researchers in the diverse fields impinging on all aspects of computation. Major conferences have been held in Park City (1995), Rio de Janeiro (1997), and Oxford (1999).

The next major FoCM conference will take place at the Institute for Mathematics and its Application (IMA) in Minneapolis in the summer of 2002. More information about FoCM can be obtained from its website at www.focm.net.

The conference in Oxford, England, on July 18-28, 1999, was attended by over 300 scientists. Workshops were held on fourteen subjects dealing with diverse research topics from computational mathematics. In addition, eighteen plenary lectures, concerned with various computational issues, were given by some of the world's foremost researchers. This volume presents thirteen papers from these plenary speakers. Some of these papers are a survey of state of the art in an important area of computational mathematics, others present new material. The range of the topics: from complexity theory to the computation of partial differential equations, from optimization to computational geometry to stochastic systems, is an illustration of the wide sweep of contemporary computational mathematics and the intricate web of its interaction with pure mathematics and application areas.

We hope that this volume will be of great interest to both researchers in computational mathematics and the non-experts who want a taste of the state of the art in this important discipline.

This is an opportunity to thank all those who have made the Oxford FoCM'99 conference such a success: local organisers and the administrative staff of our hosts, Oxford University Computing Laboratory, workshop organisers and a number of organisations that have helped us in funding and publicity: UK Engineering and Physical Sciences Research Council, International Mathematical Union, European Mathematical Society, American Mathematical Society, National Science Foundation, Institute of Mathematics and its Applications and Society for Industrial and Applied Mathematics. Yet, the greatest thanks are to the participants, who have made Oxford in July 1999 into such a vibrant and exciting epicentre of computational mathematics.

Singularities and computation of minimizers for variational problems

J.M. Ball

Mathematical Institute
University of Oxford
24-29 St. Giles'
Oxford OX1 3LB, U.K.
Email ball@maths.ox.ac.uk

Abstract

Various issues are addressed related to the computation of minimizers for variational problems. Special attention is paid (i) to problems with singular minimizers, which natural numerical schemes may fail to detect, and the role of the choice of function space for such problems, and (ii) to problems for which there is no minimizer, which lead to difficult numerical questions such as the computation of microstructure for elastic materials that undergo phase transformations involving a change of shape.

1 Introduction

In this article I give a brief tour of some topics related to the computation of minimizers for integrals of the calculus of variations. In this I take the point of view not of a numerical analyst, which I am not, but of an applied mathematician for whom questions of computation have arisen not just because of the need to understand phenomena inaccessible to contemporary analysis, but also because they are naturally motivated by attempts to apply analysis to variational problems.

I will concentrate on two specific issues. The first is that minimizers of variational problems may have singularities, *but natural numerical schemes may fail to detect them.* Connected with this is the surprising *Lavrentiev phenomenon*, according to which minimizers in different function spaces may be different. The second is that minimizers may not exist, in which case the question naturally arises as to what the behaviour of numerical schemes designed to compute such minimizers will be. In this case the predictive power of the variational problem may still

1

be retained, for example as a explanatory mechanism for the formation of *microstructure* in materials. A key tool here is the elusive concept of quasiconvexity, which helps to describe the passage from microscales to macroscales.

As a motivating example, consider (nonlinear) elasticity theory. For a homogeneous elastic body the total elastic energy is given by the integral

$$I(y) = \int_\Omega W(Dy)\, dx,$$

where W is the stored-energy function of the material. Here Ω is a bounded open subset of $\mathbf{R}^3$, with Lipschitz boundary $\partial\Omega$, that the body occupies in a reference configuration, and $y : \Omega \to \mathbf{R}^3$ denotes a typical deformation with gradient

$$Dy(x) = \left(\frac{\partial y_i}{\partial x_j} \right).$$

Thus for each x, $Dy(x) \in M^{3\times 3}$, where $M^{m\times n} = \{$real $m \times n$ matrices$\}$. In this case the singularities of minimizers could potentially be related to various kinds of fracture or its onset, dislocations, or phase boundaries, while microstructure arises in materials undergoing phase transformations, for which the minimum of I subject to suitable boundary conditions may not be attained.

2 Singular minimizers and the Lavrentiev phenomenon

2.1 The Lavrentiev phenomenon and repulsion property

Consider first the simple problem due to Manià [45] of minimizing the integral

$$I(u) = \int_0^1 (u^3 - x)^2 u_x^6 \, dx \tag{2.1}$$

among absolutely continuous functions u satisfying the end conditions

$$u(0) = 0, u(1) = 1. \tag{2.2}$$

The unique minimizer of this problem is easily seen to be

$$u^*(x) = x^{\frac{1}{3}}.$$

In fact $I(u^*) = 0$, and if $\bar{u}$ were any other function satisfying the end conditions (2.2) with $I(\bar{u}) = 0$ then $\bar{u}_x(x) = 0$ for $x \in E$ and $\bar{u}(x) =$

$u^*(x)$ for $x \in (0,1) \backslash E$, where E has positive one-dimensional Lebesgue measure. Thus

$$0 = \int_0^1 (u_x^* - \bar{u}_x)\, dx = \int_E \frac{1}{3} x^{-\frac{2}{3}}\, dx > 0,$$

a contradiction.

Consider now a very natural finite-element scheme for computing the minimizer. Take a uniform mesh subdividing $[0,1]$ into N subintervals of length $h = 1/N$ and minimize I among continuous functions satisfying the end conditions (2.2) which are affine on each element $(i/N, (i+1)/N)$. For any such function v_h the integral $I(v_h)$ can be computed exactly (due to the explicit form of the integrand), so that questions of quadrature can in the first instance be ignored. For each h there is at least one minimizer u_h^* to this discrete problem. What is the behaviour of u_h^* as $h \to 0$? Remarkably, u_h^* converges as $h \to 0$, but not to the minimizer u^*! In fact the limit u_0 is a monotone increasing function that is smooth in $(0,1)$ but has infinite slope at the end-points $x = 0, 1$.

This behaviour is hard to credit at a first glance. An illuminating initial calculation is to compute $I(u_h)$ for the function

$$u_h(x) = \begin{cases} h^{-\frac{2}{3}} x & \text{if } x \in (0,h) \\ u^*(x) & \text{if } x \in (h,1) \end{cases}$$

in which u^* is altered only on the first element. Surely $\lim_{h \to 0} I(u_h) = I(u^*) = 0$? But no,

$$\int_0^1 (u_h^3 - x)^2 u_{hx}^6\, dx = \frac{8}{105} h^{-1},$$

which tends to $+\infty$ as $h \to 0$!

In fact it can be shown (see Ball & Knowles [12]) that if $1 \le p \le \infty$ and

$$\mathcal{A}_p = \{ v \in W^{1,p}(0,1) : v(0) = 0, v(1) = 1 \}$$

(so that $\mathcal{A}_1$ is the admissible class of functions considered above) then

$$\inf_{\mathcal{A}_\infty} I = \min_{\mathcal{A}_{3/2}} I > \min_{\mathcal{A}_1} I = 0.$$

The fact that the infimum of I in different function spaces can be different is known as the *Lavrentiev phenomenon* (see Lavrentiev [37] for the original example). The initial calculation above has the following generalization, let us call it the *repulsion property*, that if $u^{(j)} \in \mathcal{A}_{3/2}$ and $u^{(j)} \to u^*$ a.e. in $(0,1)$ then $I(u^{(j)}) \to \infty$.

In the Manià example the integrand $f(x, u, p) = (u^3 - x)^2 p^6$ is convex in p, but not strictly convex. However, as was shown by Ball & Mizel [13, 14], the Lavrentiev phenomenon and the repulsion property can hold for elliptic integrands, i.e. those for which $f_{pp}(x, u, p) \geq \mu > 0$ for all x, u, p. Such an example is given by the problem [14] of minimizing

$$I(u) = \int_{-1}^{1} [(x^4 - u^6)^2 u_x^{28} + \varepsilon u_x^2] \, dx \qquad (2.3)$$

in $\mathcal{A}_p = \{v \in W^{1,p}(-1, 1) : v(-1) = -1, v(1) = 1\}$. Note that the integrand $f(x, u, p) = (x^4 - u^6)^2 p^{28} + \varepsilon p^2$ satisfies $f_{pp} \geq 2\varepsilon > 0$. Here, for sufficiently small $\varepsilon > 0$, there is an absolute minimizer u^* of I in $\mathcal{A}_1$ that is a smooth solution of the Euler-Lagrange equation in $[-1, 0) \cup (0, 1]$ but has derivative $+\infty$ at $x = 0$, where $u^*(x) \sim |x|^{\frac{2}{3}} \operatorname{sign} x$. The Lavrentiev phenomenon holds in the form

$$\inf_{\mathcal{A}_\infty} I = \min_{\mathcal{A}_3} I > \inf_{\mathcal{A}_1} I = I(u^*). \qquad (2.4)$$

As indicated in (2.4), I attains its infimum in $\mathcal{A}_3$, and every such minimizer is a smooth solution of the Euler-Lagrange equation in the whole interval $[-1, 1]$. The repulsion property also holds in the form that if $u^{(j)} \in \mathcal{A}_3$ with $u^{(j)} \to u^*$ a.e. then $I(u^{(j)}) \to \infty$.

2.2 Computation of singular minimizers

What are possible numerical methods for detecting such singular minimizers? Consider the problem of minimizing

$$I(u) = \int_a^b f(x, u, u_x) \, dx$$

in

$$\mathcal{A}_1 = \{u \in W^{1,1}(a, b) : u(a) = \alpha, u(b) = \beta\},$$

where α, β are given constants.

A first method proposed by Ball & Knowles [12] consists in decoupling u from its derivative. Thus given $\varepsilon > 0$ we minimize

$$I(u, v) = \int_a^b f(x, u, v) \, dx$$

among piecewise affine functions u in $\mathcal{A}_1$ on a uniform mesh of size h, and functions v that are piecewise constant on the same grid, subject to the constraint

$$\int_a^b \varphi(u_x - v)\, dx \le \varepsilon,$$

where $\varphi \ge 0$ is a suitable even continuous function satisfying (i) $\varphi(p) \ge |v|^s$ for all $v \in \mathbf{R}$, where $1 \le s < \infty$, (ii) $\varphi(p_1 + p_2) \le C(\varphi(p_1) + \varphi(p_2))$ for all $p_1, p_2 \in \mathbf{R}$. For example, one can take $\varphi(p) = |p|$. Then under suitable growth and convexity hypotheses on f (in particular guaranteeing that the infimum of $I(u)$ in $\mathcal{A}_1$ is attained) it can be shown that minimizers $\{u_{h,\varepsilon}, v_{h,\varepsilon}\}$, with $h < \gamma(\varepsilon)$ for a suitable function γ, converge to minimizers $\{u^*, u_x^*\}$ of I in $\mathcal{A}_1$, possibly after extraction of a subsequence, and that

$$\lim_{h,\varepsilon \to 0,\ 0 < h < \gamma(\varepsilon)} I(u_{h,\varepsilon}, v_{h,\varepsilon}) = \inf_{u \in \mathcal{A}_1} I(u) = I(u^*).$$

The effect of numerical quadrature is also studied in [12]; in fact for the Manià example a direct numerical minimization of (2.1) among piecewise affine functions in $\mathcal{A}_1$ using the trapezoidal rule succeeds in finding u^*, but this is a freak resulting from the special form of the integrand and in particular its degeneracy when $u^3 = x$.

A second idea (see Li [43]) is the *truncation method*, in which $f = f(x, u, p)$ is replaced by a truncated integrand $f_M(x, u, p)$ satisfying $f_M(x, u, p) = f(x, u, p)$ whenever $|p| \le M$, and $f_M \to f$ monotonically as $M \to \infty$. If f_M has suitable mild growth properties as $|p| \to \infty$, ensuring in particular that the truncated integral

$$I_M(u) = \int_a^b f_M(x, u, u_x)\, dx$$

does not have the Lavrentiev phenomenon, then we can first minimize I_M, and then let $M \to \infty$. Although this method works theoretically, it has practical drawbacks in that it may not be easy to find an absolute minimizer of I_M (for example, in the case when I attains a minimum among smooth functions, as for the integral (2.3), there is the danger of finding this minimizer instead, since it is a local minimizer of I_M for all sufficiently large M).

A more interesting method (see Li [41]) is that of *element removal*. Here we use piecewise affine approximations, but for each h minimize

$$\int_{[a,b]\setminus E} f(x, u, u_x)\, dx$$

where E consists of a (controlled) small number of elements (in practice these turn out to be elements where u_x^* is large). As for the method in [12] the number of unknowns is increased, in this case by variables tracking which elements are removed.

A potentially promising method, which as far as I am aware has not been studied, is that of using nonconforming elements. It seems possible that this could lead to ways of detecting minimizers in a continuous scale of Sobolev spaces.

There is a growing literature on singular minimizers of one-dimensional variational problems. These singular minimizers *do not in general satisfy the Euler-Lagrange equation in weak or integrated form on the whole interval* $[a, b]$. For elliptic integrands the Euler-Lagrange equation is satisfied on the complement of a closed set of Lebesgue measure zero. This is part of the content of the Tonelli partial regularity theorem [60]II p.359. This theorem is shown to be optimal in [14] (where a slightly improved version can be found), and in Davie [29]. The singularities of minimizers can also be studied in the (x, u) plane; it was shown by Ball & Nadirashvili [16] that under natural hypotheses on f there is a *universal singular set* $\mathcal{D}_f$ to which all points $(x, u(x))$ with $|u_x(x)| = \infty$ for minimizers u of I in $\mathcal{A}_1$ for *any* a, b, α, β must belong, and that $\mathcal{D}_f$ is of first category in $\mathbf{R}^2$. Later Sychev [58] proved that $\mathcal{D}_f$ has two-dimensional Lebesgue measure zero.

There is an important philosophical consequence of the above discussion. Suppose for a moment that one of the variational integrals (2.1), (2.3) represented the energy of some physical system (of course this is not the case, but I will give a physical example later). Since minimizers in different function spaces can be different, in other words different function spaces lead to different predictions, it follows that *the function space is part of the model.* This conclusion seems to me inescapable, but is an uncomfortable one in the sense that little attention is traditionally paid to function spaces when deriving mathematical models (an interesting exception is quantum mechanics, in which the underlying Hilbert space is introduced at the foundational level). If we accept it, then the next question is where the function space (for example, for a model of continuum physics) should come from? To the extent that it is made explicit, common practice is to adopt a pragmatic attitude to this question, making a partly phenomenological choice based on the experimentally observed singularities and the form of natural expressions such as energy that appear in the theory. A more satisfactory approach would be to derive the function space (as well as the governing equations) from a more detailed (e.g. atomistic) model. The only example that I am familiar with where this is done is in the paper of Braides, Dal Maso & Garroni [18], where a one-dimensional model for softening phenomena in fracture mechanics is derived from a (primitive) atomistic model complete with a function space (the space BV of functions of bounded variation).

As was pointed out to me by J.F. Traub at the Oxford FoCM conference, the issue of the choice of a function space arises in continuous complexity theory, for example in analysing the complexity of the problem of computing the integral of a function (see, for example, Traub & Werschutz [61]), where some *a priori* hypothesis has to be made about the regularity of the function to be integrated. It would be interesting to analyse the complexity of computation of integrals of the calculus of variations, and of the problem of minimization of such integrals, in the light of the Lavrentiev phenomenon.

3 Nonlinear elasticity

As described in the introduction, the total elastic energy of a homogeneous elastic body has the form

$$I(y) = \int_{\Omega} W(Dy)\,dx \tag{3.1}$$

Consider the problem of minimizing I among deformations y satisfying the boundary condition

$$y|_{\partial\Omega_1} = \bar{y}, \tag{3.2}$$

where $\partial\Omega_1 \subset \partial\Omega$ has positive area and where $\bar{y} : \partial\Omega_1 \to \mathbf{R}^3$ is a given measurable mapping. No condition is specified on the remainder of the boundary $\partial\Omega\backslash\partial\Omega_1$, where minimization leads to a natural boundary condition corresponding to zero applied traction.

To be physically meaningful, the deformation y should be invertible on Ω. (This is another requirement on the function space that we could ask to be the consequence of a derivation of (3.1) from a more detailed model.) In particular this leads to the requirement that

$$\det Dy(x) > 0 \quad \text{for a.e. } x \in \Omega. \tag{3.3}$$

To guarantee (3.3) for deformations of finite energy, suppose that W :

$M_+^{3\times 3} \to \mathbf{R}$, where $M_+^{3\times 3} = \{A \in M^{3\times 3} : \det A > 0\}$, with

$$W(A) \to \infty \text{ as } \det A \to 0^+. \tag{3.4}$$

If $\partial\Omega_1 \neq \partial\Omega$ there is the possibility of self-contact of the boundary (for a treatment of this see Ciarlet & Nečas [24]), and for this reason y need not be invertible on the closure $\bar{\Omega}$ of Ω. However, even if $\partial\Omega_1 = \partial\Omega$ the requirement of invertibility leads to difficulties. Consider for example the mapping $u : D \to \mathbf{R}^2$, where D is the unit disk of $\mathbf{R}^2$, given in plane polar coordinates by $(r, \theta) \mapsto (\frac{1}{\sqrt{2}}r, 2\theta)$. It is easily seen that u is Lipschitz with $\det Du(x) = 1$ a.e., but u is not locally invertible at 0. Thus for mappings in Sobolev spaces local invertibility does not follow from (3.3). This difficulty can be overcome, and global invertibility established, by an appropriate use of degree theory (see [4], Šverák [55], Fonseca & Gangbo [33]).

The problem of numerical minimization of I via finite elements leads naturally to the

Open question. If $y \in W^{1,p}$ is invertible, can y be approximated in $W^{1,p}$ by piecewise affine invertible mappings?

Here $W^{1,p} = W^{1,p}(\Omega; \mathbf{R}^n)$, $\Omega \subset \mathbf{R}^n$, $n \geq 2$. This question is also of considerable theoretical interest, and I first heard of it from L.C. Evans [31] in the context of his attempts to prove a version of his partial regularity theorem [32] for quasiconvex integrals that would be valid for elastic energies satisfying (3.4). He remarked to me that the existing literature on simplicial approximation (see e.g. [46]) did not cover the case of mappings in Sobolev spaces, since the techniques used relied on composition of mappings, and mappings in Sobolev spaces are not closed under composition.

Consider the simplest case $n = 2$. For a continuous $y \in W^{1,p}$ with $\det Dy(x) > 0$ a.e. a natural algorithm is to triangulate Ω with a regular mesh of size h and define the approximating mapping y^h to be that piecewise affine mapping coinciding with y at the mesh points. Unfortunately this fails because even if $y \in W^{1,\infty}$ with $\det Dy(x) \geq \sigma > 0$ there can be, for arbitrarily small h, triangles on which $\det Dy^h$ is negative. An algorithm is needed for choosing a sequence of finer and finer meshes without this undesirable behaviour.

The existence theory for minimizers of (3.1) has been reviewed in many places (see, for example, [7, 8], Ciarlet [23], Dacorogna [28], Šilhavý [63], Pedregal [51]). For the existence of minimizers it is necessary to impose growth and convexity conditions on the stored-energy function

W. The natural convexity condition is that of *quasiconvexity*. In fact for the variational problem of minimizing

$$I(y) = \int_\Omega f(Dy)\,dx$$

subject to

$$y|_{\partial\Omega_1} = \bar{y},$$

where $\Omega \subset \mathbf{R}^n$ is a bounded Lipschitz domain, $y : \Omega \to \mathbf{R}^m$ and $f : M^{m\times n} \to \mathbf{R}$ is continuous and satisfies suitable growth conditions, it is now understood (Morrey [47], Acerbi & Fusco [1]) that for the existence of minimizers it is sufficient (and, up to the addition of lower-order terms, necessary, see Ball & Murat [15]) that f be quasiconvex.

Definition 3.1 *Let $f : M^{m\times n} \to \mathbf{R} \cup \{+\infty\}$ be continuous. Then f is quasiconvex if*

$$\int_\Omega f(Dv)\,dx \geq \int_\Omega f(A)\,dx$$

for all $A \in M^{m\times n}$ and all $v \in Ax + C_0^\infty(\Omega; \mathbf{R}^m)$.

(This condition seems to, but does not, depend on Ω.) No tractable necessary and sufficient conditions are known for a function f to be quasiconvex. If f is quasiconvex then f is *rank-one convex*, that is the mapping $t \mapsto W(A + t\lambda \otimes \mu)$ is convex for all $A \in M^{m\times n}, \lambda \in \mathbf{R}^m, \mu \in \mathbf{R}^n$, but it was shown by Šverák [56] that the converse is false for $n \geq 2, m \geq 3$. Based on Šverák's example, Kristensen [36] proved the striking result that for the same dimensions there is no local necessary and sufficient condition for quasiconvexity.

Unfortunately, the known existence theorems for quasiconvex integrands do not apply to elasticity, because they assume growth conditions incompatible with (3.4). For this reason it is at present necessary (see the references cited above), to make the stronger convexity hypothesis that W be *polyconvex*, namely

$$W(A) = g(A, \operatorname{cof} A, \det A)$$

for some convex g, where $\operatorname{cof} A$ denotes the matrix of 2×2 subdeterminants of A. Then the existence of an absolute minimzer is assured, provided the growth condition

$$W(A) \geq c_0(|A|^2 + |\operatorname{cof} A|^{\frac{3}{2}}) - c_1 \tag{3.5}$$

holds, where $c_0 > 0, c_1$ are constants (see Müller, Qi & Yan [49] for this improved version of a result of [3]). However essentially nothing is known about the smoothness of absolute minimizers y^* for strictly polyconvex or strictly quasiconvex W. It is not even known if the usual weak form of the Euler-Lagrange equation is satisfied (though certain weak forms may be obtained, see [5, 2], Bauman, Owen & Phillips [17]), or if

$$\det Dy^*(x) \geq \sigma > 0 \text{ for a.e. } x \in \Omega. \tag{3.6}$$

Of course a proof of smoothness and of (3.6) would lead to a justification of standard finite-element minimization schemes for (3.1), (3.2). Otherwise the above Open Problem makes the construction of a scheme generating invertible approximate minimizers problematic. Another approach is to tolerate a small set on which the approximate minimizers fail to satisfy (3.3). This approach is taken in Li [42], who applies the element removal method to find a sequence of such (possibly noninvertible) approximate minimizers converging, at least theoretically, to a minimizer of (3.1),(3.2).

Under the hypotheses of the existence theorem it is not known whether the Lavrentiev phenomenon can hold. However if the growth condition (3.5) is slightly weakened then there is a physically interesting example involving *cavitation*.

As an illustrative example of cavitation consider the problem of minimizing

$$I(y) = \int_{B(0,1)} W(Dy) \, dx$$

subject to the pure displacement boundary condition

$$y|_{\partial B(0,1)} = \lambda x, \ \lambda > 0,$$

where $B(0,1)$ denotes the unit ball in $\mathbf{R}^3$, and where

$$W(A) = |A|^2 + h(\det A),$$

with $h : (0,\infty) \to \mathbf{R}$ smooth and satisfying $h'' > 0, \lim_{\delta \to \infty} \frac{h(\delta)}{\delta} = \lim_{\delta \to 0+} h(\delta) = \infty$. Note that (3.5) does not hold, but that W is polyconvex. Hence, since polyconvexity implies quasiconvexity, the minimizer of I among smooth (or even $W^{1,3}$) y is given by

$$\tilde{y}_\lambda(x) \equiv \lambda x.$$

But among radial maps

$$y(x) = r_\lambda(|x|)\frac{x}{|x|}$$

we have nontrivial minimizers for $\lambda > \lambda_{cr}$ for some critical value λ_{cr}. These radial minimizers satisfy $r_\lambda(0) > 0$. Thus a hole is formed at the origin. Furthermore we have the Lavrentiev phenomenon in the form

$$\inf_{\mathcal{A}_1} I < \inf_{\mathcal{A}_3} I = I(\tilde{y}_\lambda),$$

where $\mathcal{A}_p = \{y \in W^{1,p}(B(0,1); \mathbf{R}^3) : y|_{\partial B(0,1)} = \lambda x\}$. In fact cavitation is a common failure mechanism in polymers. See Lazzeri & Bucknall [38] for some striking images of almost radial cavitation of roughly spherical rubber particles imbedded in a matrix of nylon-6; such rubber-toughened plastics are used, for example, in car bumpers. See [7] for further remarks about cavitation and function spaces.

4 Computation of microstructure

4.1 Nonattainment of minimum energy and microstructure

Consider a single crystal of a material (for example, some metallic alloy) that can undergo a phase transformation involving a change of shape at some critical temperature $\theta = \theta_c$ from a higher symmetry *austenite* phase to a lower symmetry *martensite* phase. The crystal is assumed to be elastic with stored-energy function $W_\theta(Dy)$ that depends on the temperature θ. If W attains a finite minimum, then by adding a suitable function of θ there is no loss of generality in assuming that

$$\min_A W_\theta(A) = 0.$$

Consider the corresponding set of energy-minimizing gradients

$$K_\theta = \{A \in M^{3\times 3} : W_\theta(A) = 0\}.$$

Since the stored-energy function must satisfy the frame-indifference condition

$$W_\theta(QA) = W_\theta(A) \quad \text{for all } Q \in SO(3)$$

it follows that $K_\theta = SO(3)K_\theta$. Let $U = U^T > 0$ be the linear transformation describing the change of shape at $\theta = \theta_c$ relative to undistorted austenite. Then at $\theta = \theta_c$

$$K_{\theta_c} = SO(3) \cup \bigcup_{i=1}^{N} SO(3)U_i,$$

where the U_i are the distinct matrices RUR^T for R belonging to the symmetry group $\mathcal{S}$ of the material (assumed to be a subgroup of $SO(3)$).

The energy well $SO(3)$ corresponds to the austenite, while each energy well $SO(3)U_i$ corresponds to one of the N variants of the martensite. If, as is often the case, the austenite is stable for temperatures $\theta > \theta_c$ then for these temperatures $K_\theta = \alpha(\theta)SO(3)$, where $\alpha(\cdot)$ accounts for thermal expansion, while, for $\theta < \theta_c$, K_θ is given by the N martensitic variants, with $U = U(\theta)$.

An important example is provided by a cubic to tetragonal transformation. Here $N = 3$, and the three variants correspond to the matrices

$$U_1 = \operatorname{diag}(\eta_2, \eta_1, \eta_1),$$
$$U_2 = \operatorname{diag}(\eta_1, \eta_2, \eta_1),$$
$$U_3 = \operatorname{diag}(\eta_1, \eta_1, \eta_2),$$

where η_1, η_2 are lattice parameters.

Interfaces between variants are described by rank-one connections between the corresponding energy wells $SO(3)U_i, SO(3)U_j$, i.e. by pairs of matrices R_iU_i, R_jU_j with

$$R_iU_i - R_jU_j = \lambda \otimes \mu,$$

where $i \neq j$ and $R_i, R_j \in SO(3)$, and where μ is the interface normal. Such rank-one connections exist between any pair of the three tetragonal wells in a cubic to tetragonal transformation. Given such a rank-one connection, the function $t \mapsto W_\theta(R_iU_i + t\lambda \otimes \mu)$ has a double-well form, and therefore is not convex. Hence W_θ is not rank-one convex, and so not quasiconvex either, leading to the expectation that the minimum of

$$I_\theta(y) = \int_\Omega W_\theta(Dy)\,dx \tag{4.1}$$

in $W^{1,1}(\Omega; \mathbf{R}^3)$ subject to the boundary condition

$$y|_{\partial\Omega_1} = \bar{y}, \tag{4.2}$$

is in general *not attained*. In fact this has been proved in certain cases (see Ball & James [11]Theorem 7.1, Ball & Carstensen [9]Theorems 3.1, 3.2). Minimizing sequences $y^{(j)}$ typically have gradients $Dy^{(j)}$ that oscillate more and more finely as j increases, generating in the limit an infinitely fine *microstructure*.

Real microstructures are not of course infinitely fine, and their limited fineness can be modelled by introducing interfacial energy. For example, a crude way of doing this is to change the energy functional to

$$I_\theta^\varepsilon(y) = \int_\Omega [W_\theta(Dy) + \frac{1}{2}\varepsilon^2|D^2y|^2]\,dx, \tag{4.3}$$

for some small $\varepsilon > 0$.

I refer the reader to Ball & James [10, 11], Luskin [44], Müller [48], Pedregal [51] for further details concerning the physical model and its analysis.

4.2 What should we compute and how?

According to the elasticity model described above, minimizing sequences for the total elastic energy may develop infinitely fine microstructure. Thus, however fine a finite-element mesh is used, numerical minimization of the energy can be expected to yield oscillations in Dy at a length-scale comparable with the mesh size. For a model such as (4.3) incorporating interfacial energy a very fine mesh is still needed to capture the details of the microstructure, which in real materials can have a length-scale of as little as a few atomic spacings. Thus numerical minimization of the energy is computationally highly intensive.

A second difficulty is that, because oscillations can develop at the level of the mesh, and the problem has preferred crystallographic directions, the computations will in general be sensitive to mesh orientation.

A third difficulty is that the discretized energy has a huge number of local minimizers. Consider, for example, the one-dimensional problem of minimizing

$$I(u) = \int_0^1 [(u_x^2 - 1)^2 + u^2]\, dx$$

in $W^{1,1}(0,1)$ (so that there are no end conditions). Then

$$\inf_{W^{1,1}} I = 0$$

but the infimum is not attained. Consider the discretized problem of minimizing I among piecewise affine functions on a mesh of size $h = \frac{1}{N^2}$. Let u_h be a minimizer and set $I(u_h) = E_h$. Then (Carstensen [19]) there exists a family $\mathcal{K}$ consisting of N^N local minimizers of the same discretized problem, such that

(i) $\mathcal{K}$ is a subset of the ball in $L^2(0,1)$ with centre u_h, radius $5h$,
(ii) $I(v) < (1 + 24\sqrt{h})E_h$ for each $v \in \mathcal{K}$,
(iii) if v_0, v_1 are distinct points of $\mathcal{K}$ then $\sup_{t\in[0,1]} I(v(t)) > \frac{13}{h}E_h$ for any continuous path $v : [0,1] \to L^2(0,1)$ with $v(0) = v_0, v(1) = v_1$.

Because of the exponential number of local minimizers, and the relatively high energy thresholds between them, local descent methods will typically fail to detect a global minimizer of the discretized problem.

But what should we compute? For the problem (4.1),(4.2) (similar issues arise for problems incorporating small interfacial energy) one possible answer is the set of possible *Young measures* $(\nu_x)_{x \in \Omega}$ corresponding to sequences of deformation gradients $Dy^{(j)}$ for minimizing sequences $y^{(j)}$ (for the definition and properties of these measures see, for example, [6, 59, 48, 50, 62]). For each $x \in \Omega$, ν_x is a probability measure on $M^{3 \times 3}$. Let y be the *macroscopic deformation* given by the weak limit of the minimizing sequence $y^{(j)}$ in an appropriate Sobolev space determined by the growth of W. The Young measure determines the corresponding *macroscopic deformation gradient* Dy through the formula $Dy(x) = \bar{\nu}_x$, where $\bar{\nu}_x = \int_{M^{3 \times 3}} A \, d\nu_x(A)$ denotes the centre of mass of ν_x.

In principle y can be computed by minimizing

$$I_\theta^{\mathrm{qc}}(y) = \int_\Omega W_\theta^{\mathrm{qc}}(Dy) \, dx, \qquad (4.4)$$

subject to (4.2), where W_θ^{qc} is the *quasiconvex envelope* of W_θ, that is the supremum of all quasiconvex functions that are less than W_θ. This is the content of the relaxation theorem of Dacorogna [27], though as for the existence theorems assuming quasiconvexity, the theorem does not strictly speaking apply to elasticity because the growth hypotheses are inconsistent with the property $W_\theta(A) \to \infty$ as $\det A \to 0^+$. Thus quasiconvexification describes the passage from microscopic to macroscopic stored-energy functions for these materials.

The idea of minimizing (4.4) to compute y is attractive, but a very serious drawback is that the lack of a suitable characterization of quasiconvexity means that it is only in rare cases (see, for example, Kohn [35], Pipkin [52, 53], LeDret & Raoult [39]) that W_θ^{qc} is known. Further, to compute W_θ^{qc} numerically leads to a problem of similar difficulty to the original one.

Despite all these difficulties there have been a number of interesting computations of martensitic microstructure, though it is fair to say that there is a long way to go before the computer can be used as an effective predictive tool in these problems. Some key references are Carstensen and Plecháč [21, 20, 22], Collins, Kinderlehrer & Luskin [25], Collins & Luskin [26], Dolzmann [30], Killough [34], Li & Luskin [40] (these last two papers concerning computations of needle-like martensitic microstructures), and the review article of Luskin [44].

As an illustration of what can be achieved, in Figure 1 are shown the results of some computations due to P. Plecháč [54]. These computations were carried out for a two-dimensional version of I_θ^ε (see (4.3)) with

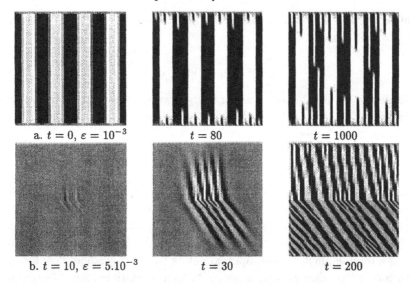

a. $t = 0$, $\varepsilon = 10^{-3}$	$t = 80$	$t = 1000$
b. $t = 10$, $\varepsilon = 5.10^{-3}$	$t = 30$	$t = 200$

Fig. 1. Microstructure evolution a. for a single crystal b. for two adjacent grains.

$W = W_\theta : M^{2\times 2} \to \mathbf{R}$ given by

$$W(A) = \kappa_1(C_{11} - (1 + \delta^2))^2 + \kappa_2(C_{22} - 1)^2 + \kappa_3(C_{12}^2 - \delta^2)^2, \quad (4.5)$$

where $U = A^T A$. In this case we have that

$$K = \{A \in M^{2\times 2} : W(A) = 0\} = SO(2)V_1 \cup SO(2)V_2,$$

where

$$V_1 = \begin{pmatrix} 1 & 0 \\ -\delta & 1 \end{pmatrix}, \quad V_2 = \begin{pmatrix} 1 & 0 \\ \delta & 1 \end{pmatrix}.$$

The two energy wells are rank-one connected with

$$V_2 - V_1 = \lambda \otimes \mu,$$

where $\lambda = (0, 2\delta), \mu = (1, 0)$. In the computations, $\Omega = (0,1)^2$, the boundary condition was taken to be linear with gradient $\frac{1}{2}(V_1 + V_2)$, that is

$$y(x) = x \text{ for } x \in \partial\Omega,$$

and the constants in (4.5) had the values $\kappa_i = 1, \delta = 0.5$. For these boundary conditions with $\varepsilon = 0$ the minimum of the energy (4.3) is not attained, and the Young measure of any minimizing sequence is given by

$$\nu_x = \frac{1}{2}(\delta_{V_1} + \delta_{V_2}).$$

Similarly, for small ε we expect a finely layered configuration to evolve in which Dy mostly takes the values V_1, V_2. The minimization algorithm was to solve the evolution equation

$$\Delta y_t = -\operatorname{div} \sigma(Dy) + \varepsilon^2 \Delta^2 y,$$

where $\sigma(A) = DW(A)$. (This equation, used in the calculations as a numerical solver, can also be thought of as a viscoelastic model for the evolution of y that neglects inertia; however it should be noted that the damping term Δy_t is not frame-indifferent.) The time-discretization was a fully implicit scheme, approximating the H^{-1}-gradient flow for I_θ^ε by a corresponding minimization problem at each time step. The mesh was uniform with 500×500 elements. This algorithm is of course not guaranteed to tend as time $t \to \infty$ to a global minimizer. In Figure 1a are shown stages in the evolution starting from sinusoidal initial data, with $\varepsilon = 10^{-3}$, the shading varying from white when $Dy(x) \in SO(2)V_1$ to black when $Dy(x) \in SO(2)V_2$. Note how the refinement of the layering is via the initiation of new layers at the boundary; a similar effect was observed earlier in the computations of Swart (see Swart & Holmes [57]).

For the computation in Figure 1b the crystal consists of two adjacent grains, corresponding to the stored-energy function $W = W(x, A)$ given by

$$W(x, A) = \begin{cases} W(A) & x \in \Omega_1 = (0,1) \times (1/2, 1) \\ W(AR_{\pi/4}) & x \in \Omega_2 = (0,1) \times (0, 1/2) \end{cases},$$

where $R_{\pi/4}$ denotes a planar rotation through the angle $\pi/4$, and $\varepsilon = 5.10^{-3}$. The initial data was taken to be the identity with a small smooth perturbation in a neighbourhood of the centre of the square. Two simple laminates are formed meeting at the grain boundary.

Acknowledgments

I am grateful to Petr Plecháč for permission to include his computations and to him, Carsten Carstensen, Craig Evans, Matt Killough, Bob Kohn and Joe Traub for useful comments that helped improve the paper and lecture at the 1999 Oxford FoCM conference on which it was based. This work was supported by EC TMR Contract ERBFMRX CT 98-229 on 'Phase transformations in crystalline solids'.

Bibliography

[1]   Acerbi, E. and Fusco, N. (1984). Semicontinuity problems in the calculus of variations. *Arch. Rat. Mech. Anal.* **86**, 125–145.

[2]   Ball, J.M. (2000). Some open problems in elasticity. To appear.

[3]   Ball, J.M. (1977). Convexity conditions and existence theorems in nonlinear elasticity. *Arch. Rat. Mech. Anal.* **63**, 337–403.

[4]   Ball, J.M. (1981). Global invertibility of Sobolev functions and the interpenetration of matter. *Proc. Royal Soc. Edinburgh* **88A**, 315–328.

[5]   Ball, J.M. (1983). Minimizers and the Euler-Lagrange equations, in *Proc. ISIMM conference, Paris* (Springer-Verlag, Berlin).

[6]   Ball, J.M. (1989). A version of the fundamental theorem for Young measures. in *Proceedings of Conference on 'Partial differential equations and continuum models of phase transitions'*, eds M. Rascle, D. Serre, and M. Slemrod, Springer Lecture Notes in Physics 359 (Springer-Verlag, Berlin) 3–16.

[7]   Ball, J.M. (1996). Nonlinear elasticity and materials science; a survey of some recent developments, in *Nonlinear Mathematics and Its Applications*, ed. P.J. Aston (Cambridge University Press, Cambridge), 93–119.

[8]   Ball, J.M. (1998). The calculus of variations and materials science. *Quart. Appl. Math.* **56**, 719–740.

[9]   Ball, J.M. and Carstensen, C. (1999). Compatibility conditions for microstructures and the austenite-martensite transition. *Materials Science & Engineering A* **273–275**, 231–236.

[10] Ball, J.M. and James, R.D. (1987). Fine phase mixtures as minimizers of energy. *Arch. Rat. Mech. Anal.* **100**, 13–52.

[11] Ball, J.M. and James, R.D. (1992). Proposed experimental tests of a theory of fine microstructure, and the two-well problem. *Phil. Trans. Roy. Soc. London A* **338**, 389–450.

[12] Ball, J.M. and Knowles, G. (1986). A numerical method for detecting singular minimizers. *Numerische Mathematik* **92**, 193–204.

[13] Ball, J.M. and Mizel, V.J. (1984). Singular minimizers for regular one-dimensional problems in the calculus of variations. *Bull. Amer. Math. Soc.* **11**, 143–146.

[14] Ball, J.M. and Mizel, V.J. (1985). One-dimensional variational problems whose minimizers do not satisfy the Euler-Lagrange equations. *Arch. Rat. Mech. Anal.* **90**, 325–388.

[15] Ball, J.M. and Murat, F. (1984). $W^{1,p}$-quasiconvexity and variational problems for multiple integrals. *J. Functional Analysis* **58**, 225–253.

[16] Ball, J.M. and Nadirashvili, N. (1993). Universal singular sets for one-dimensional variational problems. *Calculus of Variations and Partial Differential Equations*, **1**, 429–438.

[17] Bauman, P., Owen, N.C. and Phillips, D. (1991). Maximum principles and a priori estimates for a class of problems from nonlinear elasticity. *Annales de l'Institut Henri Poincaré - Analyse non linéaire* **8**, 119–157.

[18]   Braides, A., Dal Maso, G. and Garroni, A. (1999). Variational formulation for softening phenomena in fracture mechanics: the one-dimensional case. *Arch. Rat. Mech. Anal.* **146**, 23–58.

[19]   Carstensen, C. (1996). Numerical analysis of nonconvex minimization problems allowing microstructures. *Zeitschrift für Angewandte Mathematik und Mechanik* **76(S2)**, 437–438.

[20]   Carstensen, C. and Plecháč, P. (1997). Adaptive algorithms for scalar non-convex variational problems. *Appl. Numer. Maths* **26**, 203–216.

[21]   Carstensen, C. and Plecháč, P. (1997). Numerical solution of the scalar double-well problem allowing microstructure. *Maths Comp.* **6**, 997–1026.

[22]   Carstensen, C. and Plecháč, P. (2000). Numerical analysis of compatible phase transitions in. *SIAM J. Numer. Anal.* **37**, 2061–2081.

[23]   Ciarlet, P.G. (1988). *Mathematical Elasticity, Vol.I: Three-Dimensional Elasticity* (North-Holland, Amsterdam).

[24]   Ciarlet, P.G. and Nečas, J. (1985). Unilateral problems in nonlinear three-dimensional elasticity. *Arch. Rat. Mech. Anal.* **87**, 319–338.

[25]   Collins, C., Kinderlehrer, D. and Luskin, M. (1991). Numerical approximation of the solution of a variational problem with a double well potential. *SIAM J. Num. Anal.* **28**, 321–332.

[26]   Collins, C. and Luskin, M. (1991). Optimal-order error-estimates for the finite-element approximation of the solution of a nonconvex variational problem. *Maths Comp.* **57**, 621–637.

[27]   Dacorogna, B. (1982). Quasiconvexity and relaxation of non convex variational problems. *J. Funct. Anal.* **46**, 102–118.

[28]   Dacorogna, B. (1989). *Direct Methods in the Calculus of Variations* (Springer, New York).

[29]   Davie, A.M. (1988). Singular minimizers in the calculus of variations in one dimension. *Arch. Rat. Mech. Anal.* **101**, 161–177.

[30]   Dolzmann, G. (1999). Numerical computation of rank-one convex envelopes. *SIAM J. Numer. Anal.* **36**, 1621–1635.

[31]   Evans, L.C.. Private communication.

[32]   Evans, L.C. (1986). Quasiconvexity and partial regularity in the calculus of variations. *Arch. Rat. Mech. Anal.* **95**, 227–268.

[33]   Fonseca, I. and Gangbo, W. (1995). Local invertibility of Sobolev functions. *SIAM J. Math. Anal.* **26**, 280–304.

[34]   Killough, M. (1998). *A diffuse interface approach to the development of microstructure in martensite.* Courant Institute, New York University Ph.D. thesis.

[35]   Kohn, R.V. (1991). The relaxation of a double-well energy. *Continuum Mechanics and Thermodynamics* **3**, 193–236.

[36]   Kristensen, J. (1999). On the non-locality of quasiconvexity. *Ann. Inst. Henri Poincaré–Analyse Nonlinéaire* **16**, 1–13.

[37]   Lavrentiev, M. (1926). Sur quelques problèmes du calcul des variations. *Ann. Mat. Pura Appl.* **4**, 7–28.

[38]   Lazzeri, A. and Bucknall, C.B. (1995). Applications of a dilatational yielding model to rubber-toughened polymers. *Polymer* **36**, 2895–2902.

[39]   LeDret, H. and Raoult, A. (1995). The quasiconvex envelope of the Saint Venant-Kirchhoff stored energy function. *Proc. Royal Soc. Edinburgh* **125A**, 1179–1192.

[40]   Li, B. and Luskin, M. (1999). Theory and computation for the microstructure near the interface between twinned layers and a pure

variant of martensite. *Materials Science & Engineering A* **273**, 237–240.

[41] Li, Z.-P. (1992). Element removal method for singular minimizers in variational problems involving Lavrentiev phenomenon. *Proc. R. Soc. Lond. A* **439**, 131–137.

[42] Li, Z.-P. (1995). Element removal method for singular minimizers in problems of hyperelasticity. *Math. Models and Methods in Appl. Sci.* **5**, 387–399.

[43] Li, Z.-P. (1995). A numerical method for computing singular minimizers. *Numer. Mathematik* **71**, 317–330.

[44] Luskin, M. (1996). On the computation of crystalline microstructure. *Acta Numerica* **5**, 191–258.

[45] Manià, B. (1934). Sopra un esempio di Lavrentieff. *Bull. Un. Mat. Ital.* **13**, 36–41.

[46] Moise, E.E. (1977). *Geometric Topology in Dimensions 2 and 3* Graduate Texts in Mathematics **47** (Springer-Verlag, New York).

[47] Morrey, C.B. (1952). Quasi-convexity and the lower semicontinuity of multiple integrals. *Pacific J. Math.* **2**, 25–53.

[48] Müller, S. (1999). Variational methods for microstructure and phase transitions, in *Calculus of Variations and Geometric Evolution Problems*, Lecture Notes in Maths **1713** (Springer-Verlag, Berlin), 85–210.

[49] Müller, S., Qi, T. and Yan, B.S. (1994). On a new class of elastic deformations not allowing for cavitation. *Ann. Inst. Henri Poincaré,Analyse Nonlinéaire* **11**, 217–243.

[50] Pedregal, P. (1991). *Parametrized measures and variational principles*, Progress in nonlinear differential equations and their applications **30** (Birkhäuser, Basel).

[51] Pedregal, P. (2000). *Variational Methods in Nonlinear elasticity* (SIAM, Philadelphia).

[52] Pipkin, A.C. (1986). The relaxed energy density for isotropic elastic membranes. *IMA J. Applied Maths* **36**, 85–99.

[53] Pipkin, A.C. (1991). Elastic materials with two preferred states. *Quarterly J. Mech. Appl. Math.* **44**, 1–15.

[54] Plecháč, P. (1998). Computation of microstructure with interfacial energies, in *Proceedings ENUMATH 97, (Heidelberg)* (World Scientific, Singapore).

[55] Šverák, V. (1988). Regularity properties of deformations with finite energy. *Arch. Rat. Mech. Anal.* **100**, 105–127.

[56] Šverák, V. (1992). Rank-one convexity does not imply quasiconvexity. *Proc. Royal Soc. Edinburgh* **120A**, 185–189.

[57] Swart, P.J. and Holmes, P.J. (1992). Energy minimization and the formation of microstructure in dynamic anti-plane shear. *Arch. Rat. Mech. Anal.* **121**, 37–85.

[58] Sychev, M.A. (1994). Lebesgue measure of the universal singular set for the simplest problems in the calculus of variations. *Siberian Mathematical J.* **35**, 1220–1233.

[59] Tartar, L. (1982). The compensated compactness method applied to systems of conservation laws, in *Systems of Nonlinear Partial Differential Equations*, ed. J.M. Ball, NATO ASI Series, Vol. C111 (Reidel, Dordrecht), 263–285.

[60] Tonelli, L. (1921–23). *Fondamenti di Calcolo delle Variazioni*, Volumes I, II. (Zanichelli, Bologna).

[61] Traub, J.F. and Werschulz, A.G. (1998). *Complexity and Information,* Lezioni Lincei (Cambridge Univ. Press, Cambridge).

[62] Valadier, M. (1994). A course on Young measures. *Rend. Istit. Mat. Univ. Trieste* **26:suppl.**, 349–394.

[63] Šilhavý, M. (1997). *The Mechanics and Thermodynamics of Continuous Media* (Springer-Verlag, Berlin).

Adaptive Finite Element Methods for Flow Problems

Roland Becker, Malte Braack and Rolf Rannacher

Institut für Angewandte Mathematik
Universität Heidelberg
Im Neuenheimer Feld 293/294
D-69120 Heidelberg, Germany
Email: Rolf.Rannacher@iwr.uni-heidelberg.de
Url http://gaia.iwr.uni-heidelberg.de

Abstract

We present a general approach to error control and mesh adaptation for computing viscous flows by the Galerkin finite element method. A posteriori error estimates are derived for quantities of physical interest by duality arguments. In these estimates local cell residuals are multiplied by influence factors which are obtained from the numerical solution of a global dual problem. This provides the basis of a feed-back algorithm by which economical meshes can be constructed which are tailored to the particular needs of the computation. The performance of this method is illustrated by several flow examples.

1 Introduction

Approximating partial differential equations by discretization as in the finite element method may be considered as a *model reduction* where a conceptually *infinite* dimensional model is approximated by a *finite* dimensional one. As the result of the computation, we obtain an approximation to the desired output quantity of the simulation and besides that certain accuracy indicators like cell-residuals. Controlling the error in such an approximation of a continuous model requires to determine the influence factors for the *local* error indicators on the target quantity. Such a sensitivity analysis with respect to local perturbations of the model is common in optimal control theory and introduces the concept of a *dual* (or *adjoint*) problem.

For illustration, let $Lu = f$ be a (linear) differential problem posed in variational form and $L_h u_h = f_h$ its approximation by a Galerkin finite element method on a meshes $\mathcal{T}_h = \{K\}$ with cells K. Then, the error $e := u - u_h$ satisfies $Le = \rho$ in the variational sense with the "residual" $\rho := f - Lu_h$ as right-hand side. Now, let $J(u)$ be a quantity of physical interest derived from the solution u by applying a functional $J(\cdot)$. The goal is to control the error of the discretization with respect to this functional output in terms of certain computable cell residuals $\rho_K(u_h)$. An example is control of the total error $e_K = u - u_{h|K}$ in some cell K. By superposition, e_K splits into two components, the locally produced *truncation error* and the globally transported *pollution error*, i.e., $e_K^{tot} = e_K^{loc} + e_K^{trans}$. This asks for control of

- error propagation in space (global pollution effect),
- interaction of physical error sources (local sensitivity analysis).

Clearly, the effect of the cell residual ρ_K on the local error $e_{K'}$, at another cell K', is governed by a Green function of the continuous problem which is determined globally by the adjoint operator. Capturing this dependence by *numerical* evaluation is the general philosophy underlying our approach to error control. In this it differs essentially from the traditional method which is based on local residual-type information alone (see [22] and [1] for surveys). In practice it is mostly impossible to determine the complex error interaction by analytical means, it rather has to be detected by computation. This eventually results in a feed-back process in which error estimation and mesh adaptation goes hand-in-hand leading to economical discretization for computing the quantities of interest with high accuracy.

2 A general paradigm for a posteriori error estimation

We outline our concept of error estimation in an abstract setting following the general paradigm introduced in Johnson [18] and in Eriksson–Estep–Hansbo–Johnson [12]. For a detailed discussion of the practical aspects of this approach, we refer to [8] and [9]; a survey with applications to various problems in mechanics and physics has been given in [20]. Here, we concentrate on the aspects particularly relevant for flow computations.

Let $\hat{V}$ be a Hilbert space with inner product $(\cdot, \cdot)$ and corresponding norm $\| \cdot \|$, $V \subset \hat{V}$ a subspace with dual V^* and $A(\cdot; \cdot)$ a continuous semi-linear form on $\hat{V} \times \hat{V}$. Further, let $\hat{u} \in \hat{V}$ be a particular element representing prescribed boundary data. For a given $F \in V^*$, we seek a solution $u \in V + \hat{u}$ to the abstract variational problem

$$A(u; \varphi) = F(\varphi) \qquad \forall \varphi \in V. \qquad (2.1)$$

This problem is approximated by a Galerkin method using a sequence of finite dimensional subspaces $V_h \subset \hat{V}_h \subset \hat{V}$ indexed by a discretization parameter h. The discrete problems seek $u_h \in V_h + \hat{u}_h$, satisfying

$$A(u_h; \varphi_h) = F(\varphi_h) \qquad \forall \varphi_h \in V_h. \qquad (2.2)$$

The key feature of this approximation is the "Galerkin orthogonality" which in this nonlinear case is expressed as

$$A(u; \varphi_h) - A(u_h; \varphi_h) = 0, \quad \varphi_h \in V_h. \qquad (2.3)$$

By elementary calculus, there holds

$$A(u; \varphi_h) - A(u_h; \varphi_h) = \int_0^1 A'(su + (1-s)u_h; e, \varphi_h)\, ds,$$

with $A'(v; \cdot, \cdot)$ denoting the tangent form of $A(\cdot; \cdot)$ at some $v \in \hat{V}$. This leads us to introduce the bilinear form

$$L(u, u_h; \varphi, \psi) := \int_0^1 A'(su + (1-s)u_h; \varphi, \psi)\, ds,$$

which depends on the solutions u as well as u_h. Then, for the error $e = u - u_h$, there holds

$$L(u, u_h; e, \varphi_h) = \int_0^1 A'(su + (1-s)u_h; e, \varphi_h)\, ds$$
$$= A(u; \varphi_h) - A(u_h; \varphi_h) = 0, \quad \varphi_h \in V_h.$$

Suppose that the quantity $J(u)$ has to be computed, where $J(\cdot)$ is a linear functional on V. For representing the error $J(e)$, we use the solution $z \in V$ of the following so-called "dual problem":

$$L(u, u_h; \varphi, z) = J(\varphi) \qquad \forall \varphi \in V, \qquad (2.4)$$

assuming that this problem is solvable. Taking $\varphi = e$ in (2.4) and using the Galerkin orthogonality (2.3), we obtain the error representation

$$J(e) = L(u, u_h; e, z - \varphi_h) = F(z - \varphi_h) - A(u_h; z - \varphi_h), \qquad (2.5)$$

with an arbitrary $\varphi_h \in V_h$. This simple argument assumes that $e \in V$ which requires exact representation of boundary data, i.e., $\hat{u} = \hat{u}_h$. In practice this condition may not be satisfied which requires modifications in the argument using the particular structure of the problem (see the example in the next section). Since the bilinear form $L(u, u_h; \cdot, \cdot)$ contains the unknown solution u in its coefficient, the evaluation of (2.5) requires approximation. The simplest way is to replace u by u_h, yielding a perturbed dual bilinear form

$$L(u_h, u_h; \varphi, \psi) = A'(u_h; \varphi, \psi).$$

Controlling the effect of this perturbation on the accuracy of the resulting error estimator may be a delicate task and depends strongly on the particular problem considered. Our own experience with several different types of problems (including the Navier–Stokes equations) indicates that this problem is less critical as long as the continuous solution is stable. The crucial problem is the numerical solution of the linearised dual problem. To this end, we may select certain subspaces $\tilde{V}_h \subset V$ and compute an approximation $\tilde{z}_h \in \tilde{V}_h$ by solving

$$L(u_h, u_h; \varphi_h, \tilde{z}_h) = J(\varphi_h) \qquad \forall \varphi_h \in \tilde{V}_h. \tag{2.6}$$

In practice, we may require $V_h \subset \tilde{V}_h$ or even $V_h = \tilde{V}_h$. The effect of all these approximations is described in the following theorem.

Theorem 1 *Let $z \in V$ be the solution of the linearised dual problem*

$$A'(u_h; \varphi, z) = J(\varphi) \quad \forall \varphi \in V. \tag{2.7}$$

Then, there holds the a posteriori error representation

$$J(e) = F(z - \varphi_h) - A(u_h, z - \varphi_h) + R(u, u_h; e, e, z), \tag{2.8}$$

with an arbitrary $\varphi_h \in V_h$, and a remainder term

$$R(u, u_h; e, e, z) := \int_0^1 A''(u_h + se; e, e, z)\,(1 - s)\,ds.$$

Proof Setting $\varphi = e$ (assuming $e \in V$) in the dual problem (2.7), we obtain

$$J(e) = A'(u_h; e, z).$$

Further, noting that

$$A(u; z) = A(u_h; z) + A'(u_h; e, z) + \int_0^1 A''(u_h + se; e, e, z)\,(1 - s)\,ds,$$

and $A(u; z) = F(z)$, we conclude that

$$J(e) = F(z) - A(u_h; z) + R(u, u_h; e, e, z),$$

with the remainder term

$$R(u, u_h; e, e, z) := \int_0^1 A''(u_h + se; e, e, z)(1 - s) \, ds.$$

Then, using the Galerkin orthogonality (2.3), we obtain

$$J(e) = F(z - \psi_h) - A(u_h; z - \psi_h) + R(u, u_h; e, e, z),$$

for an arbitrary $\varphi_h \in V_h$.   □

Clearly, the remainder term R in (2.8) vanishes if $A(\cdot; \cdot)$ is affine-linear. We note, that Theorem 1 allows a natural generalisation to cover also the case of a nonlinear, differentiable error functional $J(\cdot)$.

2.1 The solution approach

The approximation of problem (2.1) by a Galerkin method has been described in a functional analytic setting. For actually solving it on a computer, we have to convert the *discrete* problem (2.2) into an algebraic equation. To this end, we choose a basis $\{\varphi^\nu, \nu = 1, \ldots, N = \dim V_h\}$ of the subspace $V_h \subset V$ (for instance the standard "nodal function basis" in a finite element scheme) and seek for a solution in the form

$$u_h = \sum_{\nu=1}^{N} x_\nu \varphi^\nu + \hat{u}_h.$$

Inserting this ansatz into equation (2.2) results in a nonlinear algebraic system

$$A(u_h; \varphi^\nu) = F(\varphi^\nu), \quad \nu = 1, \ldots, N. \tag{2.9}$$

for the unknown coefficient vector $x = \{x_\nu\}_{\nu=1}^N$. This may be solved for instance by a Newton iteration. In the context of adaptive discretization, this solution process is usually coupled with successive mesh refinement. The resulting *nested* algorithm reads as follows.

Let a desired error tolerance TOL or a maximum mesh complexity $N_{\max}$ be given. Starting from a coarse initial mesh $\mathcal{T}_0$, a hierarchy of successively refined meshes $\mathcal{T}_i$, $i \geq 1$, and corresponding finite element spaces $V_i \subset \hat{V}_i$ is generated as follows:

(0) *Initialisation* $i = 0$: Compute an initial approximation $u_0 \in \hat{V}_0$.

(i) *Defect correction iteration:* For $i \geq 1$, start with $u_i^{(0)} := u_{i-1} \in \hat{V}_i$.

(ii) *Iteration step:* For $j \geq 0$ evaluate the defect

$$(d_i^{(j)}, \varphi) := F(\varphi) - A(u_i^{(j)}; \varphi), \quad \varphi \in V_i. \tag{2.10}$$

Choose a suitable approximation $\tilde{A}'(u_i^{(j)}; \cdot, \cdot)$ to $A'(u_i^{(j)}; \cdot, \cdot)$ (with good stability and solubility properties) and compute a correction $v_i^{(j)} \in V_i$ from the linear equation

$$\tilde{A}'(u_i^{(j)}; v_i^{(j)}, \varphi) = (d_i^{(j)}, \varphi) \quad \forall \varphi \in V_i. \tag{2.11}$$

For this, Krylov-space or multigrid methods are employed using the hierarchy of meshes $\{\mathcal{T}_i, \dots, \mathcal{T}_0\}$. Then, update $u_i^{(j+1)} = u_i^{(j)} + \lambda_i v_i^{(j)}$, with some relaxation parameter $\lambda_i \in (0, 1]$, set $j := j+1$ and go back to (2). This process is repeated until a limit $\tilde{u}_i \in \hat{V}_i$, is reached with a certain prescribed accuracy.

(iii) *Error estimation:* Accept $u_i := \tilde{u}_i$ as the solution on mesh $\mathcal{T}_i$ and solve the discrete linearised dual problem

$$z_i \in V_i: \quad A'(u_i; \varphi, z_i) = J(\varphi) \quad \forall \varphi \in V_i, \tag{2.12}$$

and evaluate the a posteriori error estimate

$$|J(e_i)| \approx \eta(u_i). \tag{2.13}$$

For controlling the reliability of this bound, i.e. the accuracy in the determination of the dual solution z, one may check whether the change $\|z_i - z_{i-1}\|$ is sufficiently small; if this is not the case, additional global mesh refinement is advisable.

If $\eta(u_i) \leq TOL$ or $N_i \geq N_{\max}$, then stop. Otherwise, cell-wise mesh adaptation yields the new mesh $\mathcal{T}_{i+1}$. Then, set $i := i+1$ and go back to (i).

We note that the evaluation of the a posteriori error estimate (2.13) involves only the solution of *linearised* problems. Hence, the whole error estimation may amount only to a relatively small fraction of the total cost for the solution process. This has to be compared to the usually much higher cost when working on non-optimised meshes.

In using the a posteriori error estimate (2.13), it is assumed that the exact discrete solution $u_i \in V_i$ on mesh $\mathcal{T}_i$ is available. This asks for estimation of the unavoidable iteration error $\tilde{u}_i - u_i$ and its effect on the accuracy of the estimator for the discretization error. This can be achieved in the case of a Galerkin finite-element multigrid iteration by exploiting the projection properties of the combined scheme (see [3]).

3 Practical realization of error control

3.1 Evaluation of a posteriori error estimates

Next, we want to discuss how the abstract a posteriori error representation (2.8) is evaluated and used in mesh refinement in practice. To this end, we consider the usual scalar Laplace equation

$$-\Delta u = f \quad \text{in } \Omega, \quad u_{|\partial\Omega} = \hat{u}, \tag{3.1}$$

on a polygonal domain $\Omega \subset \mathbb{R}^2$ with boundary $\partial\Omega$, and some prescribed right-hand side $f \in L^2(\Omega)$ and sufficiently smooth boundary value $\hat{u}$.

Before proceeding, let us fix some notation. We will denote by $(\cdot, \cdot)_D$ the L^2 scalar product on some domain D and by $\|\cdot\|_D$ the corresponding norm; in the case $D = \Omega$, the subscript D is usually omitted. Further, we use the standard notation $L^2(D)$ for the Lebesgue space as well as $H^1(D)$ and $H_0^1(D)$ for the first-order Sobolev spaces over D.

For the following, we set $\hat{V} := H^1(\Omega)$ and $V := H_0^1(\Omega)$. Then, the variational formulation of (3.1) seeks $u \in V + \hat{u}$ satisfying

$$A(u; \varphi) := (\nabla u, \nabla \varphi) = (f, \varphi) \quad \forall \varphi \in V. \tag{3.2}$$

We consider the approximation of (3.2) by a standard Galerkin finite element method. To fix ideas, let us consider only piecewise bilinear shape functions defined on quasi regular rectangular meshes $\mathcal{T}_h = \{K\}$ consisting of non-degenerate cells K (rectangles) with edges denoted by Γ, as described in the standard finite element literature; see, e.g., Johnson [17]. The local mesh width is denoted by $h_K = \text{diam}(K)$ and $h := \max_{K \in \mathcal{T}_h} h_K$. In order to facilitate local mesh refinement and coarsening, we allow the cells in the refinement zone to have nodes which lie on faces of neighbouring cells (Fig. 3.1). The degrees of freedom corresponding to such "hanging nodes" are eliminated from the system by interpolation enforcing global conformity (i.e., continuity across inter-element boundaries) for the finite element functions.

isopar$\{1, x_1, x_2, x_1 x_2\}$

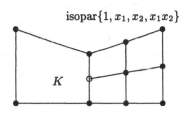

Fig. 3.1. Quadrilateral mesh patch with a "hanging node"

In this setting, the error representation (2.8) can be developed into a more concrete form. Since the form $A(\cdot;\cdot)$ is linear in this case, the nonlinear remainder term vanishes. The dual problem corresponding to a (linear) error functional $J(\cdot)$ seeks a $z \in V$ such that

$$(\nabla\varphi, \nabla z) = J(\varphi) \quad \forall\varphi \in V. \tag{3.3}$$

We do not want to assume that $\hat{u}_h = \hat{u}$, so that $e = u - u_h \notin V$. Hence, we cannot directly take $\varphi = e$ as test function in (3.3). We deal with this complication by rewriting (3.3) in the form

$$(\nabla\varphi, \nabla z) - (\varphi, \partial_n z)_{\partial\Omega} = J(\varphi), \quad \varphi \in \hat{V}. \tag{3.4}$$

Taking here $\varphi = e$ yields

$$J(e) = (\nabla e, \nabla z) - (\hat{u} - \hat{u}_h, \partial_n z)_{\partial\Omega}.$$

On the first term on the right, we can now use the Galerkin orthogonality and cell-wise integration by parts, to obtain

$$
\begin{aligned}
(\nabla e, \nabla z) &= (\nabla e, \nabla(z - \varphi_h)) \\
&= \sum_{K \in \mathcal{T}_h} \left\{ (f + \Delta u_h, z - \varphi_h)_K - (\partial_n u_h, z - \varphi_h)_{\partial K} \right\}, \\
&= \sum_{K \in \mathcal{T}_h} \left\{ (R(u_h), z - \varphi_h)_K + (r(u_h), z - \varphi_h)_{\partial K} \right\},
\end{aligned}
$$

with an arbitrary $\varphi_h \in V_h$. Here, the "equation residual" $R(u_h)$ and the "jump residual" $r(u_h)$ are cell-wise defined by

$$R(u_h)_{|K} := f + \Delta u_h, \quad r(u_h)_{|\Gamma} := \begin{cases} -\frac{1}{2}[\partial_n u_h], & \text{if } \Gamma \not\subset \partial\Omega, \\ -\partial_n u_h, & \text{if } \Gamma \subset \partial\Omega. \end{cases}$$

Here, $[\partial_n u_h]$ denotes the jump of $\partial_n u_h$ across the edge Γ. This error representation may be directly evaluated if a good approximation $\tilde{z} \sim z$ is available. This can be achieved by solving the dual problem either on a finer mesh or by patchwise higher-order interpolation of its approximation on the same mesh. For a systematic study of these alternatives, we refer to [9]. In order to derive cell error indicators which can be used in the mesh refinement process, we have to convert the error representation into an estimate. Using Hölder's inequality, we obtain the a posteriori error estimate

$$|J(e)| \leq \sum_{K \in \mathcal{T}_h} \left\{ \rho_K \omega_K + \rho_{\partial K} \omega_{\partial K} \right\} + |(\hat{u} - \hat{u}_h, \partial_n z)_{\partial\Omega}|, \tag{3.5}$$

with the cell residuals and corresponding weights defined by

$$\rho_K := \|R(u_h)\|_K, \quad \rho_{\partial K} := h_K^{-1/2}\|r(u_h)\|_{\partial K},$$

$$\omega_K := \|z - I_h z\|_K, \quad \omega_{\partial K} := h_K^{1/2}\|z - I_h z\|_{\partial K},$$

where $I_h z \in V_h$ is a suitable nodal interpolation of the dual solution z. Normally, the dual problem is solved by the same method as used for the primal problem resulting in an approximation $z_h \in V_h$. In this case, we replace the difference $z - I_h z$ in the error estimate (3.5) in terms of the local interpolation estimate

$$\max\left\{\|z - I_h z\|_K, h_K^{1/2}\|z - I_h z\|_{\partial K}\right\} \leq c_I h_K^2 \|\nabla^2 z\|_K, \qquad (3.6)$$

with an interpolation constant $c_I \sim 1$. In turn, the second derivative of the dual solution is replaced by a suitable second-order difference quotient of its computed approximation,

$$\|\nabla^2 z\|_K \approx \|\nabla_h^2 z_h\|_K. \qquad (3.7)$$

This results in the approximate a posteriori error estimate

$$|J(e)| \approx c_I \sum\nolimits_{K \in \mathcal{T}_h} h_K^2 \tilde{\rho}_K \tilde{\omega}_K + |(\hat{u} - \hat{u}_h, \partial_n z)_{\partial\Omega}|, \qquad (3.8)$$

where $\tilde{\rho}_K := \rho_K + \rho_{\partial K}$ and $\tilde{\omega}_K := \|\nabla_h^2 z_h\|_K$. In concrete applications the cell-wise parts of the boundary term in the estimator are smaller than the interior indicators. Nevertheless, they may lead to mesh refinement along the inflow boundary as can be seen in the flow examples presented below.

We recall the interpretation of this a posteriori error estimate. On each cell, we have an "equation residual" $R(u_h)$ and a "flux residual" $r(u_h)$, the latter one expressing smoothness of the discrete solution. Both residuals can easily be evaluated. They are multiplied by the weights $\tilde{\omega}_K$ which provide quantitative information about the impact of these cell-residuals on the error $J(e)$ in the target quantity (*sensitivity factors*).

3.2 Mesh adaptation strategies

Now, we want to discuss some popular strategies for mesh adaptation based on an a posteriori error estimate of the form

$$|J(e)| \leq \eta := \sum\nolimits_{K \in \mathcal{T}_h} \eta_K, \qquad (3.9)$$

with certain cell-error indicators $\eta_K = \eta_K(u_h)$ obtained on the current mesh $\mathcal{T}_h$. Suppose that a tolerance TOL for the error $J(e)$ or a maximum number $N_{\max}$ of mesh cells have been prescribed. Starting from an approximate solution $u_h \in V_h + \hat{u}_h$ obtained on the current mesh $\mathcal{T}_h$ the mesh adaptation may be organised by one of the following strategies.

- *Error balancing strategy:* Cycle through the mesh and equilibrate the local error indicators according to $\eta_K \approx \mathrm{TOL}/N$ with $N := \#\{K \in \mathcal{T}_h\}$. This leads eventually to $\eta \approx TOL$, but requires iteration with respect to N.

- *Fixed fraction strategy:* Order cells according to the size of η_K and refine a certain percentage (say 30%) of cells with largest η_K (or those which make up 30% of the estimator value η) and coarsen those cells with smallest η_K. By this strategy, we may achieve a prescribed rate of increase of N.

- *Mesh optimization strategy:* Use the (assumed) representation

$$\eta := \sum_{K \in \mathcal{T}_h} \eta_K \approx \int_\Omega h(x)^2 \Phi(x)\, dx \qquad (3.10)$$

for directly generating a formula for an optimal mesh-size distribution:

$$h_{opt}(x) = \left(\frac{W}{N_{\max}}\right)^{1/2} \Phi(x)^{-1/4}, \quad W := \int_\Omega \Phi(x)^{1/2}\, dx. \qquad (3.11)$$

Here, we think of a smoothly distributed mesh-size function $h(x)$ such that $h_{|K} \approx h_K$. The existence of such a representation with an h-independent error density function $\Phi(x) = \Phi(u(x), z(x))$ can be rigourously justified only under very restrictive conditions but is generally supported by computational experience (for details see [9]).

4 Application to incompressible fluid flow

As the first application, we consider a viscous incompressible Newtonian fluid flow modelled by the (stationary) Navier–Stokes equations

$$-\nu\Delta v + v\cdot\nabla v + \nabla p = f, \qquad \nabla\cdot v = 0, \qquad (4.1)$$

for the velocity v and the pressure p in a bounded domain $\Omega \subset \mathbb{R}^2$. Here, $\nu >$ is the normalised viscosity (density $\rho \equiv 1$), and the volume force is assumed as $f \equiv 0$. At the boundary $\partial\Omega$, the usual non-slip condition is imposed along rigid parts together with suitable inflow and free-stream outflow conditions,

$$v|_{\Gamma_{\text{rigid}}} = 0, \quad v|_{\Gamma_{\text{in}}} = \hat{v}, \quad \nu\partial_n v - pn|_{\Gamma_{\text{out}}} = 0.$$

In the following, vector functions are also denoted by normal type and no distinction is made in the notation of the corresponding inner products and norms.

As an example, we consider the flow around the cross section of a cylinder with surface S in a channel with a narrowed outlet (see Fig. 4.1). Here, a quantity of physical interest is for example the "drag coefficient" defined by

$$J_{\text{drag}}(v,p) := \frac{2}{\hat{V}^2 D} \int_S n \cdot \sigma(v,p) \psi \, ds, \qquad (4.2)$$

where $\psi := (0,1)^T$, $\sigma(v,p) = \frac{1}{2}\nu(\nabla v + \nabla v^T) + pI$ *is* the stress acting on the cylinder, D is its diameter, and $\hat{V} := \max|\hat{v}|$ is the reference velocity. In this example, the Reynolds number is $\text{Re} = \hat{V}^2 D/\nu = 50$, such that the flow is stationary.

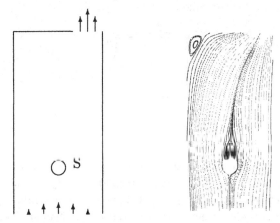

Fig. 4.1. Configuration and streamline plot of the "flow around a cylinder"

The variational formulation of (4.1) uses the function spaces

$$\hat{V} := L \times \hat{H}, \quad V := L \times H \subset \hat{V},$$

where $L := L^2(\Omega)$, $\hat{H} := H^1(\Omega)^2$, and $H := \{v \in \hat{H}, \; v_{|\Gamma_{\text{in}} \cup \Gamma_{\text{rigid}}} = 0\}$. In the special case $\Gamma_{\text{out}} = \emptyset$, we set $L := \{q \in L^2(\Omega), \; \int_\Omega q \, dx = 0\}$. For $u = \{p,v\}$, $\varphi = \{q,w\} \in \hat{V} \times \hat{V}$, we define the bilinear form

$$A(u;\varphi) := (\nabla u, \nabla w) + (v \cdot \nabla v, w) - (p, \nabla \cdot w) + (q, \nabla \cdot v)$$

and seek $u = \{p,v\} \in V + \{0,\hat{v}\}$, such that

$$A(u;\varphi) = (f,w) \quad \forall \varphi = \{q,w\} \in V. \qquad (4.3)$$

For discretising this problem, we use a finite element method based on the quadrilateral Q_1/Q_1-Stokes element with globally continuous (piece-wise isoparametric) bilinear shape functions for both unknowns, pressure and velocity. As described before, we allow "hanging" nodes while the corresponding unknowns are eliminated by linear interpolation. The corresponding finite element subspaces are denoted by

$$L_h \subset L, \quad \hat{H}_h \subset \hat{H}, \quad H_h \subset H, \quad \hat{V}_h := L_h \times \hat{H}_h, \quad V_h := L_h \times H_h,$$

and $\hat{v}_h \in \hat{H}_h$ is a suitable interpolation of the boundary function $\hat{v}$. This construction is oriented by the situation of a polygonal domain Ω for which the boundary $\partial\Omega$ is exactly matched by the mesh domain $\Omega_h := \cup\{K \in \mathcal{T}_h\}$. In the case of a general curved boundary (as in the above flow example) some standard modifications are necessary which are omitted here for the sake of brevity.

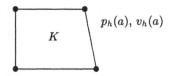

$p_h(a),\ v_h(a)$

Fig. 4.2. The Q_1/Q_1-Stokes element

In order to obtain a stable discretization of (4.3) in these spaces with "equal-order interpolation" of pressure and velocity, we use the least-squares technique proposed in [16]. Following [15] a similar approach is employed for stabilising the convective term. The Navier–Stokes system can be written in vector form for the unknown $u = \{p, v\}$ like

$$A(u) := \begin{bmatrix} -\nu\Delta v + v{\cdot}\nabla v + \nabla p \\ \nabla{\cdot}v \end{bmatrix} = \begin{bmatrix} f \\ 0 \end{bmatrix} =: F.$$

The operator $A(\cdot)$ has a differential at u which acts on elements $\varphi = \{q, w\} \in \hat{V}$ like

$$A'(u)\varphi := \begin{bmatrix} -\nu\Delta w + v{\cdot}\nabla w + w{\cdot}\nabla v + \nabla q \\ \nabla{\cdot}w \end{bmatrix}.$$

For stabilising the formulation (4.3) and also for preconditioning in the linear defect equations (2.11), we use the following simplification of $A'(u)$ (suitable for low-order finite elements) which avoids the presence of the "reaction term" $w \cdot \nabla v = (\nabla v)^T w$:

$$\tilde{A}'(u)\varphi := \begin{bmatrix} v \cdot \nabla w + \nabla q \\ 0 \end{bmatrix}.$$

With this notation, we introduce the stabilised form

$$A_\delta(u_h; \varphi_h) := A(u_h; \varphi_h) + (A(u_h) - F, S(u_h)\varphi_h)_\delta,$$

with $S(u_h) := \tilde{A}'(u_h)$ and the mesh-dependent inner product and norm

$$(v, w)_\delta := \sum_{T \in \mathcal{T}_h} \delta_K (v, w)_K, \quad \|v\|_\delta = (v, v)_\delta^{1/2}.$$

The stabilisation parameter is chosen according to

$$\delta_K = \alpha \left(\nu h_K^{-2}, \beta \, |v_h|_{K;\infty} h_K^{-1} \right)^{-1}, \quad \delta := \max_{K \in \mathcal{T}_h} \delta_K, \qquad (4.4)$$

with the heuristic choice $\alpha = \frac{1}{12}$, $\beta = \frac{1}{6}$. Now, in the discrete problems, we seek $\{p_h, v_h\} \in V_h + \hat{v}_h$, such that

$$A_\delta(u_h; \varphi_h) = F(\varphi_h) \quad \forall \varphi_h \in V_h. \qquad (4.5)$$

This approximation is fully consistent in the sense that the exact solution $u = \{p, v\}$ also satisfies (4.5). This implies again Galerkin orthogonality (2.3) for the error $e = \{e^p, e^v\} := \{p - p_h, v - v_h\}$ with respect to the form $A_\delta(\cdot; \cdot)$. We note that there are other possible choices of the stabilising operator in the form $A_\delta(\cdot; \cdot)$, e.g., $S = -\tilde{A}'(u)^*$ (see the compressible flow example below). Usually the adjustment of the mesh $\mathcal{T}_h = \{K\}$ is oriented by the size of certain local "error indicators" $\eta_K = \eta_K(v, p)$ which can cheaply be obtained from the computed solution $\{v_h, p_h\}$. Common candidates of *heuristic* indicators are

- *Vorticity:* $\eta_K := \|\nabla \times v_h\|_K$,
- *Pressure-gradient:* $\eta_K := \|\nabla p_h\|_K$,
- *"Energy-error":* $\eta_K := \|R_h\|_K + h_K^{1/2} \|[\partial_n v_h]\|_{\partial K} + \|\nabla \cdot v_h\|_K$,
  with the "equation residual" $R_h := -\nu \Delta v_h + v_h \cdot \nabla v_h + \nabla p_h$.

The vorticity and the pressure-gradient indicators measure the "smoothness" of the computed solution $\{v_h, p_h\}$ while the "energy-error" indicator additionally contains information concerning local conservation of mass and momentum.

In Fig. 4.3, we compare the results for approximating the drag coefficient $J_{\text{drag}}(v, p)$ on meshes which are constructed by using the different error indicators with that obtained on uniformly refined meshes. It turns out that all the above indicators perform equally weakly. A better result

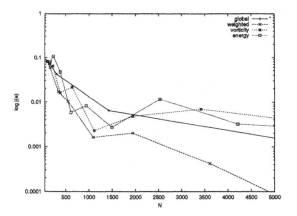

Fig. 4.3. Mesh efficiency obtained by global uniform refinement ("global" +), the weighted indicator ("weighted" ×), the vorticity indicator ("vorticity" *), and the energy indicator ("energy" □).

is obtained by using a "weighted indicator", $\eta_K := \rho_K \cdot \omega_K$ in which the three residual terms contained in the "energy-error" indicator are multiplied by weights ω_K which are obtained from the solution of a global dual problem (generalized Green function) and represent the sensitivity of the error $J_{\mathrm{drag}}(v, p) - J_{\mathrm{drag}}(v_h, p_h)$ with respect to changes of the residuals ρ_K under mesh refinement. The basis of this approach is provided by the Theorem 1 applied to the present concrete situation. Here, for simplicity, the explicit dependence of the stabilisation parameter δ on the solution is neglected.

Theorem 2 *For a (linear) functional $J(\cdot)$ let $z = \{z^p, z^v\} \in V$ be the solution of the linearised dual problem*

$$A'_\delta(u_h; \varphi, z) = J(\varphi) \quad \forall \varphi \in V. \tag{4.6}$$

Then, there holds the a posteriori error estimate

$$|J(e)| \leq \sum_{K \in \mathcal{T}_h} \left\{ \sum_{i=1}^{3} \rho_K^{(i)} \omega_K^{(i)} + \ldots \right\} + |R(u, u_h; e, e, z)|, \tag{4.7}$$

where

$$
\begin{aligned}
\rho_K^{(1)} &= \|R^p(u_h)\|_K, & \omega_K^{(1)} &= \|z^p - z_h^p\|_K, \\
\rho_K^{(2)} &= \|R^v(u_h)\|_K, & \omega_K^{(3)} &= \|z^v - z_h^v\|_K + \delta_K \|\tilde{A}'(u_h)(z - z_h)\|_K, \\
\rho_K^{(3)} &= \|r^v(u_h)\|_{\partial K}, & \omega_K^{(3)} &= \|z^v - z_h^v\|_{\partial K},
\end{aligned}
$$

with the cell and edge residuals

$$R^v(u_h)_{|K} := f + \nu \Delta v_h - v_h \cdot \nabla v_h - \nabla p_h, \quad R^p(u_h)_{|K} := \nabla \cdot v_h,$$

$$r^v(u_h)_{|\Gamma} := \begin{cases} -\frac{1}{2}[\nu \partial_n v_h - p_h n], & \text{if } \Gamma \not\subset \partial\Omega, \\ -(\nu \partial_n v_h - p_h n), & \text{if } \Gamma \subset \partial\Omega, \end{cases}$$

with $[\cdot]$ *denoting again the jump across an interior edge* Γ, *and the nodal interpolant* $z_h = I_h z \in V_h$. *The dots* " $\dots$ " *in (4.7) stand for additional terms measuring the errors in approximating the inflow data and the curved cylinder boundary which are neglected. The remainder term can be bounded by*

$$|R(u, u_h; e, e, z)| \leq 2\|e^v\|\|\nabla e^v\|\|z^v\|_\infty + \mathcal{O}(\delta\|e^v\|). \tag{4.8}$$

Proof Neglecting the errors due to the approximation of the inflow boundary data $\hat{v}$ as well as the curved cylinder boundary S, we have to evaluate the residual $\langle \rho(u_h), \varphi \rangle := F(u_h, \varphi) - A_\delta(u_h, \varphi)$ for $\varphi = \{q, w\} \in V$. By definition, there holds

$$\langle \rho(u_h), \varphi \rangle = (f, w) - \nu(\nabla v_h, \nabla w) - (v_h \cdot \nabla v_h, w) + (p_h, \nabla \cdot w)$$
$$- (q, \nabla \cdot u_h) - (A(u_h) - F, \tilde{A}'(u_h)\varphi)_\delta.$$

Splitting the integrals into their contributions from each cell $K \in \mathcal{T}_h$ and integrating cell-wise by parts yields analogously as in deriving (3.5):

$$\langle \rho(u_h), \varphi \rangle = \sum_{K \in \mathcal{T}_h} \Big\{ (R(u_h), w)_K + (r(u_h), w)_{\partial K} + (q, \nabla \cdot v_h)_K$$
$$+ \delta_K(R(u_h), v_h \cdot \nabla w + \nabla q)_K \Big\}.$$

From this, we obtain the error estimate (4.7) if we set $\varphi := z - I_h z$. It remains to estimate the remainder term $R(u, u_h; e, z)$. The first derivative of $A_\delta(\cdot; \cdot)$ has for arguments $u = \{p, v\}$, $\varphi = \{q, w\} \in V$ and $z = \{z^p, z^v\} \in V$ the explicit form

$$A'_\delta(u; \varphi, z) = (A'(u)\varphi, z) + (A'(u)\varphi, \tilde{A}'(u)z)_\delta + (A(u), \tilde{A}''(u)\varphi z)_\delta,$$

with $A'(u)\varphi$ and $\tilde{A}'(u)z$ defined as above. Further, we note that

$$\tilde{A}''(u)\varphi z := \begin{bmatrix} w \cdot \nabla z^v \\ 0 \end{bmatrix}, \quad A''(u)\psi\varphi := \begin{bmatrix} \phi \cdot \nabla w + w \cdot \nabla \phi \\ 0 \end{bmatrix}.$$

Since $\tilde{A}'''(u) \equiv 0$, the second derivative of $A_\delta(\cdot; \cdot)$ has for arguments

$u = \{p, v\}$, $\psi = \{\chi, \phi\}$, $\varphi = \{q, w\} \in V$ and $z = \{z^p, z^v\} \in V$ the form

$$
\begin{aligned}
A''_\delta(u; \psi, \varphi, z) &= (\phi \cdot \nabla w + w \cdot \nabla \phi, z^v) + (A''(u)\psi\varphi, \tilde{A}'(u)z)_\delta \\
&\quad + 2\left(A'(u)\varphi, \tilde{A}''(u)\psi z\right)_\delta \\
&= (\phi \cdot \nabla w + w \cdot \nabla \phi, z^v) + (\phi \cdot \nabla w + w \cdot \nabla \phi + \nabla q, v \cdot \nabla z^v)_\delta \\
&\quad + 2\left(-\nu\Delta w + v \cdot \nabla w + w \cdot \nabla v, \phi \cdot \nabla z^v\right)_\delta .
\end{aligned}
$$

From this, we easily infer the proposed bound for the remainder term by applying Hölder's inequality.   □

In the estimate (4.7) the additional terms representing the errors in approximating the inflow data and the curved boundary component S are neglected; they can be expected to be small compared to the other residual terms. The bounds for the dual solution $\{z^p, z^v\}$ are obtained computationally by replacing the unknown velocity v in in the derivative form $A'(v; \cdot, \cdot)$ by its approximation v_h and solving the resulting linearised problem on the same mesh. From this approximate dual solution $\tilde{z}_h$, patchwise biquadratic interpolations are taken to approximate z in evaluating the weights $I_h^{(2)} \tilde{z}_h - z_h \approx z - z_h$. This frees us from choosing any interpolation constants.

The quantitative results of Fig. 4.3 and the corresponding meshes shown in Fig. 4.4 confirm the superiority of the weighted error indicator in computing local quantities.

We note that in our computation the drag coefficient has been evaluated from the formula

$$
J_{\mathrm{drag}}(v, p) := \frac{2}{\hat{V}^2 D} \int_\Omega \left\{ \sigma(p, v)\nabla\bar{\psi} + \nabla \cdot \sigma(p, v)\bar{\psi} \right\} dx , \tag{4.9}
$$

where $\bar{\psi}$ is an extension of the directional vector $\psi := (0, 1)^T$ from S to Ω with support along S. By integration by parts, one sees that this definition is independent of the choice of $\bar{\psi}$ and that it is equivalent to the original one (4.2) as a contour integral. However, on the discrete level the two formulations differ. In fact, computational experience shows that formula (4.9) yields significantly more accurate approximations of the drag coefficient (see [13] and [4]).

5 Application to "low-Mach-number" compressible flow

As the second application, we consider low-Mach-number compressible gas flow with density variations due to temperature gradients. Such conditions often occur in chemically reactive flows and are characterised

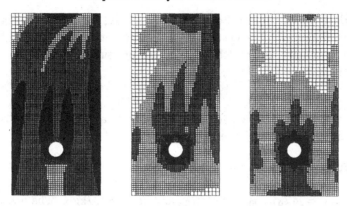

Fig. 4.4. Meshes with about 5,000 cells obtained by the vorticity indicator (left), the "energy" indicator (middle), and the weighted indicator (right).

by hydrodynamically incompressible behaviour . For the application of our approach to such problems see [10], [7], [6], and [11]. As a typical example (without chemistry), we choose a 2D benchmark "heat-driven cavity" (for details see Fig. 5.1 and [19] or [5]). Here, the flow, confined to a square box with side length $L = 1$, is driven by a temperature difference $T_h - T_c = 2\varepsilon T_0$ between the left ("hot") and the right ("cold") wall under the action of gravity g in the y-direction.

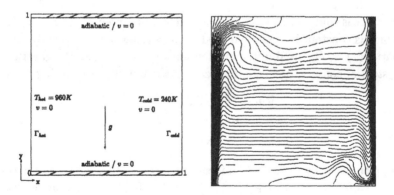

Fig. 5.1. Configuration of the heat-driven cavity problem and plot of computed temperature isolines.

For the viscosity μ Sutherland's law is used,

$$\mu(T) \;=\; \mu^* \Big(\frac{T}{T^*}\Big)^{1/3} \frac{T^* + S}{T + S},$$

with the Prandtl number $\mathrm{Pr}=0.71$, $T^* = 273\,K$, $\mu^* = 1.68\cdot10^{-5}\mathrm{kg/ms}$, and $S := 110.5\,K$. Further, the heat conductivity is $\kappa(T) = \mu(T)/\mathrm{Pr}$. In the stationary case the thermodynamic pressure is defined by

$$P_{\mathrm{th}} = P_0\Big(\int_\Omega T_0^{-1}\,dx\Big)\Big(\int_\Omega T^{-1}\,dx\Big)^{-1},$$

where $T_0 = 600\,K$ is a reference temperature and $P_0 = 101,325\,\mathrm{Pa}$. Accordingly, the Rayleigh number is determined by

$$\mathrm{Ra} = \mathrm{Pr}\,g\Big(\frac{\rho_0 L}{\mu_0}\Big)^2\frac{T_h-T_c}{T_0} \approx 10^6, \qquad \varepsilon = \frac{T_h-T_c}{T_h+T_c} = 0.6,$$

where $\mu_0 := \mu(T_0)$, $\rho_0 := P_0/RT_0$, and $R = 287\,\mathrm{J/kgK}$. The boundary conditions are "no slip" for the velocity along the whole boundary, adiabatic (Neumann) boundary conditions for the temperature along the upper and lower wall and Dirichlet conditions for the temperature along the left and right hand walls which are represented by a function $\hat{T}$.

In this benchmark, one of the quantities to be computed is the average Nusselt number along the cold wall:

$$J(u) := c\int_{\Gamma_{\mathrm{cold}}} \kappa\partial_n T\,ds, \qquad c := \frac{\mathrm{Pr}}{2\mu_0 T_0\varepsilon}.$$

The mathematical model is the set of the (stationary) compressible Navier–Stokes equations in the so-called "low-Mach-number approximation" due to the low speed of the resulting flow. Accordingly, the total pressure is split like $P(x) = P_{\mathrm{th}}+p(x)$ into a thermodynamical part P_{th} which is constant in space and used in the gas law, and a hydrodynamic part $p(x) \ll P_{\mathrm{th}}$ used in the momentum equation. Then, the governing system of conservation equations can be written in the following form:

$$\nabla\cdot v - T^{-1}v\cdot\nabla T = 0,$$
$$\rho v\cdot\nabla v + \nabla\cdot\tau + \nabla p = (\rho-\rho_0)\,g, \qquad (5.1)$$
$$\rho v\cdot\nabla T - \nabla\cdot(\kappa\nabla T) = f_T,$$

supplemented by the law of an ideal gas $\rho = P_{\mathrm{th}}/RT$. The stress tensor is given by $\tau = -\mu\{\nabla v+(\nabla v)^T-\frac{2}{3}(\nabla\cdot v)I\}$. In the present model case there are no heat sources (e.g., due to chemical reactions), i.e., $f_T=0$. The variational formulation of (5.1) uses the following semi-linear form defined for triples $u = \{p,v,T\}$, $\varphi = \{q,w,\xi\}$:

$$A(u;\varphi) := (\nabla\cdot v - T^{-1}v\cdot\nabla T, q) + (\rho v\cdot\nabla v, w) - (\tau, \nabla w) - (p, \nabla\cdot w)$$
$$- (p, \nabla\cdot w) - (\rho g, w) + (\rho v\cdot\nabla T, \xi) + (\kappa\nabla T, \nabla\xi).$$

Further, we define the functional

$$F(\varphi) := -(\rho_0 g, w).$$

The natural solution spaces are $\hat{V} := L^2(\Omega)/\mathbb{R} \times H^1(\Omega)^2 \times H^1(\Omega)$ and $V := L^2(\Omega)/\mathbb{R} \times H_0^1(\Omega)^2 \times H_0^1(\Gamma_y; \Omega)$, where $H_0^1(\Gamma_y; \Omega) := \{\xi \in H^1(\Omega), \xi_{|x \in \{0,L\}} = 0\}$. With this notation, the variational form of (5.1) seeks $u = \{p, v, T\} \in V + \hat{u}$, with $\hat{u} = \{0, 0, \hat{T}\}$, satisfying

$$A(u; \varphi) = F(\varphi) \quad \forall \varphi \in V, \tag{5.2}$$

where the density ρ is considered as a (nonlinear) coefficient determined by the temperature through the equation of state $\rho = P_{th}/RT$.

The discretization of the system (5.2) uses again the continuous Q_1-finite element for all unknowns and employs least-squares stabilisation for the velocity-pressure coupling as well as for the transport terms. We do not state the corresponding discrete equations since they have an analogous structure as already seen in the preceding section for the incompressible Navier–Stokes equations. The derivation of the related (linearised) dual problem and the resulting a posteriori error estimates follows the same line of argument. For economy reasons, we do not use the full Jacobian of the coupled system in setting up the dual problem, but only include its dominant parts. The same simplification is used in the nonlinear iteration process. For details, we refer to [7], [11], and particularly the recent paper [6].

The discrete problems seek $u_h = \{p_h, v_h, T_h\} \in V_h + \hat{u}_h$, satisfying

$$A_\delta(u_h; \varphi_h) = F(\varphi_h) \quad \forall \varphi_h \in V_h, \tag{5.3}$$

with the stabilised form

$$A_\delta(u_h; \varphi_h) := A(u_h; \varphi_h) + (A(u_h), S(u_h)\varphi_h).$$

As on the continuous level the discrete density is determined by the temperature through the equation of state $\rho_h := P_{th}/RT_h$. Here, the operator $A(u_h)$ is the generator of the form $A(u_h; \cdot)$, and the operator $S(u_h)$ in the stabilisation term is chosen according to $S(u_h) := -\tilde{A}'(u_h)^*$ where $\tilde{A}'(u_h)$ represents the differential part of $A'(u_h) = \tilde{A}'(u_h) + A_0'(u_h)$, while $A_0'(u_h)$ contains all zero-order terms. Accordingly, we have

$$S(u_h) := \begin{bmatrix} 0 & \mathrm{div} & 0 \\ \nabla & \rho_h v_h \cdot \nabla + \nabla \cdot \mu \nabla & 0 \\ -T_h^{-1} v_h \cdot \nabla & 0 & \rho_h v_h \cdot \nabla + \nabla \cdot \kappa \nabla \end{bmatrix}.$$

We introduce the following notation for the equation residuals of the solution $u_h = \{p_h, v_h, T_h\}$ of (5.3):

$$
\begin{aligned}
R^p(u_h) &= \nabla \cdot v_h - T_h^{-1} v_h \cdot \nabla T_h \,, \\
R^v(u_h) &= \rho_h v_h \cdot \nabla v_h - \nabla \cdot (\mu \nabla v_h) + \nabla p_h + (\rho_0 - \rho_h) g \,, \\
R^T(u_h) &= \rho_h v_h \cdot \nabla T_h - \nabla \cdot (\kappa \nabla T_h) - f_T \,.
\end{aligned}
$$

Then, the stabilising part in $A_\delta(\cdot\,;\cdot)$ can be written in the form

$$
\begin{aligned}
(A(u_h), S(u_h)\varphi_h)_\delta \;=\; &(R^p(u_h), \nabla \cdot w)_\delta \\
&+ (R^v(u_h), \nabla q + \rho_h v_h \cdot \nabla w + \nabla \cdot (\mu \nabla w))_\delta \\
&+ (R^T(u_h), \rho_h v_h \cdot \nabla \xi + \nabla \cdot (\kappa \nabla \xi) - T_h^{-1} v_h \cdot \nabla q)_\delta \,,
\end{aligned}
$$

for $\varphi = \{q, w, \xi\}$. These terms comprise stabilisation of the stiff pressure-velocity coupling in the low-Mach-number situation as well as stabilisation of transport in the momentum and energy equation. The parameters $\delta_K = \{\delta_K^p, \delta_K^v, \delta_K^T\}$ may be chosen differently in the three equations following rules analogous to (4.4). The stability of this discretization has been investigated in [5].

From Theorem 1, we obtain the following result by the same argument as used in Theorem 2. This analysis assumes for simplicity that viscosity μ and heat conductivity κ are determined by T_0.

Theorem 3 *For a (linear) functional $J(\cdot)$ let $z = \{z^p, z^v, z^T\} \in V$ be the solution of the linearised dual problem*

$$
A_\delta'(u_h; \varphi, z) = J(\varphi) \quad \forall \varphi \in V \,. \tag{5.4}
$$

Then, there holds the a posteriori error estimate

$$
|J(e)| \leq \sum_{K \in \mathcal{T}_h} \left\{ \sum_{i=1}^{5} \rho_K^{(i)} \omega_K^{(i)} \right\} + |R(u, u_h; e, e, z)| \,, \tag{5.5}
$$

where

$$
\begin{aligned}
\rho_K^{(1)} &= \|R^p(u_h)\|_K, & \omega_K^{(1)} &= \|z^p - z_h^p\|_K + \delta_K^p \|S^p(u_h)(z^p - z_h^p)\|_K, \\
\rho_K^{(2)} &= \|R^v(u_h)\|_K, & \omega_K^{(2)} &= \|z^v - z_h^v\|_K + \delta_K^v \|S^v(u_h)(z^v - z_h^v)\|_K, \\
\rho_K^{(3)} &= \|r^v(u_h)\|_{\partial K}, & \omega_K^{(3)} &= \|z^v - z_h^v\|_{\partial K}, \\
\rho_K^{(4)} &= \|R^T(u_h)\|_K, & \omega_K^{(4)} &= \|z^T - z_h^T\|_K + \delta_K^T \|S^T(u_h)(z^T - z_h^T)\|_K, \\
\rho_K^{(5)} &= \|r^T(u_h)\|_{\partial K}, & \omega_K^{(5)} &= \|z^T - z_h^T\|_{\partial K},
\end{aligned}
$$

with the cell and edge residuals

$$R^p(u_h)_{|K} := \nabla \cdot v_h - T_h^{-1} v_h \cdot \nabla T_h \,,$$

$$R^v(u_h)_{|K} := \rho_h v_h \cdot \nabla v_h - \nabla \cdot (\mu \nabla v_h) + \nabla p_h + (\rho_0 - \rho_h)q \,,$$

$$r^v(u_h)_{|\Gamma} := \begin{cases} -\frac{1}{2}[\nu \partial_n v_h - p_h n], & \text{if } \Gamma \not\subset \partial\Omega, \\ -(\nu \partial_n v_h - p_h n), & \text{if } \Gamma \subset \partial\Omega, \end{cases}$$

$$R^T(u_h)_{|K} := \rho_h \cdot \nabla T_h - \nabla \cdot (\kappa \nabla T_h) \,,$$

$$r^T(u_h)_{|\Gamma} := \begin{cases} -\frac{1}{2}[\kappa \partial_n T_h], & \text{if } \Gamma \not\subset \partial\Omega, \\ -\kappa \partial_n T_h, & \text{if } \Gamma \subset \partial\Omega, \end{cases}$$

with $[\cdot]$ denoting again the jump across an interior edge Γ, and the nodal interpolant $z_h = I_h z \in V_h$. The remainder term can be bounded analogously as in Theorem 2 for incompressible flow.

In Fig. 5.2 and Table 1.1, we demonstrate the mesh efficiency of the "weighted" error indicator compared to a heuristic "energy-error" indicator similar to the one discussed in Section 4 for the incompressible flow case. A collection of refined meshes produced by the two error indicators is shown in Fig. 5.3. The superiority of the "weighted" indicator is particularly evident if higher solution accuracy is required. As usual in problems of such complex structure, the a posteriori bound $\eta(u_h)$ tends to over-estimate the true error but it appears to be "reliable" and the "effectivity index" $I_{\text{eff}} := \eta/e$ stays of the order $\mathcal{O}(1)$ as $N \to \infty$.

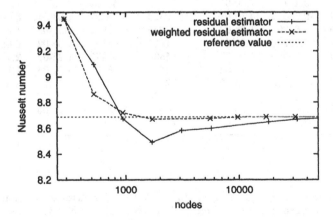

Fig. 5.2. Results of computing the Nusselt number by using the "residual indicator" (symbol "+") and the "weighted residual indicator" (symbol "×").

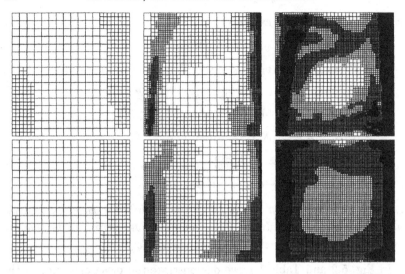

Fig. 5.3. Sequences of refined meshes obtained by the "energy-error" indicator (upper row, $N = 524, 5656, 58678$) and the weighted error estimator (lower row, $N = 523, 5530, 56077$).

Table 1.1. *Results for the computation of the Nusselt number by the "energy-error" indicator (left) and the weighted error indicator (right).*

N	$\langle Nu \rangle_c$	e	N	$\langle Nu \rangle_c$	e	η
289	-9.44783	7.6e-01	289	-9.44783	7.6e-01	7.9e-01
524	-9.09552	4.1e-01	523	-8.86487	1.8e-01	3.4e-01
945	-8.67201	1.5e-02	945	-8.71941	3.3e-02	1.5e-01
1708	-8.49286	1.9e-01	1717	-8.66898	1.8e-02	6.6e-02
3108	-8.58359	1.0e-01	5530	-8.67477	1.2e-02	2.1e-02
5656	-8.59982	8.7e-02	9728	-8.68364	3.0e-03	1.1e-02
18204	-8.64775	3.9e-02	17319	-8.68744	8.5e-04	6.3e-03
32676	-8.66867	1.8e-02	31466	-8.68653	6.9e-05	3.7e-03
58678	-8.67791	8.7e-03	56077	-8.68653	6.7e-05	2.1e-03
79292	-8.67922	7.4e-03	78854	-8.68675	1.5e-04	1.5e-03

References

[1] Ainsworth, M. and Oden, J.T. (1997). A posteriori error estimation in finite element analysis. *Comput. Methods Appl. Mech. Engrg.*, **142**, 1–88.

[2] Becker, R. (1995). An adaptive finite element method for the incompressible Navier–Stokes equations on time-dependent domains, Ph. D. Thesis, Tech. Rep. 95-44, SFB 359, Univ. Heidelberg.

[3] Becker, R. (1998). An adaptive finite element method for the Stokes equations including control of the iteration error, in *Proc. ENUMATH-97*, ed. H. G. Bock et al. (World Scient. Publ., Singapore), 609–620.

[4] Becker, R. (1998). Weighted error estimators for the incompressible Navier–Stokes equations Tech. Rep. 98-20, SFB 359, Univ. Heidelberg.

[5] Becker, R. and Braack, M. (2000). Fast computation of compressible flows at low Mach number with finite elements. Tech. Rep. SFB 359, Univ. Heidelberg.

[6] Becker, R., Braack, M., Rannacher, R. and Waguet, C. (1999). Fast and reliable solution of the Navier–Stokes equations including chemistry. *Computing & Visualization in Science* **2**, 107–122.

[7] Becker, R., Braack, M. and Rannacher, R. (1999). Numerical simulation of laminar flames at low Mach number with adaptive finite elements. *Combustion Theory and Modelling* **3**, 503–534.

[8] Becker, R. and Rannacher, R. (1998). Weighted a posteriori error control in FE methods, in *Proceedings of ENUMATH-97*, ed. H. G. Bock, et al., (World Scient. Publ., Singapore), 621–637.

[9] Becker, R. and Rannacher, R. (1996). A feed-back approach to error control in finite element methods: basic analysis and examples. *East-West J. Numer. Maths* **4**, 237–264.

[10] Braack, M. (1998). An adaptive finite element method for reactive flow problems, Dissertation, Univ. Heidelberg, 1998.

[11] Braack, M. and Rannacher, R. (1999), Adaptive finite element methods for low-Mach-number flows with chemical reactions, Lecture Series 1999-03, in *30th Computational Fluid Dynamics*, ed. H. Deconinck (von Karman Institute for Fluid Dynamics, Delgium).

[12] Eriksson, K., Estep, D., Hansbo, P. and Johnson, C. (1995). Introduction to adaptive methods for differential equations, *Acta Numerica* **4**, pp. 105–158,.

[13] Giles, M., Larsson, M., Levenstam, M. and Süli, E. (1997). Adaptive error control for finite element approximations of the lift and drag coefficients in viscous flow. Tech. Rep. NA-76/06, Oxford University Computing Laboratory, to appear in *SIAM J. Numer. Anal.*

[14] Houston, P., Rannacher, R. and Süli, E. (1999). A posteriori error analysis for stabilised finite element approximation of transport problems, to appear in *Compu. Meth. Appl. Sci. Engrg.*

[15] Hughes, T. J. R. and Brooks, A. N. (1982). Streamline upwind/Petrov Galerkin formulations for convection dominated flows with particular emphasis on the incompressible Navier–Stokes equation. *Comp. Math. Appl. Mech. Eng.* **32**, 199–259.

[16] Hughes, T. J. R., Franca, L. P. and Balestra, M. (1986). A new finite element formulation for computational fluid dynamics: V. Circumvent the Babuska–Brezzi condition: A stable Petrov–Galerkin formulation for the Stokes problem accommodating equal order interpolation, *Comp. Meth. Appl. Mech. Eng.* **59**, 89–99.

[17] Johnson, C. (1987). *Numerical Solution of Partial Differential Equations by the Finite Element Method* (Cambridge University Press, Cambridge).

[18] Johnson, C. (1993). A new paradigm for adaptive finite element methods, in *Proc. MAFELAP Conf., Brunel Univ. 1993* (John Wiley,

London).

[19] Le Quere, P. and Paillere,H. (2000). Modelling an simulation of natural convection flows with large temperature differences: a benchmark problem for low Mach number solvers, to appear in *Int. J. Thermal Sciences.*

[20] Rannacher, R. (1999). Error control in finite element computations, in *Proceedings of Summer School on Error Control and Adaptivity in Scientific Computing,* ed. H. Bulgak and C. Zenger, (Kluwer, Dordrecht), 247–278.

[21] Schäfer, M. and Turek, S. (1996). The benchmark problem "flow around a cylinder", in *Flow Simulation with High-Performance Computers,* ed. E.H. Hirschel, Notes Comput. Fluid Mech. (Vieweg, Stuttgart).

[22] Verfürth, R. (1996). *A Review of A Posteriori Error Estimation and Adaptive Mesh-Refinement Techniques* (Wiley/Teubner, New York–Stuttgart).

Newton's method and some complexity aspects of the zero-finding problem

Jean-Pierre Dedieu

MIP, Département de Mathématique
Université Paul Sabatier
31062 Toulouse Cedex 04, France
Email: dedieu@cict.fr

Abstract

Newton's iteration is a classical numerical method to find a zero of a nonlinear system of equations. In this paper we discuss recent advances in this subject: homogeneous and multihomogeneous systems, overdetermined and underdetermined systems are considered. We also discuss some complexity aspects of continuation methods using Newton's method.

1 Introduction

Newton's iteration is a classical numerical method to solve a system of nonlinear equations

$$f : \mathbf{E} \to \mathbf{F}$$

with $\mathbf{E}$ and $\mathbf{F}$ two real or complex Banach spaces. If $x \in \mathbf{E}$ is an approximation of a zero of this system then, Newton's method updates this approximation by linearizing the equation $f(y) = 0$ around x so that

$$f(x) + Df(x)(y - x) = 0.$$

When $Df(x)$ is an isomorphism we obtain the classical Newton's iteration

$$y = N_f(x) = x - Df(x)^{-1}f(x).$$

The idea of improving an estimate of the solution (here x) by addition of a correction term (here $-Df(x)^{-1}f(x)$) has been used by many cultures for millenia. The method appears in a general setting in Newton's tract *De analysi per aequationes numero terminorum infinitas* (1669)

45

where Newton considers polynomial equations and uses a linearization technique. The case of Kepler's equation (a nonpolynomial equation)

$$x - e\sin(x) = M$$

is described in *Philosophiae Naturalis Principia Mathematica* published in 1687. Two other names are associated to this method: Joseph Raphson and Thomas Simpson. In 1690 Raphson published *Analysis aequationum universalis* in which he presents a new method for solving polynomial equations. It was the same method as in Newton's tract but presented differently. Then came Simpson who described in his *Essays in Mathematicks*, 1740, "a new method for the solution of equations" using the "method of fluxions" i.e. the derivatives. More details are given by T. Ypma in [43], or in H. Goldstine's book, [19].

Three hundred and thirty years after Newton's tract we are still studying this powerful method. The reason is the wide range of problems we have to consider. Systems of nonlinear equations do not necessarily have the same number of equations and of unknowns, they sometimes have natural domains in spaces which are not necessarily vector spaces. In this paper we will emphasize these two problems: we will consider well-determined systems of equations ($\dim \mathbf{E} = \dim \mathbf{F}$) and also underdetermined systems ($\dim \mathbf{E} > \dim \mathbf{F}$) and overdetermined systems ($\dim \mathbf{E} < \dim \mathbf{F}$). We will also consider affine, homogeneous and multihomogeneous systems which associated zero sets are naturally defined in vector spaces, projective spaces and products of projective spaces respectively.

In the second part of this paper we will consider complexity aspects of the zero finding problem for a system of nonlinear equations when using Newton's method. The problem is to find a vector x in the attraction basin for this method i.e. such that Newton's sequence $x_k = N_f^{(k)}(x)$ converges to a zero for the nonlinear system f. We will consider two different strategies. One of them consists in testing many different starting points x until one of them is contained in a basin. This process is well understood for univariate polynomials: we will present recent results exploring this subject.

Another strategy uses an homotopy argument: if x_0 is an approximate zero for f (i.e. if Newton's sequence starting at x_0 converges quadratically to a zero for f) it is also an approximate zero for a nearby system g. Thus, using a predictor-corrector strategy (Newton's method is here used as a corrector operator) we are able to follow a curve of zeros given a curve of systems. We will present upper bounds and also lower bounds related to such a construction.

2 Newton's method: the affine case

For a system $f : \mathbf{E} \to \mathbf{F}$ with $\mathbf{E}$ and $\mathbf{F}$ two real or complex Banach spaces and when $Df(x)$ is an isomorphism Newton's map at x is defined by $N_f(x) = x - Df(x)^{-1}f(x)$. In such a case, fixed points for N_f correspond to zeros for f. Newton's sequence at x_0 is defined by $x_k = N_f^{(k)}(x_0)$. If this sequence converges to ζ then ζ is a fixed point for N_f and consequently a zero for f.

When $Df(x)$ is not an isomorphism the linearized equation $f(x) + Df(x)(y - x) = 0$ cannot be solved by premultiplying it by $Df(x)^{-1}$. In this case we use the Moore–Penrose inverse $Df(x)^\dagger$ instead of the classical inverse, and we define Newton's map by

$$y = N_f(x) = x - Df(x)^\dagger f(x).$$

Let $\mathbf{E}$ and $\mathbf{F}$ be two real Euclidean or complex Hermitian spaces. We recall that the Moore–Penrose inverse of a linear operator

$$A : \mathbf{E} \to \mathbf{F}$$

is the composition of two maps $A^\dagger = B \circ \Pi_{\mathrm{im}\,A}$ where $\Pi_{\mathrm{im}\,A}$ is the orthogonal projection in $\mathbf{F}$ onto $\mathrm{im}\,A$ and B is the inverse of the restriction

$$A|_{(\ker A)^\perp} : (\ker A)^\perp \to \mathrm{im}\,A.$$

We have $A^\dagger = (A^*A)^{-1}A^*$ when A is injective, $A^\dagger = A^*(AA^*)^{-1}$ when A is surjective, where A^* denotes the adjoint of A. Notice that $A^\dagger A = \Pi_{(\ker A)^\perp}$ and $AA^\dagger = \Pi_{\mathrm{im}\,A}$ in any case. When A is surjective, then $A^\dagger b$ is the minimum norm solution of the equation $Ax = b$ while when A is injective it is the least square solution of this equation i.e.

$$\|A(A^\dagger b) - b\|^2 = \min_{x \in \mathbf{E}} \|Ax - b\|^2.$$

The Moore–Penrose inverse can also be defined similarly when A is a bounded linear operator with closed range between two real or complex Hilbert spaces, see D. Luenberger, [25], Section 6.11, or A. Ben Israel and T. Greville, [4].

This Newton's map defined via the Moore–Penrose inverse of the derivative has been introduced for the first time by K. Gauss, 1809, [18], to solve the nonlinear least square problem. For this reason it is often

called the Newton–Gauss method. Gauss was considering overdetermined systems of equations and looking for least square solutions. The
Newton–Gauss method is obtained in taking the least-square solution of
the linearized equation $f(x) + Df(x)(y - x) = 0$. The first author to consider Moore–Penrose–Newton in a very general setting is A. Ben Israel
[3], Newton's method for underdetermined systems of equations is also
studied in Allgower and Georg [1], Beyn [2], Shub and Smale [36]. The
case of overdetermined systems is investigated in Dennis and Schnabel
[13], Dedieu and Shub [9], and the case of analytic maps with constant
rank derivatives in Dedieu and Kim [8].

For well-determined systems fixed points for the Newton map correspond to zeros. This is not necessarily the case for the Moore–Penrose–
Newton map. We have see [9] or [8] Proposition 1:

Proposition 1 *Fixed points for N_f correspond to least square solutions
for f: $N_f(\zeta) = \zeta$ if and only if $DF(\zeta) = 0$ with $F(x) = \frac{1}{2}\|f(x)\|^2$. When
$Df(\zeta)$ is onto then $N_f(\zeta) = \zeta$ if and only if $f(\zeta) = 0$.*

2.1 Convergence properties: the well-determined case

Let us first consider the well-determined case and a zero $\zeta \in \mathbf{E}$ of f with
$Df(\zeta)$ an isomorphism. An easy computation shows that the derivative
of the Newton map at ζ is zero so that Newton's sequence $x_k = N_f^{(k)}(x_0)$
converges quadratically for any x_0 in a certain ball about ζ. For this
reason we define with Smale (see [5] Section 8.1):

Definition 1 *Say that x_0 is an approximate zero of f if the sequence
$x_{k+1} = N_f(x_k)$ is defined for any $k \geq 0$ and there is a ζ such that
$f(\zeta) = 0$ with*

$$\|x_k - \zeta\| \leq \left(\frac{1}{2}\right)^{2^k - 1} \|x_0 - \zeta\|.$$

Call ζ the associated zero.

Approximate zeros may be described in two different ways: Kantorovich's Theorem and Smale's alpha-theory. Ostrowski in 1936 [29],
then Kantorovich in 1949 [21] pioneered such studies . Alpha-theory
has been introduced by Kim in [22] for one variable polynomial equations and by Smale [40] for analytic maps between Banach spaces. In
the following, the system f is supposed to be C^2 between two Banach
spaces:

Theorem 2 *Suppose that $f(\zeta) = 0$ and that $Df(\zeta)^{-1}$ exists. Let $r > 0$ be given and*

$$A = \max_{\|x-\zeta\|\leq r} \|Df(\zeta)^{-1}D^2 f(x)\|.$$

If $2Ar \leq 1$ then any x_0 with $\|x_0 - \zeta\| \leq r$ is an approximate zero with associated zero ζ.

The hypothesis f is C^2 may be weakened, we can suppose that f is C^1 and its derivative Lipschitz around ζ. We give below a sophisticated version of such a theorem due to Wang Xinghua [42]:

Theorem 3 *(Wang) Let f be some map between two Banach spaces. Suppose that $f(\zeta) = 0$, f has a continuous derivative in the open ball $B(\zeta, r)$, $Df(\zeta)$ is an isomorphism and $Df(\zeta)^{-1}Df$ satisfies the following condition: there exists a positive, integrable and nondecreasing function L such that*

$$\|Df(\zeta)^{-1}(Df(x) - Df(x_t))\| \leq \int_{t\|x-\zeta\|}^{\|x-\zeta\|} L(u)du,$$

for any x, $\|x - \zeta\| < r$, and $0 \leq t \leq 1$, where $x_t = \zeta + t(x - \zeta)$. Let r satisfy

$$\frac{\int_0^r u L(u)du}{r(1 - \int_0^r L(u)du)} \leq \frac{1}{2}.$$

Then any x_0 with $\|x_0 - \zeta\| \leq r$ is an approximate zero with associated zero ζ.

When f is an analytic map between two Banach spaces, a stronger regularity assumption than previously, we are able to describe approximate zeros in terms of the invariant *gamma* defined below:

Definition 2 *For any $x \in \mathbf{E}$ let us define*

$$\gamma(f, x) = \sup_{k \geq 2} \left\| Df(x)^{-1} \frac{D^k f(x)}{k!} \right\|^{\frac{1}{k-1}}.$$

We give to this invariant the value ∞ when $Df(x)$ is not an isomorphism.

The following *gamma-theorem* is taken from [5] Section 8.1:

Theorem 4 *(Smale) Suppose that $f(\zeta) = 0$ and that $Df(\zeta)^{-1}$ exists. If*

$$\|x - \zeta\| \leq \frac{3 - \sqrt{7}}{2\gamma(f, \zeta)}$$

then x is an approximate zero of f with approximate zero ζ.

2.2 Convergence properties: the underdetermined case

In this section we confine our discussion to a smooth map $f : \mathbf{E} \to \mathbf{F}$ where $\dim \mathbf{E} \geq \dim \mathbf{F}$. Let us denote by $\mathbf{V}$ the zero set of f. Convergence properties of Newton's method are studied in Allgower and Georg [1], Beyn [2], Dedieu and Kim [8], Dedieu and Shub [9], Deuflhard and Heindl [14], Shub and Smale [36]. Let us introduce the following invariant (see [36]):

Definition 3 *For any $x \in \mathbf{E}$ let us define*

$$\gamma(f, x) = \sup_{k \geq 2} \left\| Df(x)^\dagger \frac{D^k f(x)}{k!} \right\|^{\frac{1}{k-1}}.$$

We give to this invariant the value ∞ when $Df(x)$ is not surjective.

The structure of $\mathbf{V}$ around a given point $\zeta \in \mathbf{V}$ and the convergence properties of Newton's sequence are described below in the following theorem taken from Dedieu and Shub [11]. Similar results are given in Beyn [2] and Shub–Smale [36].

Suppose that f is an analytic function $f : \mathbf{E} \to \mathbf{F}$ such that $f(\zeta) = 0$ and $Df(\zeta)$ is surjective. Let us denote $H_1 = \zeta + \ker \ Df(\zeta)$, $H_2 = \zeta + \ker \ Df(\zeta)^\perp$, $B_{\delta, H_i} = \{x \in H_i \ : \ \|x - \zeta\| < \delta\}$ and $B_\delta = B_{\delta_1, H_1} \times B_{\delta_2, H_2}$.

Theorem 5 *There are universal constants c_1, $c_2 > 0$ such that*

(i) *$\mathbf{V} \cap B_{\frac{c_1}{\gamma(f, \zeta)}}$ is a smooth manifold which can be given as the graph of an analytic function*

$$h : B_{\frac{c_1}{\gamma(f, \zeta)}, H_1} \to B_{\frac{c_1}{\gamma(f, \zeta)}, H_2}$$

with $\|Dh(x)\| \leq c_2 \|\zeta - x\| \gamma(f, \zeta)$.

(ii) *If $x \in B_{\frac{c_1}{\gamma(f, \zeta)}}$ then the Newton sequence*

$$x_0 = x, \quad x_{k+1} = x_k - Df(x_k)^\dagger f(x_k)$$

is defined for all $k \geq 0$ and satisfies

$$d(x_k, \mathbf{V}) \leq \left(\frac{1}{2}\right)^{2^k - 1} d(x, \mathbf{V}).$$

Moreover, x_k converges to a unique point in $\mathbf{V}$ which we denote by $M_f(x)$. It satisfies

$$d(x, \mathbf{V}) \leq d(x, M_f(x)) \leq 2d(x, \mathbf{V}).$$

(iii) *Let denote $T_\zeta V$ the tangent space at ζ to V. If $\dot{x} \in T_\zeta V = H_1 - \zeta$ and $\|\dot{x}\| \leq \frac{c_1}{\gamma(f,\zeta)}$ then*
3.1. $d(\zeta + \dot{x}, \mathbf{V}) \leq \frac{c_2 \gamma(f,\zeta)}{2} \|\dot{x}\|^2$
3.2. $d(\zeta + \dot{x}, M_f(\zeta + \dot{x})) \leq c_2 \gamma(f, \zeta) \|\dot{x}\|^2$
Moreover c_1 and c_2 may be chosen such that c_2 is arbitrarily close to 1.

Let us suppose that $\dim \mathbf{E} = n$ and $\dim \mathbf{F} = m$. When 0 is a regular value for f then $\mathbf{V}$ is a smooth submanifold with $\dim \mathbf{V} = n - m$. According to Theorem 5 the limit of Newton's iteration $M_f(x)$ is defined in a neighborhood of this submanifold. There is an invariant foliation defined by Newton's method which consists in the set of initial values x such that $M_f(x) = \zeta$. We denote this set by $\mathbf{W}_\zeta$. The following is proved in Beyn [2] and Shub–Smale [36]

Theorem 6 *Suppose 0 is a regular value of f. Then*

(i) *The union over $\zeta \in \mathbf{V}$ of the sets $\mathbf{W}_\zeta$ is a neighborhood of $\mathbf{V}$.*
(ii) *$\mathbf{W}_\zeta$ intersected by a small neighborhood of $\mathbf{V}$ is a submanifold in $\mathbf{E}$ and $\dim \mathbf{W}_\zeta = m$.*
(iii) *the tangent spaces $T_\zeta V$ and $T_\zeta W_\zeta$ at ζ to $\mathbf{V}$ and $\mathbf{W}_\zeta$ are orthogonal: $T_\zeta V = \ker Df(\zeta)$ and $T_\zeta W_\zeta = \ker Df(\zeta)^\perp$.*

2.3 Convergence properties: the overdetermined case

In this section we consider a smooth map $f : \mathbf{E} \to \mathbf{F}$ where $\dim \mathbf{E} \leq \dim \mathbf{F}$. Let us denote by $\mathbf{V}$ the zero set of f. Because f is an overdetermined system of equations, generically $\mathbf{V}$ is empty. For this reason it is convenient to consider the set of least-square solutions of this system $\mathbf{V}_{\mathbf{ls}}$: $\zeta \in \mathbf{V}_{\mathbf{ls}}$ if and only if it is a stationary point for the residue function $F(x) = \frac{1}{2}\|f(x)\|^2$ i.e. $DF(\zeta) = 0$. We already have noticed in Proposition 1 that fixed points for Newton's method are precisely the least-square solutions of the considered system. Contrarily to the underdetermined or well-determined case a least square-solution is not necessarily an attractive fixed point and the convergence of Newton's sequence is not necessarily quadratic. These properties are analysed in Ben Israel [3], Dedieu–Kim [8], Dedieu–Shub [9], Dennis and Schnabel [13].

Before stating these convergence results we have to introduce a new invariant adapted to the injective-overdetermined case:

Definition 4 *For any $x \in \mathbf{E}$ let us define*

(i) $\gamma_1(f, x) = \sup\limits_{k \geq 2} \left(\|Df(x)^\dagger\| \dfrac{\|D^k f(x)\|}{k!} \right)^{\frac{1}{k-1}}$,

(ii) $\beta_1(f, x) = \|Df(x)^\dagger\| \|f(x)\|$,

(iii) $\alpha_1(f, x) = \beta_1(f, x)\gamma_1(f, x)$.

We give to these invariants the value ∞ when $Df(x)$ is not injective.

The convergence to a solution of our system is quadratic and the size of the quadratic attraction basin given by γ_1 as in Theorems 5 and 7.

Theorem 7 *(Dedieu–Shub) Let x and $\zeta \in \mathbf{E}$ be such that $f(\zeta) = 0$, $Df(\zeta)$ injective and*

$$v = \|x - \zeta\|\gamma_1(f, \zeta) \leq \frac{3 - \sqrt{7}}{2}.$$

Then Newton's sequence $x_k = N_f^k(x)$ satisfies

$$\|x_k - \zeta\| \leq \left(\frac{1}{2}\right)^{2^k - 1} \|x - \zeta\|.$$

The situation is completely different if we consider a least square solution instead. The following shows the geometric convergence to least-square solutions:

Theorem 8 *(Dedieu–Shub) Let us denote $\psi(v) = 1 - 4v + 2v^2$. Let x and $\zeta \in \mathbf{E}$ satisfying $Df(\zeta)^\dagger f(\zeta) = 0$, $Df(\zeta)$ injective and*

$$v = \|x - \zeta\|\gamma_1(f, \zeta) < 1 - \frac{\sqrt{2}}{2}.$$

If

$$\lambda = \frac{1}{\psi(v)}(v + \sqrt{2}(2 - v)\alpha_1(f, \zeta)) < 1$$

then Newton's sequence satisfies

$$\|x_k - \zeta\| \le \lambda^k \|x - \zeta\|.$$

The condition $\lambda < 1$ appearing in Theorem 8 is satisfied in a neighborhood of ζ if $\alpha_1(f, \zeta) < \frac{1}{2\sqrt{2}}$. More precisely we have

Theorem 9 *(Dedieu–Shub) If $\zeta \in V_{ls}$ is a least-square zero of f, if $Df(\zeta)$ is injective and if ζ is an attractive fixed point for Newton's method (i.e. all the eigenvalues of $DN_f(\zeta)$ have a modulus less than 1) then ζ is necessarily a strict local minimum for the residue function.*

The converse of this result is false. A strict local minimum for the residue function, thus necessarily a fixed point for Newton's map, may be a repelling point for this Newton's map. An example is given in [9]: we consider

$$f(x) = \begin{pmatrix} x \\ x^2 + a \end{pmatrix}, \ f : \mathbb{R} \to \mathbb{R}^2$$

where $a \in \mathbb{R}$ is given. Here $x = 0$ is a stationary point of the residue function

$$2F(x) = x^2 + (x^2 + a)^2 = x^4 + (2a + 1)x^2 + a^2.$$

When $a = 0$ then $x = 0$ is a zero of f, when $a \ne 0$ then $f(0) \ne 0$. Newton's iterate is given by

$$N_f(x) = x - \frac{2x^3 + (2a + 1)x}{1 + 4x^2}$$

so that $DN_f(0) = -2a$. The conclusion is clear: when $a = 0$ then $DN_f(0) = 0$ and Newton's sequence converges quadratically to $x = 0$. When $a \ne 0$ and $|a| < 1/2$ then $N_f^k(x)$ converges linearly to $x = 0$, when $|a| > 1/2$ then $x = 0$ is a repulsive point for Newton's iteration. At $a = 1/2$, $DN_f(0) = -1$ and N_f goes through a period doubling bifurcation. There is now a period two attracting orbit for N_f near 0, so Newton's method fails to converge to a fixed point near 0. By a equal 6, N_f appears by computer experiment to have gone through a whole period doubling cascade.

3 Newton's method: the homogeneous and multihomogeneous cases

3.1 Definition of Newton's method in the homogeneous case

Let $\mathbf{E}$ be complex or real vector spaces and $\mathbf{F} = \mathbb{C}^m$ or $\mathbb{R}^m$. Then $f : \mathbf{E} \to \mathbf{F}$ is homogeneous of degree $(d) = (d_1, \ldots, d_m)$ if and only if the ith coordinate function satisfies

$$f_i(\lambda x) = \lambda^{d_i} f_i(x)$$

for $x \in \mathbf{E}$ and λ in $\mathbb{R}$ or $\mathbb{C}$ as the case may be. When the system f is homogeneous the solution variety $\mathbf{V} = \{\zeta \in \mathbf{E} : f(\zeta) = 0\}$ consists in straight lines through the origin which define a corresponding set (still denoted by $\mathbf{V}$) in the projective space $\mathbf{P}(\mathbf{E})$. We will assume that $\mathbf{E}, \mathbf{F}$ are complex and finite dimensional vector spaces and that $\mathbf{E}$ has a Hermitian product $\langle \, , \, \rangle$. For the case where $\mathbf{E}, \mathbf{F}$ are real we would replace the Hermitian product by an inner product. Also, we denote $\mathbf{E}^{\star} = \mathbf{E} \setminus \{0\}$. Then $\mathbf{P}(\mathbf{E})$ is the quotient of $\mathbf{E}^{\star}$ by the action of $\mathbb{C}^{\star} = \mathbb{C} \setminus \{0\}$. For $x \in \mathbf{E}^{\star}$, we let $x^{\perp}$ be the Hermitian complement of x in $\mathbf{E}$. $x^{\perp}$ is a natural representative of the tangent space $T_x \mathbf{P}(\mathbf{E})$. We are now ready to define the projective Newton iteration for f. We denote this map as $N_f : \mathbf{P}(\mathbf{E}) \hookleftarrow$.

Definition 5 *(Shub [32])* $N_f(x) = f(x) - (Df(x)|_{x^{\perp}})^{\dagger} f(x)$.

The homogeneous (or projective) Newton's method is defined in $\mathbf{E}$ but is invariant under the natural identifications which define the projective space:

Proposition 10 N_f *is well-defined i.e. if* $y = \lambda x$ *for* $x \in \mathbf{E}^{\star}$ *and* $\lambda \in \mathbb{C}^{\star}$ *then* $N_f(y) = \lambda N_f(x)$.

It is now time to give an example. Let $f : \mathbb{C} \to \mathbb{C}$. To this map we associate its d−homogeneous counterpart $F : \mathbb{C}^2 \to \mathbb{C}$ defined by $F(z_0, z_1) = z_0^d f(z_1/z_0)$ so that $f(z) = F(1, z)$ and F is homogeneous with degree d.

Projective Newton method applied to F is given by

$$(Z_0, Z_1) = N_F(z_0, z_1) = (z_0, z_1) - (DF(z_0, z_1)|_{(z_0, z_1)^{\perp}})^{-1} F(z_0, z_1).$$

The pair (Z_0, Z_1) is solution of the linear system

$$\begin{pmatrix} \frac{\partial F}{\partial z_0} & \frac{\partial F}{\partial z_1} \\ \bar{z}_0 & \bar{z}_1 \end{pmatrix} \begin{pmatrix} Z_0 - z_0 \\ Z_1 - z_1 \end{pmatrix} = \begin{pmatrix} -F(z) \\ 0 \end{pmatrix}$$

so that, with $z_0 = 1$ and $z_1 = z$ we obtain

$$Z_0 = 1 - \frac{\bar{z}f(z)}{d\bar{z}f(z) - (1 + |z|^2)f'(z)}, \quad Z_1 = z + \frac{f(z)}{d\bar{z}f(z) - (1 + |z|^2)f'(z)}.$$

Since (Z_0, Z_1) and $(1, Z_1/Z_0)$ define the same line through the origin in $\mathbb{C}^2$ we may take $N_F(1, z) = (1, Z_1/Z_0) = (1, Z)$ with

$$Z = z - \frac{f(z)}{f'(z)} \frac{1}{1 - (d-1)\dfrac{\bar{z}f(z)}{(1 + |z|^2)f'(z)}}.$$

3.2 Definition of Newton's method in the multihomogeneous case

The case of multihomogeneous maps is treated similarly. It appears for the first time in Dedieu–Shub [10]. Let us first define the word multihomogeneous. Let $\mathbf{E}_1, \ldots, \mathbf{E}_k$ be complex or real vector spaces and $\mathbf{F} = \mathbb{C}^m$ or $\mathbb{R}^m$. Let $\mathbf{E} = \mathbf{E}_1 \times \ldots \times \mathbf{E}_k$ and $((d)) = ((d_1), \ldots, (d_k))$, $(d_i) = (d_{1i}, \ldots, d_{ki})$ for $i = 1, \ldots, m$. Then $f : \mathbf{E} \to \mathbf{F}$ is multihomogeneous of degree $((d))$ if and only if the ith coordinate function satisfies

$$f_i(\lambda_1 x_1, \ldots, \lambda_k x_k) = \prod_{j=1}^{k} \lambda_j^{d_{ji}} f_i(x_1, \ldots, x_k)$$

for $(x_1, \ldots, x_k) \in \mathbf{E}$ and $(\lambda_1, \ldots, \lambda_k)$ a scalar k-tuple, i.e. $(\lambda_1, \ldots, \lambda_k) \in \mathbf{G} = \mathbb{C}^k$ or $\mathbb{R}^k$ as the case may be.

Here are two examples of multihomogeneous functions. Let $\mathcal{H}_d$ be the space of homogeneous polynomials of degree d defined on $\mathbb{C}^n$. Let $(d) = (d_1, \ldots, d_m)$ and $\mathcal{H}_{(d)} = \prod_{i=1}^{m} \mathcal{H}_{d_i}$. So elements of $\mathcal{H}_{(d)}$ represent polynomial functions $f : \mathbb{C}^n \to \mathbb{C}^m$ where $f = (f_1, \ldots, f_m)$ and f_i is homogeneous of degree d_i. The evaluation map

$$ev : \mathcal{H}_{(d)} \times \mathbb{C}^n \to \mathbb{C}^m, \quad ev(f, x) = f(x),$$

is multihomogeneous. Each coordinate function of ev is linear in f and homogeneous of degree d_i in x.

A second example is given by the generalized eigenvalue problem. Let $A, B : \mathbb{C}^n \to \mathbb{C}^n$ be linear operators. Then

$$F_{(A,B)} : \mathbb{C}^2 \times \mathbb{C}^n \to \mathbb{C}^n, \quad F_{(A,B)}(\alpha, \beta, x) = (\alpha B - \beta A)x,$$

is bilinear, i.e. it is linear in (α, β) and linear in x. The generalized eigenvalue problem is to find the zeros of $F_{(A,B)}$.

We denote

$$\mathbf{E}^\star = (\mathbf{E}_1 \setminus \{0\}) \times \ldots \times (\mathbf{E}_k \setminus \{0\}).$$

If $\lambda = (\lambda_1, \ldots, \lambda_k) \in \mathbf{G} = \mathbb{C}^k$ we define

$$\times \lambda : \mathbf{E} \to \mathbf{E}$$

by

$$\times \lambda x = (\lambda_1 x_1, \ldots, \lambda_k x_k).$$

Then $\mathbf{P}(\mathbf{E}_1) \times \ldots \times \mathbf{P}(\mathbf{E}_k)$ is the quotient of $\mathbf{E}^\star$ by the action of $\mathbf{G}^\star = (\mathbb{C} \setminus \{0\}) \times \ldots \times (\mathbb{C} \setminus \{0\})$, k times. For $x \in \mathbf{E}^\star$, $x = (x_1, \ldots, x_k)$, we let $x_i^\perp$ be the Hermitian complement of x_i in $\mathbf{E}_i$, and

$$x^\perp = \prod_{i=1}^{k} x_i^\perp \subset \mathbf{E}.$$

For each i, $x_i^\perp$ is a natural representative of the tangent space $T_{x_i} \mathbf{P}(\mathbf{E}_i)$ and hence $x^\perp$ is a natural representative of the tangent space

$$T_x \left(\prod_{i=1}^{k} \mathbf{P}(\mathbf{E}_i) \right) = \prod_{i=1}^{k} T_{x_i} (\mathbf{P}(\mathbf{E}_i)).$$

When a map $f : \mathbf{E} \to \mathbf{F}$ is multihomogeneous its zero set $f^{-1}(0)$ is stable under $\times \lambda$ action, $\lambda \in \mathbf{G}$, and it defines a corresponding set in the product of projective spaces $\mathbf{P}(\mathbf{E}_1) \times \ldots \times \mathbf{P}(\mathbf{E}_k)$.

We are now ready to define the multihomogeneous projective Newton iteration for f. We denote this map as $N_f : \prod_i \mathbf{P}(\mathbf{E}_i) \hookleftarrow$.

Definition 6 *(Dedieu–Shub [10])*

$$N_f(x) = f(x) - (Df(x)|_{x^\perp})^\dagger f(x).$$

N_f is of course naturally defined on $\mathbf{E}$ but it is invariant under the action of $\times \lambda$ as is proved by the following

Proposition 11 N_f *is well-defined i.e. if* $y = \times \lambda x$ *for* $x \in \mathbf{E}^\star$ *and* $\lambda \in \mathbf{G}^\star$ *then* $N_f(y) = \times \lambda N_f(x)$.

3.3 Convergence properties

Since Newton's method for multihomogeneous maps is defined in a product of projective spaces, we will describe its metric properties in terms of the Riemannian distance on this space. We recall that for $i = 1, \ldots, k$

the Riemannian distance in $\mathbf{P}(\mathbf{E}_i)$ is given by the angle between the two considered vectors

$$d_R(x_i, y_i) = \arccos \frac{|\langle x_i, y_i \rangle_i|}{\|x_i\|_i \|y_i\|_i},$$

and in $\mathbf{P}(\mathbf{E}_1) \times \ldots \times \mathbf{P}(\mathbf{E}_k)$ by

$$d_R(x, y) = \left(\sum_{i=1}^{k} d_R(x_i, y_i)^2 \right)^{1/2}$$

where $x = (x_1, \ldots, x_k)$ and $y = (y_1, \ldots, y_k) \in \mathbf{E}^\star$. Here and throughout we identify $x_i \in \mathbf{E}_i \setminus \{0\}$ and $x \in \mathbf{E}^\star$ with their equivalence classes in $\mathbf{P}(\mathbf{E}_i)$ and $\mathbf{P}(\mathbf{E}_1) \times \ldots \times \mathbf{P}(\mathbf{E}_k)$ respectively.

We now define an Hermitian structure on $\mathbf{E}$ depending on x by

$$\langle v, w \rangle_x = \sum_{i=1}^{k} \frac{\langle v_i, w_i \rangle_i}{\langle x_i, x_i \rangle_i}$$

for $x \in \mathbf{E}^\star$, v and $w \in \mathbf{E}$.

3.3.1 Well-determined and underdetermined multihomogeneous systems

Definition 7 $\gamma_m(f, x) = \max \left(1, \sup_{k>2} \left\| (Df(x)|_{m\perp})^\dagger \frac{D^k f(x)}{k!} \right\|_x^{\frac{1}{k-1}} \right)$,

In the definition of $\gamma_m(f, x)$, $\| \ \|_x$ is the operator norm with respect to $\langle \ , \ \rangle_x$. The invariant $\gamma_m(f, x)$ is defined for $x \in \mathbf{E}^\star$ but, in fact, it depends on the class of x in $\mathbf{P}(\mathbf{E}_1) \times \ldots \times \mathbf{P}(\mathbf{E}_k)$. This is proved by the following ([10])

Proposition 12 *For any $x \in \mathbf{E}^\star$ and $\lambda \in \mathbf{G}^\star$ we have $\gamma_m(f, x) = \gamma_m(f, \times \lambda x)$.*

Convergence theorems for the homogeneous or multihomogeneous Newton's sequence are given by Blum, Cucker, Shub and S. Smale [5], Malajovich [27] and Dedieu and Shub [10]. The following is taken from this last paper. We consider the case of underdetermined systems with surjective derivatives

Theorem 13 *There is a universal constant $c > 0$ with the following property : let $\zeta \in \mathbf{E}^\star$ be a zero of f with $Df(\zeta)$ onto and $x \in \mathbf{E}^\star$. If*

$$\|x - \zeta\|_\zeta \gamma_m(f, \zeta) \leq c$$

then multihomogeneous Newton's sequence $x_k = N_f^{(k)}(x)$ converges to a zero $\zeta' \in \mathbf{E}^\star$ of f and

$$d_R(\zeta', x_k) \le \sigma \left(\frac{1}{2}\right)^{2^k-1} \|x_1 - x\|_x$$

with

$$\sigma = \sum_{i=0}^{\infty} \left(\frac{1}{2}\right)^{2^i-1} = 1.6328\ldots$$

3.3.2 Overdetermined multihomogeneous systems

In the case of overdetermined systems with injective derivatives a similar result occurs. The following is taken from Dedieu and Shub [9].

Definition 8

$$\gamma_{m,1}(f,x) = \max\left(1, \sup_{k\ge 2}\left\|(Df(x)|_{x^\perp})^\dagger\right\|_x \left\|\frac{D^k f(x)}{k!}\right\|_x^{\frac{1}{k-1}}\right).$$

Theorem 14 *There is a universal constant $c > 0$ approximately equal to .15872 with the following properties: let $\zeta \in \mathbf{E}^*$ be a zero of f with $Df(\zeta)|_{\zeta^\perp}$ injective and $x \in \mathbf{E}^*$ such that*

$$d_P(x,\zeta)\gamma_{m,1}(f,\zeta) \le c.$$

Then the multihomogeneous Newton sequence $x_0 = x$, $x_{k+1} = N_f(x_k)$, converges to ζ and, for each $k \ge 1$,

$$d_P(\zeta, x_k) \le \left(\frac{1}{2}\right)^{2^k-1} d_P(\zeta, x).$$

4 How to find good starting points?

In the previous section we have emphasized the convergence properties of Newton's method: sufficiently close to a zero it converges very rapidly. The problem is how to find good starting points: either a point in the attraction basin of a zero or an approximate zero.

4.1 Sutherland's strategy for one variable polynomials

In a recent paper [20] Hubbard, Schleicher and Sutherland prove the following:

Theorem 15 *For every $d \geq 2$, there is a set S_d consisting of at most $1.11d\log^2 d$ points in $\mathbb{C}$ with the property that for every polynomial f of degree d, normalized so that all its roots are in the open unit disk, and for each of its roots, there is a point $x \in S_d$ such that Newton's sequence $x_k = N_f^{(k)}(x)$ converges to the choosen root.*

Notice that the set S_d is independent of the normalized polynomial of degree d and has a very small cardinality. Such a set can be constructed explicitly: take $n_1 = \lceil \alpha \log d \rceil$ circles centered at the origin with radii

$$R_\nu = R \left(\frac{d-1}{d} \right)^{\frac{\nu - \frac{1}{2}}{n_1}},$$

$\nu = 1 \ldots n_1$, and on each circle $n_2 = \lceil \beta d \log d \rceil$ equally spaced points. The set S_d consists in the union of these points. Different values for the constants R, α and β are possible, see [20] for more details. Here we take $R = 1 + \sqrt{2}$, $\alpha = 0.2663$ and $\beta = 4.1627$.

Sutherland [41], 1989, proved a version of this result in his thesis: for a polynomial of degee d, $11d(d-1)$ points on a single circle are sufficient. Later, in 1993, Hubbard and Schleicher improved the result to $\alpha d \log d$ points on $\beta \log d$ circles improving the order by using several circles instead of just one. It is also proved that at least half of the roots can be found using $2.0d$ initial points. Earlier papers are Smale [39], Manning [26], Friedman [15], [16] and [17]. See also the survey by Victor Pan [30].

4.2 Homotopy methods

Let us first consider the case of well-determined polynomial systems $f : \mathbb{C}^n \to \mathbb{C}^n$, $f = (f_1, \ldots, f_n)$ with degree $f_i = d_i$. We denote by $\mathcal{P}_{(d)}$, $(d) = (d_1, \ldots, d_n)$, the space of such systems.

Consider any curve $F : I \to V \setminus \Sigma'$, $I = [0, 1]$, with

$$V = \{ (f, \zeta) \in \mathcal{P}_{(d)} \times \mathbb{C}^n \ : \ f(\zeta) = 0 \}$$

and Σ' being the critical set for V: $(f, \zeta) \in \Sigma'$ when $(f, \zeta) \in V$ and rank $Df(\zeta) < n$. Write $F(t) = (f_t, \zeta_t)$ so that $f_t(\zeta_t) = 0$ and $Df_t(\zeta_t)$ is nonsingular. If the curve f_t is smooth then by the implicit function theorem ζ_t is also smooth.

We suppose that the curve f_t and ζ_0 are given. Our objective is to "follow" the curve ζ_t and to "reach" a zero ζ_1 of f_1 which is our target system. Instead of computing ζ_t for each t, we start from a subdivision

$$0 = t_0 < t_1 < \ldots < t_k = 1$$

and we construct a finite sequence x_i by

$$x_0 = \zeta_0, \quad x_{i+1} = N_{f_{t_{i+1}}}(x_i).$$

Our objective is to compute a partition t_i so that, for each i, x_i is an approximate zero for f_{t_i} i.e. Newton's sequence associated with f_{t_i} and starting at x_i converges quadratically to ζ_{t_i}. This is necessarily the case when all the distances between two consecutive points in the partition are small, so that k is big. Thus k is a good measure for the complexity of this method.

A good introduction to this subject is given by the book of Allgower and Georg [1].

Complexity aspects of path following methods are present in Shub–Smale [33] for one variable polynomials, Kim–Sutherland [24]: this covers the path following method for finding all the roots for functions of one complex variable. Then Shub–Smale [34] and Blum–Cucker–Shub–Smale [5] for homogeneous polynomial systems $f : \mathbb{C}^{n+1} \to \mathbb{C}^n$, Shub–Smale [34] for path following in Banach spaces, Shub–Smale [36] for underdetermined systems $f : \mathbb{R}^n \to \mathbb{R}^m$: in these papers the complexity is described in terms of the condition number of the path and the degree of the system. The case of multihomogeneous underdetermined systems is studied by Dedieu and Shub in [10], sparse systems by Dedieu [7] in terms of a sparse condition number. Probabilistic aspects are investigated by Shub and Smale in [35], [36], [37].

The case of overdetermined systems is more complicated because the restriction to V of the first projection i.e $(f, \zeta) \in V \to f$ is not onto. To find a path connecting a starting system to a target system is not necessarily easy. The complexity aspect for overdetermined system is investigated by Dedieu–Shub in [9].

All these papers present upper bounds for the complexity of path following methods in terms of the degree and a condition number. In [12] Dedieu and Smale give lower bounds for this complexity in the context of homogeneous underdetermined systems $f : \mathbb{C}^{n+1} \to \mathbb{C}^m$.

4.3 Upper bounds

We present here one of these theorems, due to Shub and Smale [34], see also [5], section 14.3.

We use $\mathcal{H}_{(d)}$ to denote the set of homogeneous polynomial systems $f : \mathbb{C}^{n+1} \to \mathbb{C}^n$, $f = (f_1, \ldots, f_n)$, each f_i a homogeneous polynomial of degree exactly d_i and $(d) = (d_1, \ldots, d_n)$. For homogeneous polynomials $f_i, g_i : \mathbb{C}^{n+1} \to \mathbb{C}$ of degree d_i let

$$(f_i, g_i)_{d_i} = \sum_{|\alpha|=d_i} a_\alpha \bar{b}_\alpha \binom{d_i}{\alpha}^{-1}$$

where $f_i(z) = \sum_{|\alpha|=d_i} a_\alpha z^\alpha$, $g_i(z) = \sum_{|\alpha|=d_i} b_\alpha z^\alpha$, $\binom{d_i}{\alpha} = d_i!/\alpha_0! \ldots \alpha_n!$ and $z^\alpha = z_0^{\alpha_0} \ldots z_n^{\alpha_n}$. This induces an Hermitian inner product on $\mathcal{H}_{(d)}$: for $f, g \in \mathcal{H}_{(d)}$

$$(f, g) = \sum_{i=1}^n (f_i, g_i)_{d_i}.$$

This Hermitian inner product is unitarily invariant under the action of the unitary group in $\mathbb{C}^{n+1}$.

Definition 9 *Let $f \in \mathcal{H}_{(d)}$ and $\zeta \in \mathbb{C}^{n+1}$, $\zeta \neq 0$, with $f(\zeta) = 0$. The condition number of f at ζ is equal to*

$$\mu(f, \zeta) = \|f\| \, \| \left(Df(\zeta)|_{\zeta^\perp}\right)^{-1} \mathrm{Diag}(\|\zeta\|^{d_i - 1})\|$$

or ∞ when $Df(\zeta)|_{\zeta^\perp}$ is singular. Here $\mathrm{Diag}(\|\zeta\|^{d_i-1})$ denotes the $n \times n$ diagonal matrix with diagonal entries $\|\zeta\|^{d_i-1}$.

It is not particularly enlightening to define the condition number $\mu(f, \zeta)$ by this mysterious formula. The story is the following: let

$$V = \{(f, \zeta) \in \mathbf{P}(\mathcal{H}_{(d)}) \times \mathbf{P}(\mathbb{C}^{n+1}) \; : \; f(\zeta) = 0\}.$$

We denote here by $\mathbf{P}(\mathbf{E})$ the projective space of lines through the origin in E. V is a variety and (f, ζ) is in the smooth part of $\mathbf{V}$ if and only if $Df(\zeta)|_{\zeta^\perp}$ is nonsingular. In that case, according to the implicit function theorem, V is locally around (f, ζ) the graph of a smooth map S defined in a neighborhood of $f \in \mathbf{P}(\mathcal{H}_{(d)})$ and taking its values in a neighborhood of $\zeta \in \mathbf{P}(\mathbb{C}^{n+1})$. S is the map "solution", it associates to a system $\tilde{f}$ close to f a zero $\tilde{\zeta}$ of $\tilde{f}$ close to ζ. The derivative of this map

$$DS(f) : T_f \mathbf{P}(\mathcal{H}_{(d)}) \to T_\zeta \mathbf{P}(\mathbb{C}^{n+1})$$

gives the first order variations of the zero ζ in terms of the first order variations of the system f. The condition number is the norm of this linear operator. For more details on condition numbers the reader is refered to Blum–Cucker–Shub–Smale [5], Chap. 10 and 12 or to Dedieu [6].

Notice that the condition number defined above is invariant under scaling for both f and ζ: it is defined on the product of projective spaces $\mathbf{P}(\mathcal{H}_{(d)}) \times \mathbf{P}(\mathbb{C}^{n+1})$. The condition number $\mu(f, \zeta)$ has a strong geometric interpretation: it measures the distance to ill-posed problems in the fiber

$$V_\zeta = \{g \in \mathbf{P}(\mathcal{H}_{(d)}) : (g, \zeta) \in V\}.$$

Ill-posed problems are defined by

$$\Sigma' = \{(\tilde{f}, \tilde{\zeta}) \in V : \text{rank } D\tilde{f}(\tilde{\zeta}) < n\},$$

and the distance in $\mathbf{P}(\mathcal{H}_{(d)})$ is $d_P = sind_R$, d_R the Riemannian distance in $\mathbf{P}(\mathcal{H}_{(d)})$. The following is due to Shub and Smale [34]

Theorem 16 (Condition Number Theorem) Let $(f, \zeta) \in V$ be given. Let us define $\rho(f, \zeta) = d_P(f, V_\zeta \cap \Sigma')$ and the following normalized condition number:

$$\mu_{\text{norm}}(f, \zeta) = \|f\| \, \| \left(Df(\zeta)|_{\zeta^\perp}\right)^{-1} \text{Diag}(d_i^{1/2}\|\zeta\|^{d_i-1})\|.$$

Then

$$\mu_{\text{norm}}(f, \zeta) = \frac{1}{\rho(f, \zeta)}.$$

For a smooth curve $F : I \to \mathbf{V}$, $F(t) = (f_t, \zeta_t)$, we define its condition number by

$$\mu(F) = \max_{t \in I} \mu(f_t, \zeta_t).$$

Let L_f denote the length of f_t, $t \in I$, computed in the projective space $\mathbf{P}(\mathbb{C}^{n+1})$ for its canonical Riemannian metric, and $D = \max_{1 \leq i \leq n} d_i$. We have [5] Chap. 14, Th. 4,

Theorem 17 (Shub–Smale) There is a partition $0 = t_0 < t_1 < \ldots < t_k = 1$ with

$$k = \left\lceil \frac{8}{3 - \sqrt{7}} D^2 \mu(F)^2 L_f \right\rceil$$

and such that the sequence x_i defined by

$$x_0 = \zeta_0, \quad x_{i+1} = N_{f_{t_{i+1}}}(x_i)$$

consists in approximate zeros of f_{t_i} with associated zero ζ_{t_i}.

4.4 Lower bounds

The question of lower bounds for continuation methods is investigated in Dedieu–Smale [12]. A Newton continuation method sequence (NCM sequence) is a sequence of pairs

$$(f_i, \zeta_i) \in (\mathcal{H}_d)^* \times (\mathbb{C}^{n+1})^*, \quad 0 \leq i \leq k,$$

satisfying the following conditions :

$$f_i(\zeta_i) = 0, \quad 0 \leq i \leq k,$$

and

$$\alpha_h(f_{i+1}, \zeta_i) \leq \alpha_0, \text{ with associated zero } \zeta_{i+1}, \quad 0 \leq i \leq k - 1.$$

This last condition implies that the projective Newton's sequence

$$x_0 = \zeta_i, \quad x_{p+1} = N_{f_{i+1}}(x_p), \quad p \geq 0,$$

converges quadratically towards ζ_{i+1}. Let us be more precise:

Definition 10

(i) $\gamma_h(f, x) = \|x\| \max_{k \geq 2} \left\| (Df(x)|_{x^\perp})^{-1} \dfrac{D^k f(x)}{k!} \right\|^{\frac{1}{k-1}},$

(ii) $\beta_h(f, x) = \|x\|^{-1} \|(Df(x)|_{x^\perp})^{-1} f(x)\|,$

(III) $\alpha_h(f, x) = \beta_h(f, x)\, \gamma_h(f, x).$

These three quantities are invariant under scaling, they are defined on the product of projective spaces: $\mathbf{P}(\mathcal{H}_{(d)}) \times \mathbf{P}(\mathbb{C}^{n+1})$. We can see easily that γ_h from Definition 23 and γ_m from Definition 15 are the same numbers. When $Df(x)|_{x^\perp}$ is singular, we take

$$\alpha_h(f, x) = \beta_h(f, x) = \gamma_h(f, x) = \infty.$$

The following theorem, due to Shub and Smale [37] and Malajovich [27] justifies our definition of a NPC sequence:

Theorem 18 *There is a universal constant $\alpha_0 > 0$ with the following property : for any homogeneous system $f \in \mathcal{H}_{(d)}$ and $x \in (\mathbb{C}^{n+1})^*$, if $\alpha_h(f, x) \leq \alpha_0$, then the projective Newton sequence*

$$x_0 = x, \quad x_{k+1} = N_f(x_k)$$

is defined and satisfies

$$\|x_{k+1} - x_k\|/\|x_k\| \leq \left(\frac{1}{2} \right)^{2^k - 1} \beta_h(f, x)$$

for any $k \geq 0$. This sequence converges to a zero $\zeta \in (\mathbb{C}^{n+1})^\star$ of f and

$$d_R(\zeta, x_k) \leq \sigma \left(\frac{1}{2}\right)^{2^k - 1} \beta_h(f, x)$$

with

$$\sigma = \sum_{i=0}^{\infty} \left(\frac{1}{2}\right)^{2^i - 1} = 1.6328\ldots$$

The complexity of an NCM sequence (f_i, ζ_i), $0 \leq i \leq k$, is measured by k. A first lower bound for k is given by the degree:

Theorem 19 *(Dedieu–Smale) For any NCM sequence (f_i, ζ_i), $0 \leq i \leq k$, one has*

$$k \geq c \max(1, \frac{D-1}{2}) d_R(\zeta_0, \zeta_k),$$

where $c > 0$ is a universal constant and $d_R(\zeta_0, \zeta_k)$ the Riemannian distance in $\mathbf{P}(\mathbb{C}^{n+1})$ between ζ_0 and ζ_k.

This bound is sharp and this complexity is obtained for the family of systems defined by

$$f_{t,j}(z_0, z_1, \ldots, z_n) = z_0^{d_j - 1}(z_j - \zeta_{t,j} z_0), \ 1 \leq j \leq n,$$

where $\zeta_t = (1 - t)(1, 0, \ldots, 0) + t(1, a_1, \ldots, a_n)$, $a \in \mathbb{C}^n$ given.

There is no such lower bound for systems $f : \mathbb{C}^n \to \mathbb{C}^n$ and affine Newton's method. The proof of Theorem 25 uses the compacity of the product of projective spaces $\mathbf{P}(\mathcal{H}_{(d)}) \times \mathbf{P}(\mathbb{C}^{n+1})$: in this context $\gamma_h(f, \zeta)$, when $f(\zeta) = 0$, is bounded from below by $(D - 1)/2$.

A second lower bound involves the distance of the vectors ζ_i to the set Σ_{f_i} of singular points for f_i. Contrarily to Theorem 25, this lower bound as an affine counterpart.

Definition 11 *For any $f \in \mathcal{H}_d$ let us define*

$$\Sigma_f = \{x \in (\mathbb{C}^{n+1})^\star \ : \ rank \ Df(x) < n\}.$$

The following shows that the complexity of a NCM sequence increase with the proximity to singular points:

Theorem 20 *(Dedieu–Smale) Let $\epsilon > 0$ be given. For any NCM sequence (f_i, ζ_i), $0 \leq i \leq k$, such that*

$$d_R(\zeta_i, \Sigma_{f_i}) \leq \epsilon, \ \ 1 \leq i \leq k,$$

we have

$$k \geq c\epsilon^{-1} d_R(\zeta_0, \zeta_k),$$

with c a universal constant.

Acknowledgments

The author gratefully acknowledges very useful conversations with Myong-Hi Kim, Mike Shub and Scott Sutherland about this paper. He is also grateful to Steve Smale for many valuable discussions about the complexity aspects of the zero finding problem.

References

[1] Allgower, E. and K. Georg (1990). *Numerical Continuation Methods* (Springer, New York).

[2] Beyn W.-J. (1993). On smoothness and invariance properties of the Gauss–Newton method, *Num. Funct. Anal. Opt.* 14, 243–252.

[3] Ben Israel A. (1966). A Newton–Raphson method for the solution of systems of equations, *J. Math. Anal. Appl.* 15, 243–252.

[4] Ben Israel, A. and Greville, T. (1974). *Generalized Inverses Theory and Applications* (Wiley, New York).

[5] Blum, L., Cucker, F., Shub, M. and Smale, S. (1997). *Complexity and Real Computation* (Springer, New York).

[6] Dedieu, J.-P. (1996). Approximate solutions of numerical problems, condition number analysis and condition number theorem, in *The Mathematics of Numerical Analysis*, ed. J. Renegar, M. Shub and S. Smale, Lectures in Applied Mathematics 32, AMS, 263–283.

[7] Dedieu, J.-P. (1997). Condition number analysis for sparse polynomial systems, in *Fondations of Computationnal Mathematics*, ed. F. Cucker and M. Shub (Springer, New York).

[8] Dedieu, J.-P. and Kim, M.-H. (1999). Newton's method for analytic systems of equations with constant rank derivatives, submitted to *Maths of Comput.*

[9] Dedieu, J.-P. and Shub, M. (2000). Newton's method for overdetermined systems of equations, to appear in *Maths of Comput.*

[10] Dedieu, J.-P. and Shub, M. (2000). Multihomogeneous Newton's method, to appear in *Maths of Comput.*

[11] Dedieu, J.-P. and Shub, M. (1998). Newton and predictor-corrector method for overdetermined systems of equations, IBM Tech. Rep.

[12] Dedieu, J.-P. and Smale, S. (1998). Some lower bounds for the complexity of continuation methods, *J. of Complexity* 14, 454–465.

[13] Dennis, J., and Schnabel, R. (1983). *Numerical Methods for Unconstrained Optimization and Nonlinear Equation* (Prentice Hall, New York).

[14] Deuflhard P. and Heindl, G. (1979). Affine invariant convergence theorems for Newton's method and extensions to related methods, *SIAM J. Numer. Anal.* **16**, 1–10.

[15] Friedman, J. (1989). On the convergence of Newton's method, 27th Annual Symposium on Foundations of Computer Science.

[16] Friedman, J. (1990). A density theorem for purely iterative zero finding methods, *SIAM J. Computing* **19**, 124–132.

[17] Friedman, J. (1990). Random polynomials and approximate zeros of Newton's method, *SIAM J. Computing* **19**, 1068–1099.

[18] Gauss, K.F. (1809). *Theoria Motus Corporum Coelestian*, Werke, 7, 240–254.

[19] Goldstine, H. (1977). *A History of Numerical Analysis from the 16th Through the 19th Century* (Springer, New York).

[20] Hubbard, J., Schleicher, D. and Sutherland, S. (1998). How to really find roots of polynomials by Newton's method, preprint.

[21] Kantorovich, L. (1949). Sur la méthode de Newton, *Travaux de l'Institut des Mathématiques Steklov* **XXVIII**, 104–144.

[22] Kim, M.-H. (1986). Computational Complexity of the Euler Type Algorithm for the Roots of Polynomials, City University of New York Ph.D. thesis.

[23] Kim, M.-H (1988). On approximate zeros and rootfinding algorithms for a complex polynomial, *Maths of Comput.* **51**, 707–719.

[24] Kim, M.-H and Sutherland, S. (1994). Polynomial root-finding algorithms and branched covers, *SIAM J. Computing* **23**, 415–436.

[25] Luenberger, D. (1969). *Optimization by Vector Space Methods* (Wiley, New York).

[26] Manning, A. (1992). How to be sure of solving a complex polynomial using Newton's method, *Bol. Soc. Bras. Mat.* **22**, 157–177.

[27] Malajovich, G. (1994). On generalized Newton's methods, *Theoretical Comp. Sci.* **133**, 65–84.

[28] Ortega, J., and Rheinboldt, W. (1968). *Numerical Solutions of Nonlinear Problems* (SIAM, Philadelphia).

[29] Ostrowski, A. (1976). *Solutions of Equations in Euclidean and Banach Spaces* (Academic Press, New York).

[30] Pan, V. (1997). Solving a polynomial equation: some history and recent progress, *SIAM Review* **39**, 187–220.

[31] Shub, M. (1986). Some remarks on dynamical systems and numerical analysis, in *Dynamical Systems and Partial Differential Equations, Proceedings of VII ELAM*, ed. L. Lara-Carrero and J. Lewowicz (Equinoccio, Universidad Simon Bolivar, Caracas).

[32] Shub, M. (1993). Some remarks on Bézout's theorem and complexity, in *Proceedings of the Smalefest*, ed. M. V. Hirsch, J. E. Marsden and M. Shub (Springer, New York), 443–455.

[33] Shub, M. and Smale, S. (1986). On the geometry of polynomials and a theory of cost: Part II, *SIAM J. Computing* **15**, 145–161.

[34] Shub, M. and Smale, S. (1993). Complexity of Bézout's theorem I: Geometric aspects, *J. Am. Math. Soc.* **6**, 459–501.

[35] Shub, M. Smale, S. (1993). Complexity of Bézout's theorem II : Volumes and probabilities, in *Computational Algebraic Geometry*, ed. F. Eyssette and A. Galligo (Birkhäuser, Basel).

[36] Shub, M. and Smale, S. (1996). Complexity of Bézout's theorem IV: Probability of success, extensions, *SIAM J. Numer. Anal.* **33**, 128–148.

[37] Shub, M. and Smale, S. (1994). Complexity of Bézout's theorem V : Polynomial time, *Theoretical Computer Sci.* **133**, 141–164.

[38] Smale, S. (1985). On the efficiency of algorithms of analysis, *Bull. AMS* **13**, 87–121.

[39] Smale, S. (1986). Algorithms for solving equations, in *Proceedings of the International Congress of Mathematicians* (AMS, Providence, RI). 172–195.

[40] Smale, S. (1986). Newton's method estimates from data at one point, in *The Merging of Disciplines: New Directions in Pure, Applied and Computational Mathematics*, ed. R. Ewing, K. Gross, and C. Martin (Springer, New York).

[41] Sutherland, S. (1989). Finding Roots of Complex Polynomials with Newton's Method, Boston University Ph.D. thesis.

[42] Wang, X. (2000). convergence of newton's method and uniqueness of the solution of equations in Banach spaces, to appear in *IMA J. Num. Anal.*

[43] Ypma, T. (1995). Historical development of the Newton–Raphson method, *SIAM Review* **37**, 531–551.

Kronecker's smart, little black boxes

Marc Giusti

UMS CNRS–Polytechnique MEDICIS
Laboratoire GAGE
École Polytechnique
F-91128 Palaiseau cedex, France
Email: Marc.Giusti@gage.polytechnique.fr

Joos Heintz

Depto. de Matemáticas, Est. y Comp.
Facultad de Ciencias Universidad de Cantabria
E-39071 Santander, Spain   and
Depto. de Matemáticas
Facultad de Ciencias Exactas y Naturales
Universidad de Buenos Aires
Ciudad Universitaria, Pab. I
(1428) Buenos Aires, Argentina
Email: heintz@hall.matesco.unican.es

Abstract

This paper is devoted to the complexity analysis of certain uniformity properties owned by all known symbolic methods of parametric polynomial equation solving (geometric elimination). It is shown that *any* parametric elimination procedure which is *parsimonious* with respect to *branchings* and *divisions* must necessarily have a non-polynomial sequential time complexity, even if highly efficient data structures (as e.g. the arithmetic circuit encoding of polynomials) are used.

1 Introduction

Origins, development and interaction of modern algebraic geometry and commutative algebra may be considered as one of the most illustrative examples of historical dialectics in mathematics. Still today, and more than ever before, timeless idealism (in form of modern commuta-

Research partially supported by the following Argentinian, Spanish, German and French grants : UBA-CYT TW 80, PIP CONICET 4571/96, ANPCyT PICT 03-00000-01593, DGICYT PB96–0671–C02–02, ARG 018/98 INF (BMBF), UMS 658 MEDICIS, ECOS A99E06, HF 1999–055

tive algebra) is bravely struggling whith secular materialism (in form of complexity issues in computational algebraic geometry).

Kronecker was doubtless the creator of this eternal battle field and its first war lord. In a similar way as Gauss did for computational number theory, Kronecker laid intuitively the mathematical foundations of modern computer algebra. He introduced 1882 in [30] his famous "elimination method" for polynomial equation systems and his "parametric representation" of (equidimensional) algebraic varieties. By the way, this parametric representation was until 10 years ago rediscovered again and again. It entered in modern computer algebra as "Shape Lemma" (see e.g. [38, 8, 14, 27]). Using his elimination method in a highly skillful, but unfortunately inimitable way, Kronecker was able to state and to prove a series of fundamental results on arbitrary algebraic varieties. He was able to define in a precise way the notion of dimension and to prove a corresponding dimension theorem for arbitrary algebraic varieties over an algebraically closed field, to estimate the number of equations needed to define any algebraic variety in affine or projective space and certainly he knew already the special form of "Hilbert's Nullstellensatz".

Not everything that came to Kronecker's mind was laid down by him in a explicit and written form. Nevertheless a careful interpretation of his work suggests his deep understanding of the general structure of algebraic varieties.

A particular result, proved by Kronecker, says that any algebraic variety can be defined by finitely many equations. Later Hilbert generalized this result to his seminal "Basissatz" introducing for its proof a new, nonconstructive method, far away from the traditional elimination-type arguments used by Kronecker and other contemporary mathematicians. It took some time to convince the mathematical world that "mystics" and "magicians" are able to produce (correct) mathematical results, but finally the new discipline of commutative algebra became legitimate.

Hilbert's discovery of the Basissatz was also the starting point for a long and huge conflict which dominated a considerable part of the history of modern algebraic geometry and which did not come to an end until today.

Classical algebraic geometry is motivated by the need – or the wish – to find tools which allow to "solve" or to "reduce" (whatever this means) systems of polynomial equations. This leads to the following questions:

- are commutative algebra and its modern derivate, namely todays scheme-theoretical algebraic geometry, able to absorb classical algebraic geometry?

• is the solution to Hilbert's "Hauptproblem der Idealtheorie" (the ideal membership problem) the key to all computational issues in classical algebraic geometry?

These questions, implicitly raised by the work of Hilbert and Macaulay, look very academic. However any attempt to answer them leads to deep consequences.

Let us here outline just one possible way to answer these questions.

If we consider the "Hauptproblem der Idealtheorie" as a question of pure theoretical computer science, this problem turns out to be computationally intractable, at least in worst case. More precisely, the "Hauptproblem der Idealtheorie" turns out to be complete in exponential (memory) space (see [36, 43, 9]). On the other hand, almost all of the most fundamental problems of computational classical algebraic geometry are proved to be solvable in polynomial space (see e.g. [35, 10, 34, 33, 28]). Thus computational complexity is able to distinguish between geometry and algebra and supports the viewpoint of Kronecker (geometry) against the viewpoint of Hilbert and Macaulay (algebra).

It is well known that Kronecker's personality was highly conflictive for his time. It is less known how much posthumous rejection Kronecker's personality was able to produce. Hilbert's writings are eloquent in this point ("die Kroneckersche Verbotsdiktatur", see [11]), whereas Macaulay's attack against Kronecker's work on elimination was a rather well educated one (see [32], Preface). In this context let us also remind André Weil's "elimination of the elimination theory". Kronecker's radical spirit did not recognise limitations. His radicalism was as universal as his spirit was. Of course he was right requiring that any mathematical reflection has to end up with finitely many and practically realizable computations which settle the concrete (e.g. application) problem under consideration. However declaring the natural numbers as the only mathematical objects created by god for mankind and declaring the "rest" (real, complex numbers, infinite cardinals and ordinals) as devils work, tempting humans to play with the infinite, he demonstrated that he was not able or willing to distinguish between syntaxis and semantics. He was right to require that mathematical expressions (algebra) have always to move within finitary limits, but he was wrong to exclude reflections about infinite mathematical objects (geometry) using a finitarian language. Of course his own work ended up to be godless enough to guarantee him mathematical recognition by his worst adversaries.

Kronecker's own formulation of his elimination method in [30] was imprecise and general enough to allow different interpretations of it, computationally efficient and inefficient ones. Possibly it was in his mind to leave open the door for future complexity issues of his method.

Hilbert's and Macaulay's attacks against Kronecker's method were based on the computationally inefficient interpretation. They noticed that under this interpretation Kronecker's elimination method leads to a hyperexponential swell of intermediate expression size and used this observation for the promotion of their own, more "simple" and "mathematical" point of view. Hilbert and Macaulay's position became finally predominant in the future development of algebraic geometry and commutative algebra (see e.g. [42]). Their ideas led to the modern, computer implemented tool of Gröbner basis algorithms for the symbolic resolution (simplification) of polynomial equation systems. This tool was introduced in the sixties by B. Buchberger (and his school) and represents today the core of all current computer algebra software packages.

The discovery of effective (affine) Nullstellensätze and their application to the complexity analysis of Gröbner basis algorithms for the solution of geometrical problems, represented at the end of the eighties the turning point for a process which led finally back to Kronecker's original ideas. This process was not a conscious motion with clear goals, but rather a slow emerging of mathematical insight and of algorithmic design within the scope of Kronecker's intuitions.

The first step in this direction was even made as part of a mathematical proof and not of an algorithmic design. In its standard interpretation, Kronecker's elimination method relies on an iterated use of resultants of suitable univariate polynomials with parametric coefficients. In fact, a resultant is nothing but the constant term of the characteristic polynomial of a suitable linear map determined by the polynomials under consideration. Replacing in this version of Kronecker's elimination algorithm the occuring resultants just by the constant terms of the corresponding minimal polynomials, one obtains an enormous reduction of degree, height and also of arithmetic circuit (straight-line program) complexity for the polynomials produced during the procedure (see e.g. [4, 5, 20]). By the way, let us remark that Kronecker applies in his method the mentioned simplification, however he omits to draw these important conclusions from his argument.

Other important ingredients of the emerging new algorithmic method were the parametric representation of equidimensional algebraic varieties (a rediscovery of Kronecker's old idea) and the arithmetic circuit (straight-line program) representation of polynomials (which was neither out of the scope of Kronecker's intuition nor clearly included in the main stream of his thinking). The combination of these ingredients produced a considerable effect upon the worst case complexity of symbolic elimination procedures (see e.g. [16, 29, 18]).

Nevertheless all this progress did not suffice to allow the design of practically efficient symbolic elimination algorithms and in particular of algorithms which were able to compete in complexity aspects with their numerical counterparts. The traditional huge gap between symbolic and numeric polynomial equation solving methods remained open. Whereas numeric algorithms are very efficient with respect to the number of arithmetical operations they require, they cannot be efficiently used in case of parametric, underdetermined, overdetermined or degenerate polynomial equation systems. Symbolic algorithms are free from these restrictions but they are also too inefficient for any reasonable use in practice. A way out of this dilemma became apparent in [17, 15] by a new interpretation of Newton's classical method. Interpreting Newton's approximation algorithm as a global procedure instead as a local one, allowed its use as a tool for data compression in the Kronecker–like elimination procedure of [18] (which relied on the arithmetic circuit representation of polynomials). The point was that Newton's method is well adapted to exact symbolic computation if the correct (seminumerical) data structure is used. In this way a new algorithmic method finally emerged. This method is based on a combination of Kronecker's and Newton's ideas, and is able to distinguish dichotomically between "well behaved" and "badly behaved" polynomial equations systems. Moreover this algorithmic method is optimal for worst case (i.e. generic) systems.

Unlike Gröbner basis algorithms this new method avoids any significant computational overflow during its execution. Roughly speaking, the new algorithms are always polynomial in the output size and even polynomial in the input size, if the given polynomial equation system is "well behaved".

This view of algorithmic algebraic geometry produced also the following new insight:

Elimination polynomials are always "smart"(i.e. not easy and not hard) to evaluate. How many variables they ever may contain, their evaluation complexity is always polynomial in their degree (whereas their number of monomials may be exponential in the number of their variables).

Although this insight is beyond of the scope of Kronecker's way of thinking, his formulation of the elimination method indicates that he might have intended to leave the door open for complexity issues.

A further development of this new algorithmic method was able to demonstrate its practical efficiency (see [21] and [24]). Moreover it turns out that the parameters which dominate the complexity of the new symbolic procedures determine also the efficiency of their numerical counterparts (at least if aspects of diophantine approximation are taken into account; see [6]). The new symbolic algorithms (and their numerical counterparts) have still a worst case complexity which is exponential in the (purely syntactical) input length. One may ask whether this fact is due to the algorithms and data structures employed or whether this is due to the intrinsic nature of geometric elimination.

The new results presented in this paper address this question. In order to discuss this point, let us turn back to our interpretation of Kronecker's ideas behind geometric elimination. Kronecker's elimination method (and theory) behaves well under specialisation of the input equations. In fact, at any moment of the procedure one may consider the (parametric) input equations as given by their coefficients and these coefficients may be considered as purely algebraic objects, determined only by their algebraic relations. A given equation system may even be "generalized", i.e. the coefficients of the given input equations may be replaced by indeterminates and the discussion of the solvability of the generalized systems answers all imaginable questions about the solvability of the original system. This concept of specialisation–generalisation reveals an idea of universality (or uniformity) behind Kronecker's elimination theory. This philosophical idea of universality became one of the corner stones of modern algebraic geometry and commutative algebra. Since we are (still) unable to think in a different way, we shall consider the input equations of a parametric elimination problem as functions which may be called by their values in the variables to be eliminated (black-box representation) or simply as being given by their coefficients (formal representation).

Any elimination algorithm we are able to imagine today starts from this kind of data. In other words, the black box representation constitutes today the most general way we may think the equations of our input system to be given. Of course each evaluation of the input equations has its costs and these costs may be measured by the size of a division-free arithmetic circuit which represents the input equations.

In this paper we shall show that any sufficiently uniform elimination procedure (which avoids superfluous branchings) becomes necessarily exponential in worst case, if the input equations are given in black-box representation and if the required output is a canonical elimination polynomial (a "resolvent" in the terminology of Kronecker and Macaulay). This is a general and provable fact for any symbolic as well as for any numerical elimination procedure. We shall also show, that even in case that the input equations are not given by a black box, but by an explicit arithmetic circuit, the same conclusion holds true for any sufficiently universal and uniform elimination procedure which is able to compute efficiently Zariski closures and (generically squarefree) *parametric* greatest common divisors for circuit represented algebraic families of polynomials.

Below we shall give a precise definition of the meaning of a "sufficiently universal and uniform elimination procedure" for the case of a flat family of elimination problems. Such an elimination procedure will be called *parametric*.

If an elimination procedure is used in order to assign "coordinates" (i.e. in order to parametrise) suitable and easy-to-represent families of algebraic varieties the same conclusion holds again, without any further restriction on the input representation (see [7] and [20]). Summing up, we may say: the vision of elimination theory initiated by Kronecker hides a concept of universality and uniformity which obstructs its general efficiency. The question what happens with complexity when we drop this universality and uniformity requirement, exceeds the horizon of todays mathematical thinking and is equivalent to the question whether $P \neq NP$ holds over the complex numbers (in the sense of the BSS complexity model; see [2]).

2 Parametric elimination

2.1 Parametric elimination procedures

The procedures (algorithms) considered in this paper operate with *essentially division-free arithmetic circuits* (straight-line programs) as basic data structures for the representation of inputs and outputs. Such a circuit depends on certain input nodes, labelled by indeterminates over a given ground field k (in the sequel we shall suppose that k is infinite and perfect with algebraic closure $\bar{k}$). These indeterminates are thought to be subdivided into two disjoint sets representing the *parameters* and *variables* of the given circuit. The circuit nodes of indegree zero which are not inputs are labelled by elements of k, which are called the *scalars* of the circuit (here "indegree" means the number of incoming edges of the corresponding node). Internal nodes are labelled by arithmetic operations (addition, subtraction, multiplication and division). The internal nodes of the circuit represent polynomials in the variables of the circuit. The coefficients of these polynomials belong to the *parameter field* K, generated over the ground field k by the parameters of the circuit. This is achieved by allowing in a essentially division-free circuit only divisions which involve elements of K. Thus essentially division-free circuits do not contain divisions involving intermediate results which depend on variables. A circuit which contains only divisions by nonzero elements of k is called *totally division-free*. The output nodes of an essentially division-free circuit may occur labelled by sign marks of the form "$= 0$" or "$\neq 0$" or may remain unlabelled (by sign marks). Thus the given circuit *represents* by means of its labelled output nodes a system of polynomial equations and inequations which determine in their turn a locally closed set (i.e. an embedded affine variety) with respect to the Zariski topology of the affine space of parameter and variable instances. The unlabelled output nodes of the given circuit represent a polynomial application (in fact a morphism of algebraic varieties) which maps this locally closed set into a suitable affine space. We shall interpret the system of polynomial equations and inequations represented by the circuit as a *parametric family of systems* in the variables of the circuit. The corresponding varieties constitute an *parametric family of varieties*. The same point of view is applied to the morphism determined by the unlabelled output nodes of the circuit. We shall consider this morphism as a *parametric family of morphisms*.

To a given essentially division-free arithmetic circuit we may associate different complexity measures and models. In this paper we shall be mainly concerned with *sequential* computing *time*, measured by the *size* of the circuit. Occasionally we will also refer to *parallel time*, measured by the *depth* of the circuit. In our main complexity model is the *total* one, where we take into account *all* arithmetic operations (additions, subtractions, multiplications and possibly occuring divisions) at *unit costs*. For purely technical reasons we shall also consider two *non-scalar* complexity models, one over the ground field k and the other one over the parameter field K. In the non-scalar complexity model over

K we count only the *essential* multiplications (i.e. multiplications between intermediate results which actually involve variables and not only parameters). This means that K-linear operations (i.e. additions and multiplications by arbitrary elements of K) are *cost free*. Similarly, k-linear operations are not counted in the non-scalar model over k. For more details about complexity measures and models we refer to [3].

Given an essentially division-free arithmetic circuit as input, an *elimination problem* consists in the task of finding an essentially division-free output circuit which describes the Zariski closure of the image of the morphism determined by the input circuit. The output circuit and the corresponding algebraic variety are also called a *solution* of the given elimination problem. We say that a given parameter point fixes an *instance* of the elimination problem under consideration. In this sense a problem instance is described by an input and an output (solution) instance.

In this paper we restrict our attention to input circuits which are totally division-free and contain only output nodes labelled by "=0" and unlabelled output nodes. Mostly our output circuits will also be totally division-free and will contain only one output node, labelled by the mark "=0". This output node will always represent a canonical elimination polynomial associated to the elimination problem under consideration (see Section 2.2 for more details).

In case that our output circuit contains divisions (depending only on parameters but not on variables), we require to be able to perform these divisions for any problem instance. In order to make this requirement sound, we admit in our algorithmic model certain limit processes in the spirit of de l'Hôpital's rule (below we shall modelise these limit processes algebraically, in terms of places and valuations). The restriction we impose on the possible divisions in an output circuit represents a *first* fundamental geometric *uniformity requirement* for our algorithmic model.

An algorithm which solves a given elimination problem may be considered as a (geometric) elimination procedure. However this simple minded notion is too restrictive for our purpose of showing lower complexity bounds for elimination problems. It is thinkable that there exists for every individual elimination problem an efficient ad hoc algorithm, but that there is no universal way to find and to represent all these ad hoc procedures. Therefore, a *geometric elimination procedure* in the sense of this paper will satisfy certain uniformity and universality requirements which we are going to explain now.

We modelise our elimination procedures by families of *arithmetic networks* (also called arithmetic-boolean circuits) which solve entire classes of elimination problems of *arbitrary* input size (see [12], [13]). In this sense we shall require the *universality* of our geometric elimination procedures. Moreover, we require that our elimination procedures should be essentially division-free.

In a universal geometric elimination procedure, branchings and divisions by intermediate results (that involve only parameters, but not variables) cannot be avoided. From our elimination procedures we shall require to be *parsimonious* with respect to *branchings* (and divisions). In particular we shall require that our elimination procedures do not introduce branchings and divisions for the solution of a given elimination problem when traditional algorithms do not demand this (an example of such a situation is given by the flat families of elimination problems we are going to consider in the sequel). This restriction represents a *second* fundamental *uniformity requirement* for our algorithmic model.

We call a universal elimination procedure *parametric* if it satisfies our first and second uniformity requirement, i.e. if the procedure does not contain branchings which otherwise could be avoided and if all possibly occurring divisions can be performed on all problem instances, in the way we have explained before. In this paper we shall only consider parametric elimination procedures.

We call a parametric elimination procedure *geometrically robust* if it produces for any input instance an output circuit which depends only on the mathematical objects "input equation system" and "input morphism" but not on their circuit representation. We shall apply this notion only to elimination problems given by (geometrically or scheme-theoretically) flat families of algebraic varieties. This means informally that a parametric elimination procedure is geometrically robust if it produces for flat families of problem instances "continuous" solutions.

Of course, our notion of geometric robustness depends on the (geometric or scheme-theoretical) context, i.e. it is not the same for schemes or varieties. In Section 2.2 we shall explain our idea of geometric robustness in the typical situation of flat families of algebraic varieties given by reduced complete intersections.

Traditionally, the size of a system of polynomial equations (and inequations) is measured in purely extrinsic, syntactic terms (e.g. number of parameters and variables, degree of the input polynomials, size and depth of the input circuit etc). However, there exists a new generation of symbolic and numeric algorithms which take also into account intrinsic, semantic (e.g. geometric or arithmetic) invariants of the input equation system in order to measure the complexity of elimination procedure under consideration more accurately (see e.g. [17, 15, 19] and [40, 41]).

In this paper we shall turn back to the traditional point of view. In Theorem 5 we shall show that, under our universality and uniformity restrictions, *no parametric elimination procedure which includes efficient computation of Zariski closures and of generically squarefree parametric greatest common divisors for circuit represented algebraic families of polynomials, is able to solve an arbitrary elimination problem in polynomial (sequential) time, if time is measured in terms of circuit size and input length is measured in syntactical terms only.*

Finally let us refer to the books [3], [37] and [39] as a general background for notions of algebraic complexity theory and algebraic geometry we are going to use in this paper.

2.2 Flat families of elimination problems

Let, as before, k be an infinite and perfect field with algebraic closure $\bar{k}$ and let $U_1, \ldots, U_r, X_1, \ldots, X_n, Y$ be indeterminates over k. In the sequel we shall consider $X_1, \ldots, X_n$ and Y as variables and $U_1, \ldots, U_r$ as parameters. Let $G_1, \ldots, G_n$ and F be polynomials belonging to the k-algebra $k[U_1, \ldots, U_r, X_1, \ldots, X_n]$. Suppose that the polynomials $G_1, \ldots, G_n$ form a regular sequence in $k[U_1, \ldots, U_r, X_1, \ldots, X_n]$ defining thus an equidimensional subvariety $V := \{G_1 = 0, \ldots, G_n = 0\}$ of the $(r+n)$-dimensional affine space $\mathbb{A}^r \times \mathbb{A}^n$ over the field $\bar{k}$. The algebraic variety V has dimension r. Let δ be the (geometric) degree of V as defined in [22] (this degree does not take into account multiplicities or components at infinity). Suppose furthermore that the morphism of affine varieties $\pi : V \longrightarrow \mathbb{A}^r$, induced by the canonical projection of $\mathbb{A}^r \times \mathbb{A}^n$ onto $\mathbb{A}^r$, is finite and generically unramified (this implies that π is flat and that the ideal generated by $G_1, \ldots, G_n$ in $k[U_1, \ldots, U_r, X_1, \ldots, X_n]$ is radical). Let $\tilde{\pi} : V \longrightarrow \mathbb{A}^{r+1}$ be the morphism defined by $\tilde{\pi}(z) := (\pi(z), F(z))$ for any point z of the variety V. The image of $\tilde{\pi}$ is a hypersurface of $\mathbb{A}^{r+1}$ whose minimal equation is a polynomial of $k[U_1, \ldots, U_r, Y]$ which we denote by P. Let us write $\deg P$ for the total degree of the polynomial P and $\deg_Y P$ for its partial degree in the variable Y. Observe that P is monic in Y and that $\deg P \leq \delta \deg F$ holds. Furthermore, for a Zariski dense set of points u of $\mathbb{A}^r$, we have that $\deg_Y P$ is the cardinality of the image of the restriction of F to the finite set $\pi^{-1}(u)$. The polynomial $P(U_1, \ldots, U_r, F)$ vanishes on the variety V.

Let us consider an arbitrary point $u = (u_1, \ldots, u_r)$ of $\mathbb{A}^r$. For arbitrary polynomials $A \in k[U_1, \ldots, U_r, X_1, \ldots, X_n]$ and $B \in k[U_1, \ldots, U_r, Y]$ we denote by $A^{(u)}$ and $B^{(u)}$ the polynomials $A(u_1, \ldots, u_r, X_1, \ldots, X_n)$ and $B(u_1, \ldots, u_r, Y)$ which belong to $k(u_1, \ldots, u_r)[X_1, \ldots, X_n]$ and $k(u_1, \ldots, u_r)[Y]$ respectively. Similarly we denote for an arbitrary polynomial $C \in k[U_1, \ldots, U_r]$ by $C^{(u)}$ the value $C(u_1, \ldots, u_r)$ which belongs to the field $k(u_1, \ldots, u_r)$. The polynomials $G_1^{(u)}, \ldots, G_n^{(u)}$ define a zero dimensional subvariety $V^{(u)} := \{G_1^{(u)} = 0, \ldots, G_n^{(u)} = 0\} = \pi^{-1}(u)$ of the affine space $\mathbb{A}^n$. The degree (cardinality) of $V^{(u)}$ is bounded by δ. Denote by $\tilde{\pi}^{(u)} : V^{(u)} \longrightarrow \mathbb{A}^1$ the morphisms induced by the polynomial $F^{(u)}$ on the variety $V^{(u)}$. Observe that the polynomial $P^{(u)}$ vanishes on the (finite) image of the morphism $\tilde{\pi}^{(u)}$. Observe also that the polynomial $P^{(u)}$ is not necessarily the minimal equation of the image of $\tilde{\pi}^{(u)}$).

We call the equation system $G_1 = 0, \ldots, G_n = 0$ and the polynomial F a *flat family of elimination problems depending on the parameters* $U_1, \ldots, U_r$ and we call P the associated *elimination polynomial*. An element $u \in \mathbb{A}^r$ is considered as a *parameter point* which determines a *particular problem instance* (see Section 2.1). The equation system $G_1 = 0, \ldots, G_n = 0$ together with the polynomial F is called the *general instance* of the given flat family of elimination problems and the elimination polynomial P is called the *general solution* of this flat family.

The *problem instance* determined by the parameter point $u \in \mathbb{A}^r$ is given by the equations $G_1^{(u)} = 0, \ldots, G_n^{(u)} = 0$ and the polynomial $F^{(u)}$. The polynomial $P^{(u)}$ is called *a solution* of this particular problem instance. We call two parameter points $u, u' \in \mathbb{A}^r$ *equivalent* (in symbols: $u \sim u'$) if $G_1^{(u)} = G_1^{(u')}, \ldots, G_n^{(u)} = G_n^{(u')}$ and $F^{(u)} = F^{(u')}$ holds. Observe that $u \sim u'$ implies $P^{(u)} = P^{(u')}$. We call polynomials $A \in k[U_1, \ldots, U_r, X_1, \ldots, X_n]$, $B \in k[U_1, \ldots, U_r, Y]$ and $C \in k[U_1, \ldots, U_r]$ *invariant* (with respect to $\sim$) if for any two parameter points u, u' of $\mathbb{A}^r$ with $u \sim u'$ the respective identities $A^{(u)} = A^{(u')}$, $B^{(u)} = B^{(u')}$ and $C^{(u)} = C^{(u')}$ hold.

An arithmetic circuit in $k[U_1, \ldots, U_r, Y]$ *with scalars in* $k[U_1, \ldots, U_r]$ is a totally division-free arithmetic circuit in $k[U_1, \ldots, U_r, Y]$, say β, modelised in the following way: β is given by a directed acyclic graph whose internal nodes are labelled as before by arithmetic operations. There is only one input node of β, labelled by the variable Y. The other nodes of indegree zero the circuit β may contain, are labelled by arbitrary elements of $k[U_1, \ldots, U_r]$. These elements are considered as the *scalars of β*. We call such an arithmetic circuit β *invariant* (with respect to the equivalence relation $\sim$) if all its scalars are invariant polynomials of $k[U_1, \ldots, U_r]$. Considering instead of Y the variables $X_1, \ldots, X_n$ as inputs, one may analogously define the notion of an arithmetic circuit in $k[U_1, \ldots, U_r, X_1, \ldots, X_n]$ with scalars in $k[U_1, \ldots, U_r]$ and the meaning of its invariance. However, typically we shall limit ourselves to circuits in $k[U_1, \ldots, U_r, Y]$ with scalars in $k[U_1, \ldots, U_r]$.

We are now ready to characterise in the given situation what we mean by a *geometrically robust parametric elimination procedure*. Suppose that the polynomials $G_1, \ldots, G_n$ and F are given by a totally division-free arithmetic circuit β in $k[U_1, \ldots, U_r, X_1, \ldots, X_n]$. A geometrically robust parametric elimination procedure accepts the circuit β as input and produces as output an *invariant* circuit Γ in $k[U_1, \ldots, U_r, Y]$ with scalars in $k[U_1, \ldots, U_r]$, such that Γ represents the polynomial P. Observe that in our definition of geometric robustness we did not require that β is an invariant circuit because this would be too restrictive for the modelling of concrete situations in computational elimination theory.

The invariance property required for the output circuit Γ means the following: let $u = (u_1, \ldots, u_r)$ be a parameter point of $\mathbb{A}^r$ and let $\Gamma^{(u)}$ be the arithmetic circuit in $k(u_1, \ldots, u_r)[Y]$ obtained from the circuit Γ evaluating in the point u the elements of $k[U_1, \ldots, U_r]$ which occur as scalars of Γ. Then the invariance of Γ means that the circuit $\Gamma^{(u)}$ depends only on the particular *problem instance* determined by the parameter point u but not on u itself. Said otherwise, a geometrically robust elimination procedure produces the solution of a particular problem instance in a way which is independent of the possibly different representations of the given problem instance.

By definition, a geometrically robust parametric elimination procedure produces always the *general* solution of the flat family of elimination problems under consideration. This means that for flat families, geometrically robust parametric elimination procedures do not introduce branchings in the output circuits. In Section 3.1 we shall exhibit a complexity result which may be paraphrased as follows: *within the standard philosophy of commutative algebra, none of the known (exponential time) parametric elimination procedures can be improved to a polynomial time algorithm.* For this purpose it is important to remark that the known *parametric* elimination procedures (which are without exception based on linear algebra as well as on comprehensive Gröbner basis techniques) are all geometrically robust.

The invariance property of these procedures is easily verified in the situation of a flat family of elimination problems. One has only to observe that all known elimination procedures accept the input polynomials $G_1, \ldots, G_n$ and F in their dense or sparse *coefficient representation* or as *evaluation black box* with respect to the variables $X_1, \ldots, X_n$.

Finally we are going to explain what we mean by a *(generically square-free) parametric greatest common divisor of an algebraic family of polynomials* and by the problem of the computation of this greatest common divisor (in case of a circuit represented family).

Suppose that there is given a positive number s of nonzero polynomials, say $B_1, \ldots, B_s \in k[U_1, \ldots, U_r, Y]$. Let $V := \{B_1 = 0, \ldots, B_s = 0\}$. Suppose that V is nonempty. We consider now the morphism of affine varieties $\pi : V \longrightarrow \mathbb{A}^r$, induced by the canonical projection of $\mathbb{A}^r \times \mathbb{A}^1$ onto $\mathbb{A}^r$. Let S be the Zariski closure of $\pi(V)$ and suppose that S is an *irreducible* closed subvariety of $\mathbb{A}^r$. Let us denote by $k[S]$ the coordinate ring of S. Since S is irreducible we conclude that $k[S]$ is a domain with a well defined function field which we denote by $k(S)$.

Let $b_1, \ldots, b_s \in k[S][Y]$ be the polynomials in the variable Y with coefficients in $k[S]$, induced by $B_1, \ldots, B_s$. Suppose thet there exists an index $1 \le k \le s$ with $b_k \ne 0$. Without loss of generality we may suppose that for some index $1 \le m \le s$ the polynomials $b_1, \ldots, b_m$ are exactly the non-zero elements of $b_1, \ldots, b_s$. Observe that each polynomial $b_1, \ldots, b_s$ has positive degree (in the variable Y).

We consider $b_1, \ldots, b_m$ as an *algebraic family of polynomials* (in the variable Y) and $B_1, \ldots, B_m$ as their representatives. The polynomials $b_1, \ldots, b_m$ have in $k(S)[Y]$ a well defined *normalised* (i.e. monic) greatest common divisor, which we denote by h. Let D be the degree of h (with respect to the variable Y).

We are now going to describe certain geometric requirements which will allow us to consider h as a parametric greatest common divisor of the algebraic family of polynomials $b_1, \ldots, b_m$.

Our first requirement is $D \ge 1$.

Let us fix for the moment an arbitrary point $u \in S$. The evaluation in u determines a canonical k-algebra homomorphism $\phi_u : k[S] \longrightarrow \bar{k}$. Let $\phi : k(S) \longrightarrow \bar{k} \cup \{\infty\}$ be any *place* which extends the homomorphism ϕ_u (this means that the valuation ring of ϕ contains the local ring of the point u in the variety S and that the residue class of ϕ is contained in $\bar{k}$).

We require now that the place ϕ takes only *finite* values (i.e. values of

$\bar{k}$) in the coefficients of the polynomial h (recall that these coefficients belong to the field $k(S)$).

Moreover we require that the values of the place ϕ in these coefficients *depend only on the point u* and not on the particular choice of the place ϕ extending the homomorphism ϕ_u.

In this way the place ϕ maps the polynomial h to a monic polynomial of degree D in Y with coefficients in $\bar{k}$. This polynomial depends only on the point $u \in S$ and we denote it therefore by $h(u)(Y)$. In analogy with this notation we write $b_k(u)(Y) := B_k(u)(Y)$ for $1 \le k \le m$.

Finally we require that the polynomials $b_1(u)(Y), \ldots, b_m(u)(Y)$ of $\bar{k}(Y)$ are not all zero and that their (normalized) greatest common divisor is divisible by $h(u)(Y)$.

We say that a polynomial H of $k(U_1, \ldots, U_r)[Y]$ with $deg_Y H = D$ *represents* the greatest common divisor $h \in k(S)[Y]$ if the coefficients of H with respect to the variable Y induce well–defined rational functions of the variety S and if these rational functions are exactly the coefficienst of h (with respect to the varable Y).

Suppose now that the polynomials $B_1, \ldots, B_s \in k[U_1, \ldots, U_r, Y]$ satisfy all our requirements for any point $u \in S$. Then we call h a *parametric greatest common divisor of the algebraic family of polynomials* $b_1, \ldots, b_m \in k[S][Y]$. Any polynomial $H \in k(U_1, \ldots, U_r)[Y]$ which represents h is said to *represent the parametric greatest common divisor associated to the polynomials* $B_1, \ldots, B_s$.

A monic *squarefree* polynomial $\hat{h} \in k(S)[Y]$ with the same zeroes as h in an algebraic closure of $k(S)$, is called a *generically squarefree parametric greatest common divisor of the algebraic family* $b_1, \ldots, b_m \in k[S][Y]$.

If the characteristic of the ground field k is zero, such a generically squarefree parametric greatest common divisor $\hat{h}$ always exists and has the same properties as h with respect to the places $\phi : k(S) \longrightarrow \bar{k} \cup \{\infty\}$ considered before.

If the characteristic of k is positive, this general conclusion is not true anymore. However, in the purely geometric context of the present paper, we may always arrange the polynomials $B_1, \ldots, B_s$ and their arithmetic circuits in order to assure the existence of a generically squarefree greatest common divisor. For this purpose we need that k is a perfect field.

In case that a generically squarefree greatest common divisor $\hat{h}$ exists, the notion of a representative of $\hat{h}$ is defined in the same way as for h.

Suppose now that the polynomials $B_1, \ldots, B_s$ are given by a totally division–free arithmetic circuit β_* in $k[U_1, \ldots, U_r, Y]$.

We are now going to formulate the algorithmic problem of computing the (generically square–free) parametric greatest common divisor h (or $\hat{h}$) of the algebraic family of polynomials $b_1, \ldots, b_m \in k[S][Y]$.

We consider Y as variable and $U_1, \ldots, U_r$ as parameters. We are looking for an essentially division–free arithmetic circuit Γ_* in $k(U_1, \ldots, U_r)[Y]$ which computes a representative of the (generically square–free) parametric greatest common divisor h (or $\hat{h}$), associated to the polynomials $B_1, \ldots, B_s$.

We require that the scalars of Γ_* induce well–defined rational functions of the variety S and that for any point $u \in S$ and any place $\phi : k(S) \longrightarrow \bar{k} \cup \{\infty\}$ extending the homomorphism $\phi_u : k[S] \longrightarrow \bar{k}$ the following condition is satisified:

any scalar of the arithmetic circuit Γ_* induces a rational function of the variety S, which is mapped by ϕ into a *finite* value of $\bar{k}$. This value is *uniquely* determined by the point $u \in S$.

The problem of computing the (generically square–free) parametric greatest common divisor of the algebraic family of polynomials $b_1, \ldots, b_m \in k[S][Y]$ consists now in producing from the input circuit β_* a (smallest possible) circuit Γ_* which satisfies the requirement above.

Observe that the scalars of such a circuit Γ_*, as well as the coefficients of h, are rational functions of the variety S which belong to the integral closure of $k[S]$ in the function field $k(S)$. This property of Γ_* is conserved under specialisations of the k–algebra $k[S]$.

In the proof of Theorem 5 we shall make substantial use of this observation.

For general background and details about places and valuation rings we refer to [31].

3 The intrinsic complexity of parametric elimination procedures

3.1 A particular flat family of elimination problems

Let n be a fixed natural number and let $T, U_1, \ldots, U_n, X_1, \ldots, X_n$ and Y be indeterminates over $\mathbb{Q}$. In the sequel we are going to use the following notation: for arbitrary natural numbers i and j we shall denote by $[j]_i$ the ith digit of the binary representation of j. Let P be the following polynomial of $\mathbb{Q}[T, U_1, \ldots, U_n, Y]$:

$$P(T, U_1, \ldots, U_n, Y) := \prod_{j=0}^{2^n-1} \left(Y - (j + T \prod_{i=1}^{n} U_i^{[j]_i}) \right). \qquad (3.1)$$

We observe that the dense representation of P with respect to the variable Y takes the form

$$P(T, U_1, \ldots, U_n, Y) = Y^{2^n} + A_1 Y^{2^n - 1} + \cdots + A_{2^n},$$

where $A_1, \ldots, A_{2^n}$ are suitably defined polynomials of $\mathbb{Q}[T, U_1, \ldots, U_n]$.

Let $1 \leq k \leq 2^n$. In order to determine the polynomial A_k, we observe, by expanding the right hand side of (3.1), that A_k collects the contribution of all terms of the form

$$\prod_{h=1}^{k} \left(-(j_h + T \prod_{i=1}^{n} U_i^{[j_h]_i}) \right)$$

with $0 \leq j_1 < \cdots < j_k \leq 2^n - 1$. Therefore the polynomial A_k can be expressed as follows:

$$
\begin{aligned}
A_k &= \sum_{0 \leq j_1 < \cdots < j_k \leq 2^n - 1} \prod_{h=1}^{k} \left(-(j_h + T \prod_{i=1}^{n} U_i^{[j_h]_i}) \right) \\
&= \sum_{0 \leq j_1 < \cdots < j_k \leq 2^n - 1} (-1)^k \prod_{h=1}^{k} \left(j_h + T \prod_{i=1}^{n} U_i^{[j_h]_i} \right).
\end{aligned}
$$

Observe that for $0 \leq j_1 < \cdots < j_k \leq 2^n - 1$ the expression

$$\prod_{h=1}^{k} \left(j_h + T \prod_{i=1}^{n} U_i^{[j_h]_i} \right)$$

can be rewritten as:

$$j_1 \cdots j_k + T \left(\sum_{h=1}^{k} j_1 \cdots \hat{j_h} \cdots j_k \prod_{i=1}^{n} U_i^{[j_h]_i} \right) + \text{terms of higher degree in } T.$$

Therefore, we conclude that A_k has the form:

$$
\begin{aligned}
A_k &= \sum_{0 \leq j_1 < \cdots < j_k \leq 2^n - 1} j_1 \cdots j_k \\
&\quad + T \left(\sum_{0 \leq j_1 < \cdots < j_k \leq 2^n - 1} \sum_{h=1}^{k} j_1 \cdots \hat{j_h} \cdots j_k \prod_{i=1}^{n} U_i^{[j_h]_i} \right) \qquad (3.2) \\
&\quad + \text{terms of higher degree in } T.
\end{aligned}
$$

Let us denote by L_k the coefficient of T in the representation (3.2), namely:

$$L_k := \sum_{0 \le j_1 < \cdots < j_k \le 2^n - 1} \sum_{h=1}^{k} j_1 \cdots \hat{j_h} \cdots j_k \prod_{i=1}^{n} U_i^{[j_h]_i}.$$

For later use we are now going to show the following result, for whose proof we are indebted to G. Matera.

Lemma 1 *The polynomials $L_1, \ldots, L_{2^n}$ are $\mathbb{Q}$-linearly independent in $\mathbb{Q}[U_1, \ldots, U_n]$.*

Proof Let us abbreviate $N := 2^n - 1$. We observe that for $1 \le k \le N + 1$ and $0 \le j \le N$ the coefficient $\ell_{k,j}$ of the monomial $\prod_{i=1}^{n} U_i^{[j]_i}$ occuring in the polynomial L_k can be represented as

$$\ell_{k,j} = \sum_{\substack{0 \le j_1 < \cdots < j_{k-1} \le N \\ j_r \ne j \text{ for } r=1,\ldots,k-1}} j_1 \cdots j_{k-1}.$$

Claim: *For fixed N and k, the coefficient $\ell_{k,j}$ can be written as a polynomial expression of degree exactly $k - 1$ in the index j. Moreover, this polynomial expression for $\ell_{k,j}$ has integer coefficients.*

Proof of the Claim. We proceed by induction on the index parameter k.

For $k = 1$ we have $\ell_{1,j} = 1$ for any $0 \le j \le N$ and therefore $\ell_{1,j}$ is a polynomial of degree $k - 1 = 0$ in the index j.

Let $1 \le k \le N + 1$. Assume inductively that $\ell_{k,j}$ is a polynomial of degree exactly $k - 1$ in the index j and that the coefficients of this polynomial are integers. We are now going to show that $\ell_{k+1,j}$ is a polynomial of degree exactly k in j and that the coefficients of this polynomial are integers too. Observe that

$$\ell_{k+1,j} = \sum_{\substack{0 \le j_1 < \cdots < j_k \le N \\ j_r \ne j \text{ for } r=1,\ldots,k}} j_1 \cdots j_k$$

$$= \sum_{0 \le j_1 < \cdots < j_k \le N} j_1 \cdots j_k - j \left(\sum_{\substack{0 \le j_1 < \cdots < j_{k-1} \le N \\ j_r \ne j \text{ for } r=1,\ldots,k-1}} j_1 \cdots j_{k-1} \right).$$

holds. Since the term

$$\sum_{0 \le j_1 < \cdots < j_k \le N} j_1 \cdots j_k$$

does not depend on j and since by induction hypothesis

$$\ell_{k,j} = \sum_{\substack{0 \leq j_1 < \cdots < j_{k-1} \leq N \\ j_r \neq j \text{ for } r=1,\ldots,k-1}} j_1 \cdots j_{k-1}$$

is a polynomial of degree exactly $k - 1$ in j, we conclude that $\ell_{k+1,j}$ is a polynomial of degree exactly k in j. Moreover, the coefficients of this polynomial are integers. This proves our claim.

It is now easy to finish the proof of Lemma 1. By our claim there exist for arbitrary $1 \leq k \leq N + 1$ integers $c_0^{(k)}, \cdots, c_{k-1}^{(k)}$ with $c_{k-1}^{(k)} \neq 0$ such that for any $0 \leq j \leq N$ the identity $\ell_{k,j} = c_0^{(k)} + \cdots + c_{k-1}^{(k)} j^{k-1}$ holds. Hence for arbitrary $0 \leq k \leq N$ there exist rational numbers $\lambda_1^{(k)}, \ldots, \lambda_{k+1}^{(k)}$ (not depending on j) such for any $0 \leq j \leq N$ the condition

$$j^k = \lambda_1^{(k)} \ell_{1,j} + \cdots + \lambda_{k+1}^{(k)} \ell_{k+1,j}$$

is satisfied (here we use the convention $0^0 := 1$). This implies for any index $0 \leq k \leq N$ the polynomial identity

$$\lambda_1^{(k)} L_1 + \cdots + \lambda_{k+1}^{(k)} L_{k+1} = \sum_{0 \leq j \leq N} j^k \prod_{i=1}^{n} U_i^{[j]_i}.$$

Therefore for any $0 \leq k \leq N$ the polynomial $P_k \sum_{0 \leq j \leq N} j^k \prod_{i=1}^{n} r_i^{[j]_i}$
belongs to the $\mathbb{Q}$-vector space generated by $L_1, \ldots, L_{N+1}$.

On the other hand, we deduce from the nonsingularity of the Vandermonde matrix $(j^k)_{0 \leq k,j \leq N}$ that the polynomials $P_0, \ldots, P_N$ are $\mathbb{Q}$-linearly independent. Therefore the $\mathbb{Q}$-vector space generated by $L_1, \ldots, L_{N+1}$ in $\mathbb{Q}[U_1, \ldots, U_n]$ has dimension $N + 1 = 2^n$. This implies that $L_1, \ldots, L_{2^n}$ are $\mathbb{Q}$-linearly independent. $\square$

With the notations of Section 2.2, put now $r := n + 1$, $T := U_{n+1}$ and let us consider the following polynomials of $\mathbb{Q}[T, U_1, \ldots, U_n, X_1, \ldots, X_n]$:

$$G_1 := X_1{}^2 - X_1, \ldots, G_n := X_n{}^2 - X_n \quad \text{and}$$

$$F := \sum_{i=1}^{n} 2^{i-1} X_i + T \prod_{i=1}^{n} (1 + (U_i - 1) X_i). \tag{3.3}$$

It is clear from their definition that the polynomials $G_1, \ldots, G_n$ and F can be evaluated by a (non-invariant) totally division-free arithmetic circuit β of size $O(n)$ in $\mathbb{Q}[T, U_1, \ldots, U_n, X_1, \ldots, X_n]$. Observe that the polynomials $G_1, \ldots, G_n$ do not depend on the $T, U_1, \ldots, U_n$ and that their degree is two. The polynomial F is of degree $2n + 1$. More precisely, we have $\deg_{X_1,\dots,X_n} F = n$, $\deg_{U_1,\dots,U_n} F = n$, and $\deg_T F = 1$. Although the polynomial F can be evaluated using $O(n)$ arithmetic operations, the sparse representation of F, as a polynomial in the variables $T, U_1, \dots, U_n, X_1, \dots, X_n$, contains asymptotically 2^n non-zero monomial terms.

Let us now verify that the polynomials $G_1, \dots, G_n$ and F form a flat family of elimination problems depending on the parameters $T, U_1, \dots, U_n$.

The variety $V := \{G_1 = 0, \dots, G_n = 0\}$ is nothing but the union of 2^n affine linear subspaces of $\mathbb{A}^{n+1} \times \mathbb{A}^n$, each of them of the form $\mathbb{A}^{n+1} \times \{\xi\}$, where ξ is any point of the hypercube $\{0,1\}^n$. The canonical projection $\mathbb{A}^{n+1} \times \mathbb{A}^n \to \mathbb{A}^{n+1}$ induces a morphism $\pi : V \to \mathbb{A}^{n+1}$ which glues together the canonical projections $\mathbb{A}^{n+1} \times \{\xi\} \to \mathbb{A}^{n+1}$ for any ξ in $\{0,1\}^n$. Obviously the morphism π is finite and generically unramified. In particular π has constant fibres which are all canonically isomorphic to the hypercube $\{0,1\}^n$. Let $(j_1, \dots, j_n)$ be an arbitrary point of $\{0,1\}^n$ and let $j := \sum_{1 \le i \le n} j_i 2^{i-1}$ be the integer $0 \le j < 2^n$ whose bit representation is $j_n j_{n-1} \dots j_1$. One verifies immediately the identity

$$F(T, U_1, \dots, U_n, j_1, \dots, j_n) = j + T \prod_{i=1}^{n} U_i^{j_i}.$$

Therefore for any point $(t, u_1, \dots, u_n, j_1, \dots, j_n) \in V$ with $j := \sum_1^n j_i 2^{i-1}$ we have

$$F(t, u_1, \dots, u_n, j_1, \dots, j_n) = j + T \prod_{i=1}^{n} u_i^{j_i}.$$

From this observation we deduce easily that the polynomial

$$P \in \mathbb{Q}[T, U_1, \dots, U_n, Y]$$

we are looking for (as the elimination polynomial associated to the flat family $G_1, \dots, G_n, F$) is in fact

$$P = \prod_{j=0}^{2^n-1} (Y - (j + T \prod_{i=1}^{n} U_i^{[j]_i})).$$

With the notations of the beginning of this section, this polynomial has the form

$$P = Y^{2^n} + \sum_{1 \le k \le 2^n} A_k Y^{2^n-k} \equiv Y^{2^n} + \sum_{1 \le k \le 2^n} (a_k + TL_k) Y^{2^n-k} \text{ modulo } T^2,$$

$$(3.4)$$

with $a_k := \sum_{1 \le j_1 < \ldots < j_k \le 2^n - 1} j_1 \ldots j_k$ for $1 \le k \le 2^n$.

Suppose now that there is given a geometrically robust parametric elimination procedure. This procedure produces from the input circuit β an invariant arithmetic circuit Γ in $\mathbb{Q}[T, U_1, \ldots, U_n, Y]$, with scalars in $\mathbb{Q}[T, U_1, \ldots, U_n]$, which evaluates the polynomial P. Recall that the invariance of Γ means that the scalars of Γ are invariant polynomials of $\mathbb{Q}[T, U_1, \ldots, U_n]$, say $\Omega_1, \ldots, \Omega_N$.

Let $\mathcal{L}(\Gamma)$ be the total and $L(\Gamma)$ the non-scalar size of the arithmetic circuit Γ over $\mathbb{Q}[T, U_1, \ldots, U_n]$. Without loss of generality we have $L(\Gamma) \le \mathcal{L}(\Gamma)$ and $N \le (L(\Gamma) + 3)^2$.

Let $Z_1, \ldots, Z_N$ be new indeterminates. From the graph structure of the circuit Γ we deduce that there exists for each $1 \le k \le 2^n$ a polynomial $Q_k \in \mathbb{Q}[Z_1, \ldots, Z_N]$ satisfying the condition

$$Q_k(\Omega_1, \ldots, \Omega_N) = A_k \qquad (3.5)$$

(see [3] for details). Let us now consider two arbitrary elements $u, u' \in \mathbb{A}^n$. Observe that the parameter points $(0, u)$ and $(0, u')$ of $\mathbb{A}^{n+1}$ are equivalent (in symbols: $(0, u) \sim (0, u')$). From the invariance of $\Omega_1, \ldots, \Omega_N$ we deduce therefore that $\Omega_j(0, u) = \Omega_j(0, u')$ holds for any $1 \le j \le N$. This means that the polynomials

$$\omega_1 = \Omega_1(0, U_1, \ldots, U_n), \ldots, \omega_N := \Omega_N(0, U_1, \ldots, U_n)$$

are independent from the variables $U_1, \ldots, U_n$ and therefore elements of $\mathbb{Q}$. From identity (3.4) we conclude that the same is true for $\alpha_1 := A_1(0, U_1, \ldots, U_n), \ldots, \alpha_{2^n} := A_{2^n}(0, U_1, \ldots, U_n)$. We shall abbreviate $\omega := (\omega_1, \ldots, \omega_N)$ and $\alpha := (\alpha_1, \ldots, \alpha_{2^n})$.

Let us consider the morphisms of affine spaces $\mu : \mathbb{A}^{n+1} \longrightarrow \mathbb{A}^N$ and $\psi : \mathbb{A}^N \longrightarrow \mathbb{A}^{2^n}$ given by $\mu := (\Omega_1, \ldots, \Omega_N)$ and $\psi := (Q_k)_{1 \le k \le 2^n}$. Observe that

$$\psi \circ \mu = (Q_k(\Omega_1, \ldots, \Omega_N))_{1 \le k \le 2^n} = (A_k)_{1 \le k \le 2^n}$$

holds. From our previous considerations we deduce the identities

$$(\psi \circ \mu)(0, U_1, \ldots, U_n)$$
$$= (Q_k(\Omega_1(0, U_1, \ldots, U_n), \ldots, \Omega_N(0, U_1, \ldots, U_n)))_{1 \le k \le 2^n}$$
$$= (Q_k(\omega))_{1 \le k \le 2^n}.$$

In particular we have $\psi(\omega) = \alpha$.

We analyse now the local behaviour of the morphism ψ in the point

$\omega \in \mathbb{A}^N$. Let E_ω and E_α be the tangent spaces of the points $\omega \in \mathbb{A}^N$ and $\alpha \in \mathbb{A}^{2^n}$. Let us denote the differential of the map ψ in the point ω by $(D\psi)_\omega : E_\omega \longrightarrow E_\alpha$. Taking the canonical projections of $\mathbb{A}^N$ and $\mathbb{A}^{2^n}$ as local coordinates in the points ω and α respectively, we identify E_ω with $\mathbb{A}^N$ and E_α with $\mathbb{A}^{2^n}$.

For any point $u \in \mathbb{Q}^n$ we consider the parametric algebraic curves $\gamma_u : \mathbb{A}^1 \longrightarrow \mathbb{A}^N$ and $\delta_u : \mathbb{A}^1 \longrightarrow \mathbb{A}^{2^n}$ defined as

$$\gamma_u := (\Omega_1(T, u), \dots, \Omega_N(T, u)) \text{ and } \delta_u := (A_k(T, u))_{1 \le k \le 2^n}.$$

Observe that $\psi \circ \gamma_u = \delta_u$ and that $\gamma_u(0) = \omega$, $\delta_u(0) = \alpha$ holds (independently of the point u).

Now fix $u \in \mathbb{Q}^n$ and consider

$$\gamma'_u(0) = (\frac{\partial \Omega_1}{\partial T}(0, u), \dots, \frac{\partial \Omega_N}{\partial T}(0, u)) \text{ and } \delta'_u(0) = (\frac{\partial A_k}{\partial T}(0, u))_{1 \le k \le 2^n}.$$

We have $\gamma'_u(0) \in E_\omega$ and $\delta'(0) \in E_\alpha$. From (3.4) we deduce that for any $1 \le k \le 2^n$ the identity $\frac{\partial A_k}{\partial T}(0, u)) = L_k(u)$ holds. Therefore we have $\delta'_u(0) = (L_k(u))_{1 \le k \le 2^n}$. This implies

$$(D\psi)_\omega(\gamma'_u(0)) = \delta'_u(0) = (L_k(u))_{1 \le k \le 2^n}. \tag{3.6}$$

holds. From Lemma 1 we deduce that there exist for $1 \le l \le 2^n$ rational points $u_l \in \mathbb{Q}^n$ such that the matrix $(L_k(u_l))_{1 \le k, l \le 2^n}$ is regular. From (3.6) we deduce now that the $2^n \times N$-matrix built by the row vectors $\gamma'_{u_1}(0), \dots, \gamma'_{u_{2^n}}(0)$ has rank at least 2^n. This implies the lower bound $N \ge 2^n$.

Therefore we have $2^n \le N \le (L(\Gamma) + 3)^2$ and hence the estimate $2^{\frac{n}{2}} - 3 \le L(\Gamma) \le \mathcal{L}(\Gamma)$. *We have therefore shown that any geometrically robust parametric elimination procedure applied to the flat family of elimination problems (3.3) produces a solution circuit of size at least $2^{\frac{n}{2}} - 3$, i.e. a circuit of exponential size in the length $O(n)$ of the input.*

From the previous example we deduce that the goal of a polynomial time procedure for geometric (or algebraic) elimination can not be reached just by means of an improvement of known, commutative algebra-based elimination methods.

On the other hand, our proof method is not very specific for elimination problems. This can be visualised by the following example:

Let $\mathcal{A}$ be the $\mathbb{Q}$–subalgebra of $\mathbb{Q}[T, U_1, \dots, U_n]$ generated by the coefficients of the polynomial F of (3.3) with respect to the variables $X_1, \dots, X_n$. In the same manner as above, one may show that in the non–scalar complexity model with respect to $\mathcal{A}$, any totally division–free arithmetic circuit which evaluates the polynomial F using only scalars from the $\mathbb{Q}$–algebra $\mathcal{A}$, has necessarily exponential size in n.

On the other hand, the polynomial F can be evaluated by a totally division–free arithmetic circuit in $\mathbb{Q}[T, U_1, \ldots, U_n, X_1, \ldots, X_n]$ of size $O(n)$.

Therefore we see that in the non-scalar model the (sequential time) complexity of a polynomial depends strongly on the structure of the algebra of scalars.

It was fundamental for our argumentation above that our notion of a *geometrically robust parametric elimination procedure* excludes branchings and divisions in the output program. We resume the conclusions from the complexity discussion of our example in the following form.

Theorem 2 ([23]) *For any $n \in \mathbb{N}$ there exists a flat family of elimination problems depending on $n+1$ parameters and n variables over $\mathbb{Q}$ and having input length $O(n)$, such that the following holds: any geometrically robust parametric elimination procedure which solves this problem produces an output circuit of size at least $2^{\frac{n}{2}} - 3$ (i.e. of exponential size in the input length).*

3.2 Circuit encoding of polynomials

We are now going to explain how we may encode polynomials of a certain complexity class by their values in suitable test sequences.

Let L and n be given natural numbers which we think fixed for the moment and let $X_1, \ldots, X_n$ be indeterminates over $\bar{k}$. In this section we shall only consider totally division-free arithmetic circuits in $\bar{k}[X_1, \ldots, X_n]$ and we shall only consider the non-scalar sequential complexity model over $\bar{k}$.

By $\mathcal{H} := \mathcal{H}_{L,n}$ we denote the *complexity class* of all polynomials $H \in \bar{k}[X_1, \ldots, X_n]$ which can be evaluated by a totally division-free arithmetic circuit in $\bar{k}[X_1, \ldots, X_n]$ of size at most L.

For any polynomial $H \in \mathcal{H}$ we have $\deg H \leq 2^L$. On the other hand we have e.g. $X_1^{2^L} \in \mathcal{H}$. This implies that 2^L is an *exact* degree bound for the elements of $\mathcal{H}$. Let $D := 2^L + 1$. Since the elements of $\mathcal{H}$ are contained in the D-dimensional $\bar{k}$-linear subspace of $\bar{k}[X_1, \ldots, X_n]$ consisting of the polynomials of degree at most $2^L = D - 1$, we may consider $\mathcal{H}$ as a subset of $\mathbb{A}^D$. Observe that for any $H \in \mathcal{H}$ and any $\alpha \in \bar{k}$ the element αH belongs to $\mathcal{H}$. Let $\mathcal{W} := \mathcal{W}_{L,n}$ be the Zariski closure of $\mathcal{H} := \mathcal{H}_{L,n}$ in $\mathbb{A}^D$. Since $\mathcal{H}$ is a cone over $\bar{k}$, the same is true for $\mathcal{W}$. On the other hand,

$\mathcal{H}$ is the image of a k-definable morphism of affine spaces $\mathbb{A}^{(L+3)^2} \longrightarrow \mathbb{A}^D$. Thus, in conclusion, W is an *irreducible* algebraic variety of $\mathbb{A}^D$, definable by *homogeneous* polynomials belonging to $k[X_1, \ldots, X_n]$ (see [3], Chapter 9 for details).

In the sequel we shall interpret the points of $\mathbb{A}^D$ as polynomials of $\bar{k}[X_1, \ldots, X_n]$ having degree at most $D - 1$ and viceversa. In particular we shall interpret the points of W as polynomials.

Definition 1 *Let be given a sequence of points $\gamma_1, \ldots, \gamma_m \in k^n$ and let $\gamma := (\gamma_1, \ldots, \gamma_m)$. Let us call m the length of γ.*

(i) *We call γ a correct test sequence for the polynomials of*

$$\bar{k}[X_1, \ldots, X_n]$$

of non-scalar complexity at most L (i.e. for $\mathcal{H}_{L,n}$), if for any $H \in \mathcal{W}_{L,n}$ the following implication holds:

$$H(\gamma_1) = 0, \ldots, H(\gamma_m) = 0 \implies H = 0.$$

(ii) *We call γ an identification sequence for the polynomials of*

$$\bar{k}[X_1, \ldots, X_n]$$

of non-scalar complexity at most L (i.e. for $\mathcal{H}_{L,n}$), if for any two elements H_1, H_2 of $\mathcal{W}_{L,n}$ the following implication holds:

$$H_1(\gamma_1) = H_2(\gamma_1), \ldots, H_1(\gamma_m) = H_2(\gamma_m) \implies H_1 = H_2.$$

Although the polynomials of the complexity class $\mathcal{H}_{L,n}$ may have exponential degree 2^L, there exist short identification sequences for $\mathcal{H}_{L,n}$. This is the content of the next result.

Lemma 3 ([26, 29]) *Let the notation be as before. Let $M \subseteq k$ be a finite set of cardinality at least $4LD^2$ and let $m := 6(2L + n + 1)^2$. The there exist points $\gamma_1, \ldots, \gamma_m \in M^n$ such that $\gamma := (\gamma_1, \ldots, \gamma_m)$ is an identification sequence for $\mathcal{H}_{L,n}$. Suppose that the points of M^n are equidistributed. The probability of finding by a random choice such an identification sequence is at least $(1 - 4LD^{-12(2L+n+1)^2}) > 1/2$.*

Proof From [26, 29] we deduce that there exist in M^n correct test sequences of length $m := 6(2L + n + 1)^2$ for $\mathcal{H}_{2L,n}$ and that such a correct test sequence can be found with probability of success $(1 - 4LD^{-12(2L+n+1)^2})$ by a random choice in M^n.

Let $\gamma := (\gamma_1, \ldots, \gamma_m) \in k^{m \times n}$ be a correct test sequence for $\mathcal{H}_{2L,n}$. Let $H_1, H_2 \in \mathcal{W}_{L,n}$ and suppose that

$$H_1(\gamma_1) = H_2(\gamma_1), \ldots, H_1(\gamma_m) = H_2(\gamma_m)$$

holds. Observe that $H := H_2 - H_1$ belongs to $\mathcal{W}_{2L,n}$. Therefore we have $H(\gamma_1) = 0, \ldots, H(\gamma_m) = 0$, and, since γ is a correct test sequence for $\mathcal{H}_{2L,n}$, we may conclude $H = H_2 - H_1 = 0$. $\qquad \square$

Let now $m := 6(2L + n + 1)^2$. Then, by Lemma 3, we may fix an identification sequence $\gamma := (\gamma_1, \ldots, \gamma_m) \in k^{m \times n}$ for the complexity class $\mathcal{H} = \mathcal{H}_{L,n}$. Let $S_1, \ldots, S_m$ be new indeterminates denoting the canonical coordinate functions of $\mathbb{A}^m$. We are going to consider the morphism $\sigma_{L,n}^{(\gamma)} : \mathcal{W}_{L,n} \longrightarrow \mathbb{A}^m$, defined for $H \in \mathcal{W}_{L,n}$ by $\sigma_{L,n}^{(\gamma)}(H) :=$ $(H(\gamma_1), \ldots, H(\gamma_m))$. Let $S_{L,n}^{(\gamma)}$ be the Zariski closure of the image of the morphism $\sigma_{L,n}^{(\gamma)}$ and let us abbreviate $\sigma := \sigma_{L,n}^{(\gamma)}$ and $S := S_{L,n}^{(\gamma)}$.

Since the variety $\mathcal{W}$ and the morphism $\sigma : \mathcal{W} \longrightarrow S$ are k-definable, we conclude that S is k-definable too. Moreover for any polynomial $H \in \mathcal{W}$ and any value $\alpha \in \bar{k}$ we have

$$\sigma(\alpha H) = (\alpha H(\gamma_1), \ldots, \alpha H(\gamma_m)) = \alpha \sigma(H)$$

and from this *homogeneity* property of σ we conclude that S is a *cone* over the field $\bar{k}$. Therefore S is definable by homogeneous polynomials belonging to $k[S_1, \ldots, S_m]$. Since the variety $\mathcal{W}$ is irreducible and S is the closure of the image of σ, we conclude that S is irreducible too.

By assumption $\gamma := (\gamma_1, \ldots, \gamma_m)$ is an identification sequence for the complexity class $\mathcal{H} = \mathcal{H}_{L,n}$. Therefore we conclude that $\sigma : \mathcal{W} \longrightarrow S$ is an *injective*, dominant morphism. Hence σ is *birational*.

Let $\sigma = (\sigma_1, \ldots, \sigma_m)$ where $\sigma_1, \ldots, \sigma_m$ are suitable coordinate functions of the affine variety $\mathcal{W}$. Let us consider $\mathcal{W}$ as a closed subvariety of the affine space $\mathbb{A}^D$. We recall that the points of $\mathbb{A}^D$ correspond to the polynomials of $\bar{k}[X_1, \ldots, X_n]$ of degree at most D and that the morphism σ is defined by means of the evaluation of the polynomials of $\mathcal{W}$ in the points $\gamma_1, \ldots, \gamma_m$. This implies that there exist *linear forms* in the coordinate ring of $\mathbb{A}^D$ such that $\sigma_1, \ldots, \sigma_m$ are the restrictions of these linear forms to the variety $\mathcal{W}$. From the injectivity of σ we deduce that $\mathcal{W} \cap \{\sigma_1 = 0, \ldots, \sigma_m = 0\}$ contains only the origin of $\mathbb{A}^D$. This implies that the homogeneous map σ induces a finite morphism between the projective varieties associated to the cones $\mathcal{W}$ and S. In fact, the standard proof of this classical result implies something more, namely that also the morphism $\sigma : \mathcal{W} \longrightarrow S$ is *finite* (see [39] I.5.3, Theorem

8 and proof of Theorem 7). In particular, σ is a surjective, closed map and hence a homeomorphism with respect to the Zariski topologies of $\mathcal{W}$ and $\mathcal{S}$.

Thus we have shown the following result:

Lemma 4 *Let the notation be as before and let* $m := 6(2L + n + 1)^2$. *Suppose that there is given an identification sequence* $\gamma := (\gamma_1, \ldots, \gamma_m) \in k^{m \times n}$ *for the complexity class* $\mathcal{H}_{L,n}$ *and let* $\sigma_{L,n}^{(\gamma)} : \mathcal{W}_{L,n} \longrightarrow \mathcal{S}_{L,n}^{(\gamma)}$ *be the morphism of affine varieties associated to the identification sequence* γ. *Then* $\sigma_{L,n}^{(\gamma)}$ *is a finite, bijective and birational morphism of algebraic varieties and in particular* $\sigma_{L,n}^{(\gamma)}$ *is a homeomorphism with respect to the Zariski topologies of* $\mathcal{W}_{L,n}$ *and* $\mathcal{S}_{L,n}^{(\gamma)}$

Since the varieties $\mathcal{W}_{L,n}$ and $\mathcal{S}_{L,n}^{(\gamma)}$ are irreducible, we conclude that their coordinate rings $k[\mathcal{W}_{L,n}]$ and $k[\mathcal{S}_{L,n}^{(\gamma)}]$ are domains with function fields $k(\mathcal{W}_{L,n})$ and $k(\mathcal{S}_{L,n}^{(\gamma)})$. The morphism $\sigma_{L,n}^{(\gamma)}$ induces an embedding of the coordinate ring $k[\mathcal{S}_{L,n}^{(\gamma)}]$ into $k[\mathcal{W}_{L,n}]$ (and the same is true for the corresponding function fields). Moreover, the finiteness of the morphism $\sigma_{L,n}^{(\gamma)}$ means that $k[\mathcal{W}_{L,n}]$ is an *integral* ring extension of $k[\mathcal{S}_{L,n}^{(\gamma)}]$.

Disregarding the complexity aspect, Lemma 4 says that it is possible to reconstruct the coefficients of a polynomial $H \in \mathcal{H}_{L,n}$ from the values of H in a given identification sequence γ, even if the sequence γ is short in comparison with the degree of H. Lemma 4 says further that this reconstruction is rational (i.e. it uses only arithmetical operations) and that possibly occuring divisions may always be performed by limit processes in the spirit of de l'Hôpital's rule. These processes produce only *finite* limits, because the coefficients of H are integrally dependent on the values of H in the given identification sequence γ. In algebraic terms, we may modelise these limit processes by *places* (corresponding to valuation rings) which map the function field $k(\mathcal{S}_{L,n}^{(\gamma)})$ of the variety $\mathcal{S}_{L,n}^{(\gamma)}$ into the set of values $\bar{k} \cup \{\infty\}$ and take only finite values on $k[\mathcal{S}_{L,n}^{(\gamma)}]$. The finiteness of the limits mentioned before is modelised by the requirement that any extension of such a place to the function field $k(\mathcal{W}_{L,n})$ takes finite values on $k[\mathcal{W}_{L,n}]$. This requirement is satisfied, because $k[\mathcal{W}_{L,n}]$ is integral over $k[\mathcal{S}_{L,n}^{(\gamma)}]$ (see [31] for details about places and integral extensions).

3.3 The complexity of parametric elimination procedures

In this section we suppose that the ground field k is of characteristic zero. Let n be a fixed natural number, $m(n) := 6(3n+1)^2, N(n) := (n+3)^2$ and $X_1, \ldots, X_n, Z_1, \ldots, Z_{N(n)}$ and $Y, S_1, \ldots, S_{m(n)}$ indeterminates over k.

Let us fix an identification sequence $\gamma := (\gamma_1, \ldots, \gamma_{m(n)}) \in k^{m(n) \times n}$ for the polynomials of $\bar{k}[X_1, \ldots, X_n]$ having non-scalar complexity at most n (i.e. for the complexity class $\mathcal{H}_{n,n}$).

From [3], Chapter 9, Theorem 9.9 we deduce that there exists a polynomial

$$R_n \in k[Z_1, \ldots, Z_{N(n)}, X_1, \ldots, X_n]$$

satisfying the following conditions,

- R_n can be evaluated by a totally division-free arithmetic circuit in $k[Z_1, \ldots, Z_{N(n)}, X_1, \ldots, X_n]$ of non-scalar size $N(n)$
- for any $H \in \mathcal{H}_{n,n}$ and any totally division-free arithmetic circuit β in $\bar{k}[X_1, \ldots, X_n]$, such that β has non-scalar size n, scalars $\zeta_1, \ldots, \zeta_{N(n)} \in \bar{k}$ and such that β evaluates the polynomial H, we have

$$H = R_n(\zeta_1, \ldots, \zeta_{N(n)}, X_1, \ldots, X_n).$$

Let us consider the following existential formula $\Phi_n(S_1, \ldots, S_{m(n)}, Y)$ in the free indeterminates ("free variables" in the terminology of mathematical logic) $S_1, \ldots, S_{m(n)}, Y$,

$$(\exists X_1), \ldots, (\exists X_n)(\exists Z_1), \ldots, (\exists Z_{N(n)})(\bigwedge_{1 \le i \le n} X_i^2 - X_i = 0 \, \wedge$$

$$\bigwedge_{1 \le k \le m(n)} S_k = R_n(Z_1, \ldots, Z_{N(n)}, \gamma_k) \wedge Y = R_n(Z_1, \ldots, Z_{N(n)}, X_1, \ldots, X_n)).$$

Observe that the existential formula $(\exists Y)\Phi_n(S_1, \ldots, S_{m(n)}, Y)$ describes the set $\sigma_{n,n}^{(\gamma)}(\mathcal{H}_{n,n})$. Thus, the formula $\Phi_n(S_1, \ldots, S_{m(n)}, Y)$ introduces an implicit semantical dependence between the indeterminates $S_1, \ldots, S_{m(n)}$. In the sequel we shall consider the inteterminates $S_1, \ldots, S_{m(n)}$ as parameters and Y as variable.

Let us finally remark that the formula $\Phi_n(S_1, \ldots, S_{m(n)}, Y)$ may be represented by a totally division-free circuit in

$$k[X_1, \ldots, X_n, Z_1, \ldots, Z_{N(n)}, S_1, \ldots, S_{m(n)}, Y]$$

of size $O(n^4)$.

Let us now consider an arbitrary (universal) *parametric* elimination procedure Π with associated sequential time complexity measure $\mathcal{T}$ and suppose that Π and $\mathcal{T}$ satisfy the following conditions:

(i) For any totally division-free arithmetic input network β of total size $\mathcal{L}(\beta)$ such that β represents an existential input formula Φ in the elementary language of algebraic geometry over the field k, the elimination procedure Π is able to produce a totally division-free arithmetic output network Γ with $\mathcal{L}(\Gamma) \leq \mathcal{T}(\mathcal{L}(\beta))$, such that Γ represents a quantifier-free formula which is semantically equivalent to Φ.

(ii) For any totally division-free arithmetic input network β_1 representing a constructible subset $\mathcal{X}$ of an appropriate affine space, the procedure Π is able to produce a totally division-free arithmetic output circuit Γ_1 with $\mathcal{L}(\Gamma_1) \leq \mathcal{T}(\mathcal{L}(\beta_1))$, such that Γ_1 represents a suitable polynomial equation system for the Zariski closure of the constructible set $\mathcal{X}$.

(iii) Let $U_1, \ldots, U_r$ and Y be indeterminates and let be given a positive number s of nonzero polynomials, say $B_1, \ldots, B_s \in k[U_1, \ldots, U_r, Y]$. Let $V := \{B_1 = 0, \ldots, B_s = 0\}$. Suppose that V is nonempty. Consider the morphism of affine varieties $\pi : V \longrightarrow \mathbb{A}^r$, induced by the canonical projection of $\mathbb{A}^r \times \mathbb{A}^1$ onto $\mathbb{A}^r$. Let S be the Zariski closure of $\pi(V)$ and assume that S is an irreducible closed subvariety of $\mathbb{A}^r$. Suppose that the polynomials $B_1, \ldots, B_s$ are given by a totally division-free arithmetic circuit β_2 in $k[U_1, \ldots, U_r, Y]$ and suppose that they satisfy all our requirements at the end of Section 2.2. Hence there exists a well–defined parametric greatest common divisor $h \in k(S)[Y]$ associated to the polynomials $B_1, \ldots, B_s$.

Then the procedure Π is able to produce an essentially division-free arithmetic circuit Γ_2 in $k(U_1, \ldots, U_r)[Y]$ with $\mathcal{L}(\Gamma_2) \leq \mathcal{T}(\mathcal{L}(\beta_2))$ such that Γ_2 computes a representative $H \in k(U_1, \ldots, U_r)[Y]$ of the (generically squarefree) parametric greatest common divisor $h \in k(S)[Y]$ and such that Γ_2 satisfies the requirements formulated at the end of Section 2.2 for such a circuit. Moreover, the same holds true for the generically squarefree parametric greatest common divisor associated to the polynomials $B_1, \ldots, B_s$.

In view of conditions (*ii*) and (*iii*) above, we say that the parametric elimination procedure Π *computes efficiently Zariski closures and (generically squarefree) greatest common divisors.*

The requirement that Γ_1 and Γ_2 are *circuits* (and not arithmetic networks) contains implicitly the meaning that the procedure Π is *branching parsimonious* (see Section 2). Similarly, the behaviour of the circuit Γ_2 under specialisation by places expresses the requirement that the procedure Π is *division parsimonious*.

Let us finally observe that, after a suitable adaption of the data structures, all known universal elimination procedures satisfy with respect to appropriate sequential time complexity measures our conditions $(i), (iii)$ and (iii) (see Section 3.4 for more details).

We are now ready to state the main result of this paper.

Theorem 5 *Assume that the ground field k is of characteristic zero. Let Π be an arbitrary parametric elimination procedure with associated sequential time complexity measure $\mathcal{T}$ and suppose that Π and $\mathcal{T}$ satisfy conditions $(i), (iii)$ and (iii) above. There exists a family of totally division-free arithmetic input circuits $\beta = (\beta_n)_{n \in \mathbb{N}}$ such that each β_n has total size $n^{O(1)}$ and represents a parametric family of polynomial equation systems with a well defined, canonical elimination polynomial P_n. The elimination procedure Π produces for each $n \in \mathbb{N}$ from the input circuit β_n an essentially division-free output circuit Γ_n which represents the elimination polynomial P_n. The total size of the output circuit Γ_n cannot be polynomial in the parameter n. Therefore $\mathcal{T}$ is not a polynomial function.*

Proof Let be given a parametric elimination procedure Π with associated sequential time complexity measure $\mathcal{T}$, as in the statement of the theorem. Let $n \in \mathbb{N}$. We apply first the procedure Π to any totally division-free arithmetic circuit β_n of size $O(n^4)$ which represents the formula $\Phi_n(S_1, \ldots, S_{m(n)}, Y)$. In virtue of condition (i) above, we obtain as output a totally division-free arithmetic network $\beta_1^{(n)}$ of size $\mathcal{T}(O(n^4))$ representing a quantifier-free formula which is semantically equivalent to $\Phi_n(S_1, \ldots, S_{m(n)}, Y)$. Applying now the procedure Π to the network $\beta_1^{(n)}$, we obtain in virtue of condition (ii) above a totally division-free circuit $\beta_2^{(n)}$ of size $\mathcal{T}^2(O(n^4))$ which represents a positive number s of polynomials, say $B_1, \ldots, B_s \in k[S_1, \ldots, S_{m(n)}, Y]$, such that $B_1 = 0, \ldots, B_s = 0$ forms a polynomial equation system for the Zariski closure of the constructible subset of the affine space $\mathbb{A}^{m(n)} \times \mathbb{A}^1$, defined by the formula $\Phi_n(S_1, \ldots, S_{m(n)}, Y)$. This equation system contains polynomials of $k[S_1, \ldots, S_{m(n)}]$ which determine the Zariski closed subset $\mathcal{S} := \mathcal{S}_{n,n}^{(\gamma)}$ of $\mathbb{A}^{m(n)}$ (recall that the existential formula

$$(\exists Y)\Phi_n(S_1, \ldots, S_{m(n)}, Y)$$

describes the set $\sigma_{n,n}^{(\gamma)}(\mathcal{H}_{n,n})$ and that $\mathcal{S}_{n,n}^{(\gamma)}$ is its Zariski closure).

Let $b_1, \ldots, b_s \in k[\mathcal{S}][Y]$ be the polynomials in Y with coefficients in $k[\mathcal{S}]$ induced by $B_1, \ldots, B_s$ and observe that not all polynomials $b_1, \ldots, b_s$ are zero. We consider now an arbitrary point

$$u = (u_1, \ldots, u_{m(n)}) \in \sigma_{n,n}^{(\gamma)}(\mathcal{H}_{n,n}).$$

There exists a point $\zeta = (\zeta_1, \ldots, \zeta_{N(n)}) \in \mathbb{A}^{N(n)}$ such that

$$E := R_n(\zeta, X_1, \ldots, X_n) := R_n(\zeta_1, \ldots, \zeta_{N(n)}, X_1, \ldots, X_n)$$

satisfies the condition $u = \sigma_{n,n}^{(\gamma)}(E) = (R_n(\zeta, \gamma_1), \ldots, R_n(\zeta, \gamma_{m(n)}))$.

From Lemma 4 we deduce that the polynomial $E \in \bar{k}[X_1, \ldots, X_n]$ depends only on the point u and not on the particular choice of $\zeta \in \mathbb{A}^{N(n)}$. Let us therefore write $E_u := E$. Let $P_u := \prod_{(\epsilon_1, \ldots, \epsilon_n) \in \{0,1\}^n}(Y - E_u(\epsilon_1, \ldots, \epsilon_n))$.

Interpreting now the formula $\Phi_n(S_1, \ldots, S_{m(n)}, Y)$ semantically, we see that $\Phi_n(u_1, \ldots, u_{m(n)}, Y)$ is equivalent to $P_u(Y) = 0$.

For a suitable choice of $u \in \sigma_{n,n}^{(\gamma)}(\mathcal{H}_{n,n})$ (e.g. choosing u with $H_u = \sum_{i=1}^n 2^{i-1} X_i$) we obtain a *separable* polynomial P_u.

There exists therefore a nonempty Zariski open subset $\mathcal{U} \subset \mathcal{S}$, contained in $\sigma_{n,n}^{(\gamma)}(\mathcal{H}_{n,n})$, such that for $u \in \mathcal{U}$ the greatest common divisor of the non–zero elements of $b_1(u, Y) := B_1(u, Y), \ldots, b_s(u, Y) := B_s(u, Y)$ has the same zeroes in $\bar{k}$ as the polynomial $P_u(Y)$ and such that $P_u(Y)$ is separable.

Let $h \in k(\mathcal{S})[Y]$ be the (normalised) greatest common divisor of the non–zero elements between the polynomials $b_1, \ldots, b_s$ and let $\hat{h} \in k(\mathcal{S})[Y]$ be the unique monic sparable polynomial with the same roots as h in an algebraic closure of $k(\mathcal{S})$. Without loss of generality we may assume that for any point $u \in \mathcal{U}$ the specialised polynomial $\hat{h}(u, Y)$ is a well–defined element of $\bar{k}[Y]$ with $\hat{h}(u, Y) = P_u(Y)$.

Let $\mathcal{A}$ be the integral closure of the coordinate ring $k[\mathcal{S}]$ in its fraction field $k(\mathcal{S})$. From Lemma 4 we deduce now that $\hat{h}$ and h belong to the polynomial ring $\mathcal{A}[Y]$ and that the polynomials $B_1, \ldots, B_s$ satisfy the requirements of the end of Section 2.2. In other words, the polynomial h is the parametric greatest common divisor associated to $B_1, \ldots, B_s$ and $\hat{h}$ is its generically squarefree counterpart.

Let us now consider $S_1, \ldots, S_{m(n)}$ as parameters and Y as variable.

Applying finally the procedure Π to the circuit $\beta_2^{(n)}$ we obtain by virtue of condition (iii) above an essentially division-free arithmetic circuit Γ_n^* in $k(S_1, \ldots, S_{m(n)})[Y]$ which computes a representative of the generically squarefree greatest common divisor $\hat{h} \in \mathcal{A}[Y]$. The size of Γ_n^* is $\mathcal{T}^3(O(n^4))$ and the circuit Γ_n^* satisfies the requirements of the end of Section 2.2. From condition (iii) above we deduce that all scalars of Γ_n^* are rational functions of $k(S_1, \ldots, S_{m(n)})$ which represent elements of the ring $\mathcal{A}$.

We consider now the polynomial $F := \sum_{i=1}^n 2^{i-1} X_i + T \prod_{i=1}^n (1 + (U_i - 1)X_i)$ introduced in Section 3.1. This polynomial can be represented by a totally division-free circuit of non-scalar size n over $k[U_1, \ldots, U_n, T]$.

Let $A_1 := F(U_1, \ldots, U_n, T, \gamma_1), \ldots, A_{m(n)} := F(U_1, \ldots, U_n, T, \gamma_{m(n)})$. Then $A_1, \ldots, A_{m(n)}$ are invariant polynomials of $k[U_1, \ldots, U_n, T]$. Specialising now the parameters $S_1, \ldots, S_{m(n)}$ into the polynomials $A_1, \ldots, A_{m(n)}$, we obtain a k–algebra homomorphism

$$k[\mathcal{S}] \longrightarrow k[A_1, \ldots, A_{m(n)}].$$

We consider this homomorphism as a specialization of the coordinate ring $k[\mathcal{S}]$.

Since the circuit Γ_n^* satisfies the requirements of the end of Section 2.2, the scalars of Γ_n^* become now specialised into elements of the field $k(A_1, \ldots, A_{m(n)})$. These elements belong to the integral closure of the k–algebra $k[A_1, \ldots, A_{m(n)}]$ in its fraction field $k(A_1, \ldots, A_{m(n)})$. Since $k[A_1, \ldots, A_{m(n)}]$ is a k–subalgebra of the (integrally closed) polynomial ring $k[U_1, \ldots, U_r, T]$, we conclude that the above specialisation maps the scalars of the circuit Γ_n^* into elements of $k[U_1, \ldots, U_r, T]$ which are integral over $k[A_1, \ldots, A_{m(n)}]$. This implies that the scalars of Γ_n^* are specialised into *invariant* polynomials of $k[U_1, \ldots, U_r, T]$.

Denote now by Γ_n the totally division–free invariant circuit of $k[U_1, \ldots, U_r, T, Y]$ obtained by specializing in Γ_n^* the scalars as explained before. Let $L(\Gamma_n)$ be the non–scalar size of the circuit Γ_n over $k[U_1, \ldots, U_r, T]$. Then we have $L(\Gamma_n) \geq \mathcal{L}(\Gamma_n^*) \geq \mathcal{T}^3(O(n^4))$ and Γ_n computes the image of $\hat{h}$ under the above specialisation, namely the elimination polynomial $P := \prod_{j=0}^{2^n-1}(Y - (j + T \prod_{i=1}^n U_i^{[j]_i}))$ of Section 3.1.

From the invariance of the circuit Γ_n and Theorem 2 and its proof we deduce now the estimate $L(\Gamma_n) \geq 2^{\frac{n}{2}} - 3$. On the other hand we have $\mathcal{T}^3(O(n^4)) \geq L(\Gamma_n)$. This implies that $\mathcal{T}$ cannot be a polynomial function. $\qquad\square$

3.4 State of the art in circuit-based elimination

Let us now analyse from a general non-uniform point of view how the seminumerical elimination procedure designed in [17] and [15] works on a given flat family of zero-dimensional elimination problems.

Let $U_1, \ldots, U_m, X_1, \ldots, X_n, Y$ be indeterminates over the ground field k and let $G_1, \ldots, G_n, F$ be polynomials belonging to the k-algebra

$$k[U_1, \ldots, U_m, X_1, \ldots, X_n].$$

Let $d := \max\{\deg G_1, \ldots, \deg G_n\}$ and suppose that $G_1, \ldots, G_n$ and F are given by straight-line programs in $k[U_1, \ldots, U_m X_1, \ldots, X_n]$ of length L and K respectively. Suppose that the polynomials $G_1, \ldots, G_n$ form a regular sequence in $k[U_1, \ldots, U_m, X_1, \ldots, X_n]$ defining thus an equidimensional subvariety $V = \{G_1 = 0, \ldots, G_n = 0\}$ of $\mathbb{A}^{m+n}$ of dimension m.

Assume that the morphism $\pi : V \longrightarrow \mathbb{A}^m$, induced by the canonical projection of $\mathbb{A}^{m+n}$ onto $\mathbb{A}^m$ is finite and generically unramified. Let δ be the degree of the variety V and let $D \leq \delta$ be the degree of the morphism π. Furthermore let $\tilde{\pi} : V \longrightarrow \mathbb{A}^{m+1}$ be the morphism of affine varieties defined by $\tilde{\pi}(z) := (\pi(z), F(z))$ for any point z of V. Let $P \in k[U_1, \ldots, U_m, Y]$ be the minimal polynomial of the image of $\tilde{\pi}$. The polynomial P is monic in Y and one sees immediately that $\deg P \leq \delta \deg F$ and $\deg_Y P \leq D$ holds. Let us write $\delta_* := \deg_{U_1, \ldots, U_m} P$.

Let us consider as Algorithm 1 and Algorithm 2 two non-uniform variants of the basic elimination method designed in [17] and [15].

- Algorithm 1 is represented by an arithmetic network of size $K\delta^{O(1)} + L(nd\Delta)^{O(1)}$ where Δ is the degree of the equation system $G_1 = 0, \ldots, G_n = 0$ (observe that always $\delta \leq \Delta \leq \deg G_1 \cdots \deg G_n$ holds). The output is a straight-line program Γ_1 in $k[U_1, \ldots, U_m, Y]$ of length $(K + L)(n\delta)^{O(1)}$ which represents the polynomial P.

- Algorithm 2 starts from the geometric description of a unramified parameter (and lifting) point $u = (u_1, \ldots, u_m)$ of k^m which has the additional property that the image of F restricted to the set $\{u\} \times \pi^{-1}(u)$ has cardinality $D_* = \deg_Y P$. The algorithm produces then an arithmetic circuit Γ_2 in $k[U_1, \ldots, U_m, Y]$ of length

$$O(KD^{O(1)} \log \delta_*) + \delta_*^{O(1)} = K(\delta \deg F)^{O(1)}$$

which represents the polynomial P.

We observe that $K\delta^{O(1)}$ is a characteristic quantity which appears in the length of both circuits Γ_1 and Γ_2. One may ask whether a complexity of type $K\delta^{O(1)}$ is intrinsic for the elimination problem under consideration.

In view of Bézout's Theorem and the lower complexity bound of Theorem 5, one may guess that a quantity of asymptotic order $K\delta$ may be characteristic for the intrinsic sequential time complexity of universal parametric elimination procedures.

4 Conclusions

It was fundamental for our argumentation in Section 3 that our notion of *parametric elimination procedure* restricts severely the possibility of branchings in the output circuit or network. This suggests that any polynomial time elimination algorithm (if there exists one) must have a huge topological complexity. Thus hypothetical efficiency in geometric elimination seems to imply complicated casuistics.

Let us also mention that the proof method of Section 3 contributes absolutely *nothing* to the elucidation of the fundamental thesis of algebraic complexity theory, which says that geometric elimination produces elimination polynomials which are intrinsically hard to evaluate (i.e. these polynomials need asymptotically degree–much sequential time for their evaluation). A good example for this thesis is the Pochhammer Wilkinson polynomial which can be obtained easily as the solution of a suitable elimination problem but which is conjecturally hard to evaluate (see e.g. [25, 1]) Similarly no advance is obtained by our method with respect to the question whether $P_{\mathbb{C}} \neq NP_{\mathbb{C}}$ holds in the BSS complexity model, see [2], Chapter7.

In fact our contribution consists only in the discovery of a very limiting uniformity property present in all known elimination procedures. This uniformity property inhibits the transformation of these elimination procedures into polynomial time algorithms.

In conclusion, when treating with algorithmic elimination problems, one should not expect too much from the output: as observed already by Hilbert, a strong limitation to *specific* elimination problems (with additional a priori information or additional semantical structure) is necessary in order to avoid huge difficulties and one should also renounce to compute the whole elimination object, if this is not required in advance. In particular, canonical elimination polynomials contain the complete information about the elimination problem under consideration, they are "co–versal", and this makes necessarily intricate their representation in any reasonable data–structure (for more details see [7]). A way out of this dilemma should be found in analysing which information about elimination objects is really relevant, avoiding in this way to struggle with enormous objects which encode a lot of spurious knowledge.

Acknowlegments

The authors wish to express their gratitude to Guillermo Matera and Rosa Wachenchauzer (Buenos Aires, Argentina) and to David Castro and Luis Miguel Pardo (Santander, Spain) for many fruitful discussions and ideas. They are deeply indepted to Jacques Morgenstern (Nice, France), who suggested this research and made together with the authors the first and most fundamental steps toward the complexity model and results of this paper.

References

[1] Blum, L., Cucker, F., Shub, M. and Smale, S. (1998). Algebraic settings for the problem $P \neq NP$, in *The Mathematics of Numerical Analysis*, eds J. Renegar, M. Shub & S. Smale, Lect. in Appld Maths **32** (American Maths Soc., Providence, RI), 125–144.

[2] Blum, L., Cucker, F., Shub, M. and Smale, S. (1998). *Complexity and Real Computation* (Springer, New York).

[3] Bürgisser, P., Clausen, M. and Shokrollahi, M.A. (1997). *Algebraic Complexity Theory* (Springer-Verlag, Berlin).

[4] Caniglia, L, Galligo, A. and Heintz, J. (1988). Borne simple exponentielle pour les degrés dans le théorème des zéros sur un corps de charactéristique quelconque. *C. R. Acad. Sci. Paris* **307**, 255–258.

[5] Caniglia, L, Galligo, A. and Heintz, J. (1989). Some new effectivity bounds in computational geometry. In *Applied Algebra, Algebraic Algorithms and Error Correcting Codes. Proceedings of AAECC-6*, ed. T. Mora (Springer-Verlag, Berlin), 131–152.

[6] Castro, D., Hägele, K., Morais, J.E. and Pardo, L.M. (2000). Kronecker's and Newton's approaches to solving: a first comparison, *J. Complexity*, to appear.

[7] Castro, D., Heintz, J., Giusti, M., Matera, G. and Pardo, L.M. (2000). Data structures and smooth interpolation procedures in elimination, to appear.

[8] Chistov, A.L. and Grigoriev, D.Y. (1983). Subexponential time solving systems of algebraic equations. LOMI preprint E-9-83, E-10-83 (Steklov Institute, Leningrad).

[9] Demazure, M. (1985), Le théorème de complexité de Mayr–Meyer, Tech. Rep., Ecole Polytechnique, Paris.

[10]   Dickenstein, A. Fitchas, N., Giusti, M. and Sessa, C. (1991). The membership problem for unmixed polynomial ideals is solvable in single exponential time. *Discrete Appld Maths* **33**, 73–94.

[11]   Hilbert, D. (1922), Neubegründung der Mathematik. Erste Mitteilung. *Abhandl. aus dem Math. Seminar d. Hamb. Univ.* **1**, 157–177.

[12]   von zur Gathen, J. (1986). Parallel arithmetic computations: a survey, in *Proceedings of the 12th Symposium on Mathematical Foundations of Computer Science, Bratislava, 1986*, ed. B. Rovan J. Gruska and J. Wiedermann (Springer-Verlag, Berlin), 93–112.

[13]   von zur Gathen, J. (1993). Parallel linear algebra, in *Synthesis of Parallel Algorithms*, ed. J.H. Reif (Morgan Kaufmann).

[14]   Gianni, P. and Mora, T. (1989). Algebraic solution of systems of polynomial equations using Gröbner bases, in *pplied Algebra, Algebraic Algorithms and Error Correcting Codes, Proceedings of AAECC-5* (Springer-Verlag, Berlin), 247–257.

[15]   Giusti, M., Hägele, K., Heintz, J., Morais, J.E., Montaña, J.L. and Pardo, L.M. (1997), Lower bounds for diophantine approximation. *J. Pure & Appld Maths* **117,118**, 277–317.

[16]   Giusti, M. and Heintz, J. (1993). La détermination des points isolés et de la dimension d'une variété algébrique peut se faire en temps polynomial, in *Computational Algebraic Geometry and Commutative Algebra*, ed. D. Eisenbud and L. Robbiano (Cambridge University Press, Cambridge), 216–256.

[17]   Giusti, M., Heintz, J., Morais, J.E., Morgenstern, J. and Pardo, L.M. (1998). Straight-line programs in geometric elimination theory. *J. Pure & Appld Maths* **124**, 101–146.

[18]   Giusti, M., Heintz, J., Morais, J.E. and Pardo, L.M. (1995). When polynomial equation systems can be solved fast? in *Applied Algebra, Algebraic Algorithms and Error Correcting Codes, Proceedings AAECC-11* ed. G. Cohen, H. Giusti and T. Mora (Springer-Verlag, Berlin), 205–231.

[19]   Giusti, M., Heintz, J., Morais, J.E. and Pardo, L.M. (1997). Le rôle des structures de données dans les problèmes d'élimination. *C. R. Acad. Sci. Paris* **325**, 1223–1228.

[20]   Giusti, M., Heintz, J. and Sabia, J. (1993). On the efficiency of effective Nullstellensätze. *Computational Complexity* **3**, 56–95.

[21]   Giusti, M., Lecerf, G. and Salvy, B. (2000). A Gröbner free alternative for polynomial systems solving, *J. Complexity*, to appear.

[22]   Heintz, J. (1983). Definability and fast quantifier elimination in algebraically closed fields. *Theor. Comput. Sci.* **24**, 239–277.

[23]   Heintz, J., Matera, G., Pardo, L.M. and Wachenchauzer, R. (1998). The intrinsic complexity of parametric elimination methods, *Electron. J. SADIO* **1**, 37–51.

[24]   Heintz, J., Matera, G. and Waissbein, A. (2000). On the time–space complexity of geometric elimination procedures, *Appld Algebra Engrg Comm. Comput.*, to appear.

[25]   Heintz, J. and Morgenstern, J. (1993). On the intrinsic complexity of elimination theory. *J. Complexity* **9**, 471–498.

[26]   Heintz, J. and Schnorr, C.P. (1982). Testing polynomials which are easy to compute, in *Logic and Algorithmic*, Monographie de l'Enseignement Mathématique **30**, 237–254

[27]   Kobayashi, H., Moritsugu, S. and Hogan, R.W. (1989). Solving

systems of algebraic equations, in *Symbolic and algebraic computation (Rome, 1988)* (Springer-Verlag, Berlin), 139–149.

[28] Kühnle, K. (1998). *Space Optimal Computation of Normal Forms of Polynomials*, Berichte aus der Informatik (Shaker-Verlag, Aachen).

[29] Krick, T. and Pardo, L.M. (1996). A computational method for diophantine approximation, in *Algorithms in Algebraic Geometry and Applications. Proceedings of MEGA'94*, ed. L. González-Vega and T. Recio (Birkhäuser, Basel), 193–254.

[30] Kronecker, L. (1982). Grundzüge einer arithmetischen Theorie de algebraischen Grössen, *J. Reine Angew. Math.* **92**, 1–122.

[31] Lang, S. (1984), *Algebra*, 2nd edition (Addison-Wesley, Reading, MA).

[32] Macauley, F.S. (1916). *The Algebraic Theory of Modular Systems* (Cambridge University Press, Cambridge).

[33] Matera, G. (1999). Probabilistic algorithms for geometric elimination. *Appld Algebra Engrg Comm. Comput.* **9**, 463–520.

[34] Matera, G. and Turull, J.M. (1997). The complexity space of elimination: upper bounds, in *Foundations of Computational Mathematics (Rio de Janeiro, 1997)*, ed. F. Cucker and M. Shub (Springer-Verlag, Berlin), 267–276.

[35] Mayr, E. (1989). Membership in polynomial ideals over Q is exponential space complete, in *Proceedings of the 6th Annual Symposium on Theoretical Aspects of Computer Science (STACS'89), Paderborn (FRG), 1989*, ed. B. Monien et al. (Springer-Verlag, Berlin), 400–406.

[36] Mayr, E. and Meyer, A. (1982). The complexity of the word problem for commutative semigroups. *Adv. in Maths* **46**, 305–329.

[37] Mumford, D. (1988). *The Red Book of Varieties and Schemes* (Springer-Verlag, Berlin).

[38] Narasimhan, R. (1966). *Introduction to the Theory of Analytic Spaces* (Springer-Verlag, Berlin).

[39] Shafarevich, I.R. (1994). *Basic Algebraic Geometry : Varieties in Projective Space* (Springer-Verlag, Berlin).

[40] Shub, M. and Smale, S. (1993). Complexity of Bézout's theorem I: Geometric aspects. *J. AMS* **6**, 459–501.

[41] Shub, M. and Smale, S. (1994). Complexity of Bézout's theorem V: Polynomial time. *Theor. Comp. Sci.* **133**, 141–164.

[42] van der Waerden, B.L. (1931), *Moderne Algebra II* (Springer-Verlag, Berlin).

[43] Yap, C.-K. (1991). A new lower bound construction for commutative Thue systems with applications. *J. Symbolic Comput.* **12**, 1–27.

Numerical analysis in Lie groups

Arieh Iserles

Department of Applied Mathematics and Theoretical Physics
University of Cambridge
Cambridge CB3 9EW, U.K.
Email: A.Iserles@damtp.cam.ac.uk
Url: http://www.damtp.cam.ac.uk/user/ai/

Abstract

There is growing recognition in the last few years that Lie groups and homogeneous spaces are often the right configuration space for the discretization of time-dependent differential equations. In this paper we review briefly recent advances in Lie-group calculations, concentrating mainly on approximation methods that advance a trivialised version of the differential equation in a Lie algebra in terms of either Magnus or Cayley expansions.

1 Why Lie groups?

The subject matter of this paper is as old as computational mathematics itself, yet still replete with exciting challenges: the numerical solution of a time-dependent ordinary differential system

$$y' = f(t, y), \quad t \geq t_0, \quad y(t_0) = y_0, \tag{1.1}$$

where $f \in \text{Lip}([t_0, \infty) \times \mathbb{R}^m \to \mathbb{R}^m)$. Practical methods for the equation (1.1), e.g. Runge–Kutta, multistep and extrapolation schemes, are well known to all students of numerical analysis and they are accompanied by powerful analysis and quality software. Our first goal in this paper is thus to motivate the consideration of an entirely new breed of time-stepping algorithms, of a form unfamiliar to most practitioners and, indeed, based upon a radically different approach to the entire subject area.

Important qualitative and structural features of (1.1) can be often phrased in the language of differential geometry, an issue discussed at greater length in [3]. In particular, it is often known that the configuration space of the solution is a differentiable manifold, $y(t) \in \mathcal{M} \subset \mathbb{R}^m$,

105

$t \geq t_0$. Equivalently, $f(t, y(t)) \in \mathrm{T}|_{y(t)} \mathcal{M}$, $t \geq t_0$. There are important advantages in respecting this feature under discretization. Firstly, the fact that $y(t)$ evolves on a manifold has frequently deeper physical or mathematical significance, e.g. it might correspond to the conservation of Hamiltonian energy. Secondly, conservation of various structural features of the underlying flow often leads to better reproduction of dynamics and slower error accumulation by the numerical algorithm. Thirdly, the conserved quantity is sometimes absolutely crucial to computation, e.g. we might wish to solve the *isospectral flow* $Y' = [B(Y), Y]$, $Y(0) = \mathrm{diag}\,\lambda \in \mathrm{M}^{m \times m}$, with a suitable matrix function $B(Y) \in \mathrm{SO}(m)$ chosen so that $\lim_{t \to \infty} Y(t)$ is a symmetric Toeplitz matrix, in order to compute an inverse Toeplitz eigenvalue problem [7]. In that case any error in conserved quantities, namely the departure of $\sigma(Y(t))$ from λ, voids the entire purpose of the calculation.

The new area of computational mathematics concerned with the approximation and computation of mathematical quantities while respecting their qualitative attributes is known (mainly in the context of differential equations) as *geometric integration* [3]. Conservation of Lie-group invariant, the concern of this brief survey, is a theme that pervades much of current geometric-integration research.

The manifolds under consideration in this paper are *Lie groups*, i.e. smooth manifolds equipped with a group structure which is consistent with the underlying topology [25]. Lie groups feature in a wide variety of applications, since they are a convenient framework to express symmetries intrinsic in physical and mathematical settings. Thus, it is of an independent interest to develop numerical time-stepping algorithms that respect Lie-group structure. Moreover, the availability of *Lie-group solvers* brings into the realm of geometric integration a considerably wider set of manifolds.

Most manifolds of practical relevance are homogeneous spaces: We say that $\mathcal{M}$ is a *homogeneous space*, acted upon by the Lie group $\mathcal{G}$, if there exists a map $\lambda : \mathcal{G} \times \mathcal{M} \to \mathcal{M}$ such that $\lambda(X_1 X_2, Y) = \lambda(X_1, \lambda(X_2, Y))$ for all $X_1, X_2 \in \mathcal{G}$, $Y \in \mathcal{M}$ and such that for every $Y_1, Y_2 \in \mathcal{M}$ there exists $X \in \mathcal{G}$ such that $\lambda(X, Y_1) = Y_2$ [23]. Examples of homogeneous spaces include spheres, tori, Stiefel and Grassmann manifolds, isospectral orbits and, of course, Lie groups themselves. Given a numerical method which is guaranteed to stay on a Lie group $\mathcal{G}$, it is an easy matter to amend it into a numerical method that evolves on an arbitrary homogeneous manifold $\mathcal{M}$ acted upon by $\mathcal{G}$ [22]. For example, an isospectral orbit is acted by the special orthogonal group $\mathrm{SO}(m)$ through the similarity map $\lambda(X, Y) = XYX^{-1}$. Instead of solving $Y' = [B(Y), Y]$, $Y(0) = Y_0$, we may thus solve $X' = B(XY_0X^{-1})X$, $X(0) = I$, and let $Y = XY_0X^{-1}$. This brings isospectral flows into the realm of geometric integration: it is possible to show that no classical numerical method can respect isospectral structure for $m \geq 3$ [4], while methods that evolve in SO(m) are widely available, not least in this paper.

An extensive survey of Lie-group methods is available and it goes into considerable detail in the many aspects of this fast-evolving subject [14]. Our purpose in this paper is considerably more modest, to introduce a mathematical reader to this area by highlighting a limited number of themes and focussing our narrative on principles, rather than details. To this end, we will make two important simplifying assumptions. Firstly, we consider in the sequel only *linear* ordinary differential equations. Secondly, we consider mainly two expansion methods, namely Magnus and Cayley series. It goes without saying that other methods are available and that the scope of Lie-group solvers is restricted neither to linear problems nor to ordinary differential equations [14]. Having said this, the relative simplicity of linear ODEs renders the theory more succinct and complete, makes for cleaner exposition and avoids issues which might be important in their own right but are marginal to our argument, e.g. solution of nonlinear algebraic equations in Lie groups, semidiscretization of space derivatives or treatment of boundary values.

2 Lie groups, Lie algebras and the dexpinv equation

We say that the manifold $\mathcal{G}$ is a Lie group if it is equipped with group structure and the operators $L_X Y = XY$ and $R_X Y = YX$ are continuous [23, 25]. Familiar examples are the *orthogonal group* O(m), the *special linear group* SL(m), the *special linear group* SO(m) = O(m) $\cap$ SL(m) and the *unitary group* U(m). We recall that elements of the tangent space $T_X\mathcal{G}$, $X \in \mathcal{G}$, can be expressed in the form $X\mathfrak{g}$, where $\mathfrak{g} = T_{\mathrm{Id}}\mathcal{G}$, the tangent space at identity, is a *Lie algebra*. Thus, $\mathfrak{g}$ is a linear space closed under a skew-symmetric binary operation which in a finite-dimensional case can be represented as the classical commutator of two matrices [25]. Since computation is relevant only in a finite-dimensional context, we henceforth assume that $\mathcal{G}$ is a multiplicative subgroup of the *general linear group* GL(m) of $m \times m$ nonsingular real matrices, whereby $\mathfrak{g}$ is a linear subspace of $\mathfrak{gl}(m) = \mathrm{M}^{m \times m}$, closed under the standard commutator $[X, Y] = XY - YX$.

It follows at once from the above construction that a differential system evolving in $\mathcal{G}$ can be always written in the form

$$Y' = A(t, Y)Y, \quad t \geq t_0, \qquad Y(0) = Y_0 \in \mathcal{G}, \qquad (2.1)$$

where $A : [t_0, \infty) \times \mathcal{G} \to \mathfrak{g}$ is Lipschitz. Equations of the form (2.1) feature in many applications, either directly or through the connection with homogeneous spaces that we have mentioned in the previous section:

- $O(m)$, $SO(m)$, $SE(m)$: rigid mechanics, robotics, computation of Lyapunov exponents, numerical linear algebra, isospectral flows;
- $SL(m)$: volume-preserving flows, motion by mean curvature, Riccati systems, Sturm–Liouville problems;
- $Sp(m)$: Hamiltonian problems;
- SO_{m_1, m_2}: relativity theory;
- $U(m)$, E_8: theory of superstrings.

Cf. [14] for more details and references.

Let ϕ be a map taking $\mathfrak{g}$ to $\mathcal{G}$, so that $O \in \mathfrak{g}$ is mapped to the identity of $\mathcal{G}$, and assume that ϕ is one-to-one in an open neighbourhood of the zero element of $\mathfrak{g}$. The differential equation (2.1) implies that $d\phi(X) = A(t, \phi(X))\phi(X)$ for $Y(t) = \phi(X(t))$, Provided that $t - t_0$ is sufficiently small (in a numerical method this will correspond to a sufficiently small time step), we can invert this equation and the outcome is the *left trivialisation* of (2.1),

$$X' = d\phi^{-1}(X, A(\,\cdot\,, \phi(X))). \qquad (2.2)$$

The most useful and ubiquitous map is the matrix exponential, $\phi(X) = e^X Y_0$, whereby (2.2) becomes the *dexpinv equation*

$$\begin{aligned} X' &= \operatorname{dexp}_X^{-1} A(t, e^X Y_0) \\ &= \sum_{k=0}^{\infty} \frac{B_k}{k!} \operatorname{ad}_X^k A(t, e^X Y_0), \quad t \geq t_0, \qquad X(t_0) = O, \end{aligned} \qquad (2.3)$$

where $\{B_k\}_{k \in \mathbb{Z}_+}$ are Bernoulli numbers and ad_X is the *adjoint operator* in $\mathfrak{g}$,

$$\operatorname{ad}_X^0 A = A, \qquad \operatorname{ad}_X^k A = [X, \operatorname{ad}_X^{k-1} A], \quad k \in \mathbb{N}.$$

The pedigree of the dexpinv equation extends all the way to the work of Campbell and Hausdorff, early in the Twentieth Century [11].

A Lie group $\mathcal{G}$ is, in general, a nonlinear object. Only in exceptional cases can a 'classical' numerical method be expected to evolve in $\mathcal{G}$. An exception is a 'quadratic' Lie group (cf. Section 4), whose structure is respected by certain implicit Runge–Kutta methods. However, *a Lie algebra is a linear space!* As long as we restrict ourselves to linear-space operations (matrix addition and multiplication by a scalar) and to commutation, a numerical method is assured to respect Lie-algebraic structure. Seen in this light, it is natural to replace the Lie-group equation (2.1) with its left trivialisation (2.2), advance the solution a single step with a suitable classical numerical solver, map it back to $\mathcal{G}$ and to repeat this procedure in subsequent steps.

A natural candidate for the time-stepping procedure in the Lie algebra $\mathfrak{g}$ is a *Runge–Kutta* scheme, since it affords an opportunity to discretise the underlying equation to an arbitrary order of accuracy within a one-step framework. This leads to *Runge–Kutta–Munthe-Kaas (RK-MK)* methods [20]. For example, the familiar fourth-order, four-stage explicit RK scheme with the Butcher tableau

$$
\begin{array}{c|cccc}
0 & & & & \\
\frac{1}{2} & \frac{1}{2} & & & \\
\frac{1}{2} & 0 & \frac{1}{2} & & \\
1 & 0 & 0 & 1 & \\
\hline
 & \frac{1}{6} & \frac{2}{3} & \frac{2}{3} & \frac{1}{6}
\end{array},
$$

applied to the Lie-group equation (2.1) from t_0 to $t_1 = t_0 + h$, reads

$$
\begin{aligned}
V_1 &= Y_0, & K_1 &= A(t_0, V_1)V_1, \\
V_2 &= Y_0 + \tfrac{1}{2}hK_1, & K_2 &= A(t_0 + \tfrac{1}{2}h, V_2)V_2, \\
V_3 &= Y_0 + \tfrac{1}{2}hK_2, & K_3 &= A(t_0 + \tfrac{1}{2}h, V_3)V_3, \\
V_4 &= Y_0 + hK_3, & K_4 &= A(t_0 + h, V_4)V_4, \\
\Delta &= h(\tfrac{1}{6}K_1 + \tfrac{1}{3}K_2 + \tfrac{1}{3}K_3 + \tfrac{1}{6}K_4), \\
Y_1 &= Y_0 + \Delta.
\end{aligned}
$$

In general, we cannot expect Y_1 to reside in $\mathcal{G}$. On the other hand, applying the method to (2.3) and truncating the infinite series consistently with order 4 yields

$$
\begin{aligned}
V_1 &= O, & K_1 &= A(t_0, Y_0), \\
V_2 &= \tfrac{1}{2}hK_1, & K_2 &= A(t_0 + \tfrac{1}{2}h, e^{V_2}Y_0) - \tfrac{1}{2}[V_2, A(t_0 + \tfrac{1}{2}h, e^{V_2}Y_0)], \\
V_3 &= \tfrac{1}{2}hK_2, & K_3 &= A(t_0 + \tfrac{1}{2}h, e^{V_3}Y_0) - \tfrac{1}{2}[V_3, A(t_0 + \tfrac{1}{2}h, e^{V_3}Y_0)], \\
V_4 &= hK_3, & K_4 &= A(t_0 + h, e^{V_4}Y_0) - \tfrac{1}{2}[V_4, A(t_0 + h, e^{V_4}Y_0)], \\
\Delta &= h(\tfrac{1}{6}K_1 + \tfrac{1}{3}K_2 + \tfrac{1}{3}K_3 + \tfrac{1}{6}K_4), \\
Y_1 &= e^{\Delta}Y_0.
\end{aligned}
$$

Since the method is based on operations consistent with Lie-algebraic structure, the exponential takes us back to the group and $Y_1 \in \mathcal{G}$.

RK-MK schemes constitute the most robust and versatile general-purpose family of Lie-group solvers, a close competitor being methods based on canonical coordinates of the second kind [24]. Yet, the remainder of this paper is devoted to a more specialised discretisation methods of the trivialised equation in a Lie algebra, based on expansions of certain form. Although less general, such expansions possess many appealing features and interesting mathematical structure. They are particularly effective when applied to linear Lie-group equations.

3 Magnus expansions

The origins of the *Magnus expansion* of the linear Lie-group equation

$$ Y' = A(t)Y, \quad t \geq t_0, \qquad Y(0) = Y_0 \in \mathcal{G}, \tag{3.1} $$

where $A : [t_0, \infty) \to \mathfrak{g}$ are in the work of Wilhelm Magnus [18], who showed that the solution of the dexpinv equation (2.3) can be written in the form

$$ X(t) = \int_{t_0}^{t} A(\xi)\mathrm{d}\xi - \tfrac{1}{2} \int_{t_0}^{t}\int_{t_0}^{\xi_1} [A(\xi_2), A(\xi_1)]\mathrm{d}\xi_2\mathrm{d}\xi_1 $$
$$ + \tfrac{1}{12} \int_{t_0}^{t}\int_{t_0}^{\xi_1}\int_{t_0}^{\xi_1} [A(\xi_3), [A(\xi_2), A(\xi_1)]]\mathrm{d}\xi_3\mathrm{d}\xi_2\mathrm{d}\xi_1 $$
$$ + \tfrac{1}{4} \int_{t_0}^{t}\int_{t_0}^{\xi_1}\int_{t_0}^{\xi_2} [[A(\xi_3), A(\xi_2)], A(\xi_1)]\mathrm{d}\xi_3\mathrm{d}\xi_2\mathrm{d}\xi_1 + \cdots . $$

Although Magnus carried the expansion further, to terms consisting of fourfold integrals, he neither presented a general formula nor a convergence proof. This, incidentally, did not prevent the Magnus expansion (or, at any rate, its first few terms) from being used in literally hundreds of papers in theoretical physics and quantum chemistry, mainly as a means of perturbative approximation.

Comprehensive analysis of the Magnus expansion within the context of numerical analysis has been carried out recently by Iserles and Nørsett [13]. It has led to a recursive formula for the generation of expansion coefficients, a convergence proof (much improved subsequently in [1] and [19]) and efficient means of discretising multivariate integrals in the course of numerical implementation of the expansion. The point of departure is the observation that the terms in the expansion of X'

(that is, 'stripping' the outer integral) are either the function A or scalar multiples of $\int_{t_0}^{t}[F_1(\xi), F_2(t)]d\xi$, where F_1 and F_2 have already featured earlier in the expansion. This we formalise by associating each expansion term of X' with an element of a subset of rooted planar binary trees according to the following rule:

(i) The function A is associated with the trivial tree with a single vertex;

(ii) If terms F_1 and F_2 are associated with the trees τ_1 and τ_2, respectively, then

$$\int_{t_0}^{t}[F_1(\xi), F_2(t)]d\xi \qquad \text{is associated with} \qquad \text{(tree figure)} . \qquad (3.2)$$

Intuitively speaking, vertical lines stand for integration and 'bifurcations' for commutation. We express this relationship by $\tau \to \mathcal{C}_\tau$, where τ is a binary tree and $\mathcal{C}_\tau$ a function acting in $\mathfrak{g}$. It has been proved in [13] that the coefficient in front of $\mathcal{C}_\tau$ is 1 if τ is the trivial tree, otherwise it equals

$$\alpha(\tau) = \frac{B_s}{s!}\prod_{l=1}^{s}\alpha(\tau_l) \qquad \text{where} \qquad \tau = \text{(tree figure)} . \qquad (3.3)$$

Let $\mathbb{T}_m$, $m \geq 0$ denote the set of all the trees which have been obtained after m iterations of (3.2). In other words, each $\tau \in \mathbb{T}_m$ contains exactly m nodes of degree 2 (corresponding to vertical lines, i.e. to integration) and m nodes of degree 3 (corresponding to 'bifurcating' nodes and to commutation). Thus,

$\mathbb{T}_0 :$ (figure) ;

$\mathbb{T}_1 :$ (figure) ;

$\mathbb{T}_2 :$ (figure) , (figure) ;

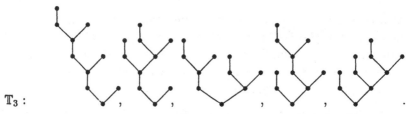

$\mathbb{T}_3$:

Because of (3.3), we have now enough information to construct a Magnus expansion recursively. A general *Magnus expansion* is thus

$$X(t) = \sum_{m=0}^{\infty} \sum_{\tau \in \mathbb{T}_m} \alpha(\tau) \int_{t_0}^{t} C_\tau(\xi) d\xi, \qquad (3.4)$$

where $C_\tau : [t_0, \infty) \to \mathfrak{g}$ is the expansion term corresponding to the tree τ. Expressed in the terminology of trees (appending a vertex to the root to signify extra integration), we have

$$X(t) \sim \quad - \tfrac{1}{2} \quad + \tfrac{1}{4} \quad + \tfrac{1}{12} \quad - \tfrac{1}{8} \quad - \tfrac{1}{24}$$

$$- \tfrac{1}{24} \quad - \tfrac{1}{24} \quad + 0 \quad + \cdots .$$

(The last tree, having a zero coefficient, can be dropped: it has been included for expository reasons only.) It is possible to prove that the expansion (3.4) converges, subject to $\int_{t_0}^{t} \|A(\xi)\| d\xi \leq 1.086868702\cdots$ [1, 19].

In practical implementation the expansion (3.4) need be truncated. An obvious truncation is

$$X_{(p)}(t) = \sum_{m=0}^{p-1} \sum_{\tau \in \mathbb{T}_m} \alpha(\tau) \int_{t_0}^{t} C_\tau(\xi) d\xi, \qquad (3.5)$$

where $p \geq 1$, and it is easy to prove that $X_{(p)}(t) = X(t) + \mathcal{O}((t - t_0)^{p+1})$

[13]. This, however, is suboptimal: it is possible to attain order p with fewer terms. We let $\mathbb{F}_m$, $m \geq 0$, denote the set of trees in $\cup_{k=0}^{m} \mathbb{T}_k$ such that

$$\tau \in \mathbb{F}_m \quad \Rightarrow \quad C_\tau(t) = \mathcal{O}((t - t_0)^m)$$

for all sufficiently smooth functions $A : [t_0, \infty) \to \mathfrak{g}$. The $\mathbb{F}_m$s can be constructed recursively: $\mathbb{F}_0 = \mathbb{T}_0$, $\mathbb{F}_1 = \emptyset$ and, if τ has been obtained from $\tau_1 \in \mathbb{F}_{m_1}$ and $\tau_2 \in \mathbb{F}_{m_2}$ according to (3.2) then

$$\tau_1 = \tau_2 \quad \Rightarrow \quad \tau \in \mathbb{F}_{m_1+m_2+2}, \qquad \tau_1 \neq \tau_2 \quad \Rightarrow \quad \tau \in \mathbb{F}_{m_1+m_2+1}$$

[15]. It clearly makes sense to replace (3.5) with

$$X_{[p]}(t) = \sum_{m=0}^{p-1} \sum_{\tau \in \mathbb{F}_m} \alpha(\tau) \int_{t_0}^{t} C_\tau(\xi) \mathrm{d}\xi, \qquad (3.6)$$

since the order is p, whilst the number of terms is in general smaller than in (3.5). The number of terms in (3.6) can be further decreased by exploiting time symmetry. A flow $\Psi_{t_0,t}$ is *time symmetric,* provided that $\Psi_{t,t_0} \circ \Psi_{t_0,t} = \mathrm{Id}$ for every (suitably small) $t \geq t_0$ [10]. It is self evident that the flow $\Psi_{t_0,t} Y_0 = Y(t) = \mathrm{e}^{X(t)} Y_0$, where Y is the solution of (3.1), is time symmetric. A considerably less trivial statement has been proved in [15], namely that $\Psi_{t_0,t} Y_0 = \mathrm{e}^{X_{[p]}(t)} Y_0$ is time symmetric [15]. This has an unexpected benefit: it is well known that time symmetry implies that $\Psi_{t_0,t} = \mathrm{e}^{\Omega_{t_0,t}}$, where the operator $\Omega_{t_0,t}$ can be expanded in *odd* powers of $t - t_0$. We deduce that $X_{[p]}(t) = X(t) + \mathcal{O}((t - t_0)^{2q+1})$ for some integer $q \geq 1$, regardless of whether p itself is odd or even. In other words, *the truncated Magnus expansion (3.6) is always of an even order of accuracy.* Therefore $p = 2q - 1$ in (3.6) delivers order $2q$, while reducing further the number of terms in the expansion.

The outcome is a family of even-order truncations of the Magnus expansion which consist of a relatively small number of expansion coefficients:

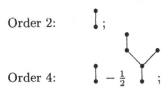

Order 6: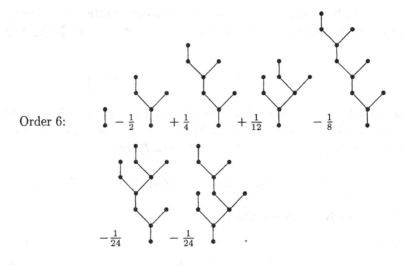

The matrix function A often consists of polynomial, trigonometric or exponential entries, or a mixture thereof. In that case the integrals $\int_{t_0}^{t} C_\tau(\xi)\mathrm{d}\xi$ can be executed in a closed form. However, in general, multivariate integrals need be replaced by quadrature. Although multivariate quadrature is notoriously expensive, the cost can be reduced a very great deal by exploiting the special structure of integrals in (3.6) and dependencies among terms in the Lie algebra $\mathfrak{g}$.

The integrals $\int_{t_0}^{t_0+h} C_\tau(\xi)\mathrm{d}\xi$, where $h > 0$ is the step size, are of the form

$$I(h) = \int_{t_0+h\mathcal{S}} \boldsymbol{L}(A(\xi_1), A(\xi_2), \dots, A(\xi_s))\mathrm{d}\boldsymbol{\xi}, \qquad (3.7)$$

where $\boldsymbol{L}$ is a *multilinear form* and $\mathcal{S}$ is a polytope of the form

$$\mathcal{S} = \{\boldsymbol{\xi} \in \mathbb{R}^s : 0 \le \xi_1 \le 1,\ 0 \le \xi_i \le \xi_{j_i},\ i = 2, 3, \dots, s\},$$

where $j_i \in \{1, 2, \dots, i - 1\}$. For example, for

$$I(h) = \int_{t_0}^{t_0+h} \int_{t_0}^{\xi_1} \int_{t_0}^{\xi_1} [A(\xi_3), [A(\xi_2), A(\xi_1)]]\mathrm{d}\boldsymbol{\xi}$$

we have $s = 3$, $\boldsymbol{L}(E_1, E_2, E_3) = [E_3, [E_2, E_1]]$ and $j_2 = j_3 = 1$.

Let $c_1, c_2, \dots, c_\nu \in [0,1]$ be distinct *collocation points* and set $A_k = hA(t_0 + c_k h)$, $k = 1, 2, \dots, \nu$. Iserles and Nørsett proposed in [13] to approximate (3.7) with the quadrature formula

$$Q(h) = \sum_{\boldsymbol{k} \in \mathcal{S}_s^\nu} b_{\boldsymbol{k}} \boldsymbol{L}(A_{k_1}, A_{k_2}, \dots, A_{k_s}), \qquad (3.8)$$

where S_s^ν is the set of all length-s words from the alphabet $\{1, 2, \ldots, \nu\}$ and

$$b_k = \int_S \prod_{i=1}^s \prod_{\substack{j=1 \\ j \neq k_i}}^\nu \frac{\xi_i - \xi_j}{c_{k_i} - c_j} d\xi, \qquad k \in S_s^\nu.$$

Note that just ν function evaluations are recycled time and again in different positions in *all* the different multilinear forms L that occur in the expansion (3.6).

The quadrature (3.8) reduces to the familiar univariate quadrature with interpolatory weights when $s = 1$ and $L(E) = E$. Remarkably, *the order of accuracy of (3.8) is exactly the same as the order of accuracy of the univariate formula!* [13]. In particular, if $c_1, c_2, \ldots, c_\nu$ are the standard Gaussian weights in $[0, 1]$ then *all* the integrals in (3.6) are computed to order 2ν in just ν function evaluations.

For example, Gaussian nodes for $\nu = 2$ are $c_1 = \frac{1}{2} - \frac{1}{6}\sqrt{3}$, $c_2 = \frac{1}{2} + \frac{1}{6}\sqrt{3}$, and

$$\int_{t_0}^{t_0+h} \int_{t_0}^{\xi_1} [A(\xi_2), A(\xi_1)] d\xi_2 d\xi_1 = \frac{\sqrt{3}}{6}[A_2, A_1] + \mathcal{O}(h^5),$$

where $A_k = hA(t_0 + c_k h)$, $k = 1, 2$. Likewise, for $\nu = 3$ we have $c_1 = \frac{1}{2} - \frac{1}{10}\sqrt{15}$, $c_2 = \frac{1}{2}$ and $c_3 = \frac{1}{2} + \frac{1}{10}\sqrt{15}$ and the quadrature formula is

$$\int_{t_0}^{t_0+h} \int_{t_0}^{\xi_1} [A(\xi_2), A(\xi_1)] d\xi_2 d\xi_1$$
$$= \frac{\sqrt{15}}{54}(2[A_1, A_2] + [A_1, A_3] + 2[A_2, A_3]) + \mathcal{O}(h^7),$$

where, again, $A_k = hA(t_0 + c_k h)$. We reiterate that the very same values of A_k are used to compute all the integrals in the expansion, therefore just ν function values are needed to approximate X to order 2ν.

Quadrature (3.8) is extremely cheap in terms of function evaluations, but the *quid pro quo* is that it requires exceedingly large number of commutator evaluations, since the cardinality of C_s^ν increases very fast with both ν and s. Naively applied, the volume of linear algebra becomes prohibitive for larger values of ν: the case $\nu = 5$, corresponding to order 10, requires 1256567 commutator evaluations in each step! However, the volume of linear algebra can be cut very substantially (in the above case, $\nu = 5$, down to 73 commutators per step) by exploiting Lie-algebra symmetries. Suppose thus that $c_1, c_2, \ldots, c_\nu$ are symmetric with respect to $\frac{1}{2}$ (e.g., Gaussian nodes) and replace $\{A_1, A_2, \dots, A_\nu\}$ by the *self-adjoint basis* $\{B_1, B_2, \dots, B_\nu\}$, where

$$\sum_{l=1}^{\nu} (c_k - \tfrac{1}{2})^{l-1} B_l = A_k, \qquad k = 1, 2, \dots, \nu.$$

We rewrite (3.8) in the new basis,

$$Q(h) = \sum_{\boldsymbol{k} \in C_s^\nu} \tilde{b}_{\boldsymbol{k}} L(B_{k_1}, B_{k_2}, \dots, B_{k_s}), \qquad (3.9)$$

where

$$\tilde{b}_{\boldsymbol{k}} = \int_S \prod_{i=1}^{s} (\xi_i - \tfrac{1}{2})^{k_i - 1} \mathrm{d}\boldsymbol{\xi}.$$

The self-adjoint representation has been introduced by Munthe-Kaas and Owren [21], who have pointed out three distinct mechanisms that reduce the number of terms that need be incorporated into (3.9).

We consider a free graded Lie algebra generated by the alphabet $\{C_1, C_2, \dots, C_\nu\}$, where each C_k is equipped with the *grade* $\omega(C_k)$. The grades propagate in a natural manner under commutation, $\omega([Z_1, Z_2]) = \omega(Z_1) + \omega(Z_2)$. We denote by $\mathcal{K}_r^\nu$ the linear space spanned by elements of grade $r \geq 1$. In the particular context of Lie-group methods, we associate the grade with the order of magnitude of a term. Thus, in (3.9) we note that, by construction, $B_k = \mathcal{O}(h^k)$, thereby we consider the generators $\{B_1, B_2, \dots, B_\nu\}$ with the grades $\omega(B_k) = k$. It follows that

$$\omega(L(B_{k_1}, B_{k_2}, \dots, B_{k_s})) = \sum_{i=1}^{s} k_i,$$

Therefore $L(B_{k_1}, B_{k_2}, \dots, B_{k_s})$ can be discarded from (3.9) whenever $\sum_{i=1}^{s} k_i$ exceeds the order of the method.

Suppose that a linear combination of integrals can be expanded in odd powers of h. It then follows from the symmetry of collocation nodes that so is the case with the corresponding linear combination of quadratures (3.9). It is a consequence of time symmetry of (3.6) that we can discard from (3.9) all the terms of even grade, since their linear combination in the discretized Magnus expansion is nil.

Finally, we note that the dimension of the linear space $\mathcal{K}_r^\nu$ is typically very small. The underlying reason is that Lie-algebraic symmetries, namely the skew-symmetry $[A, B] + [B, A] = O$ and the Jacobi identity $[A, [B, C]] + [B, [C, A]] + [C, [A, B]] = O$, allow us to express many terms as linear combinations or other terms. This is well known in the classical case $\omega(C_k) \equiv 1$ (which corresponds to (3.8)) and has been extended to general grades in [21]. In particular, for grades $\omega(B_k) = k$, we let $\lambda_1, \lambda_2, \ldots, \lambda_\nu$ be the zeros of

$$q(z) := \frac{1 - 2z + z^{\nu+1}}{1 - z},$$

whence

$$\dim \mathcal{K}_r^\nu = \frac{1}{r} \sum_{k|r} \mu(k) \sum_{i=1}^\nu \lambda_i^{r/k}, \qquad (3.10)$$

where μ is the *Möbius function*. Since a basis of $\mathcal{K}_r^\nu$ can be constructed similarly to the classical Hall basis [21], we can use it to obtain all the remaining terms in (3.9), letting r range over odd numbers in $\{1, \ldots, p\}$.

As an example, exploiting all the different economies inherent in the self-adjoint basis, we obtain the following sixth-order method,

$$X(t_0 + h) \approx B_1 + \tfrac{1}{12}B_3 - \tfrac{1}{12}[B_1, B_2] + \tfrac{1}{240}[B_2, B_3] + \tfrac{1}{360}[B_1, [B_1, B_3]]$$
$$- \tfrac{1}{240}[B_2, [B_1, B_2]] + \tfrac{1}{720}[B_1, [B_1, [B_1, B_2]]],$$

where $c_1 = \tfrac{1}{2} - \tfrac{1}{10}\sqrt{15}$, $c_2 = \tfrac{1}{2}$, $c_3 = \tfrac{1}{2} + \tfrac{1}{10}\sqrt{15}$ and

$$B_1 = A_2, \quad B_2 = \tfrac{\sqrt{15}}{3}(A_3 - A_1), \quad B_3 = \tfrac{10}{3}(A_3 - 2A_2 + A_1).$$

This, actually, is not the best sixth-order Magnus methods, since further economies can be achieved by lumping terms together [2]. Thus, we let

$$P_1 = [B_2, \tfrac{1}{12}B_1 + \tfrac{1}{240}B_3],$$
$$P_2 = [B_1, [B_1, \tfrac{1}{360}B_3 - \tfrac{1}{60}P_1]],$$
$$P_3 = \tfrac{1}{20}[B_2, P_1]$$

and this leads to the sixth-order method

$$X(t_0 + h) \approx B_1 + \tfrac{1}{12}B_3 + P_1 + P_2 + P_3.$$

We mention in passing that nonlinear versions of the Magnus method have been developed by Zanna [26, 27]. We refer to [14] for details.

4 Cayley expansions

The exponential map presents the natural trivialisation of Lie-group flows and it can be applied in all matrix Lie groups. However, its numerical evaluation is, in general, costly, in particular since we wish our approximation to the exponential to reside in a Lie group [5, 6]. In an important special case it is possible to dispose of the exponential function altogether and choose $\phi(X) = \mathrm{cay}(X) := (I - \tfrac{1}{2}X)^{-1}(I + \tfrac{1}{2}X)Y_0$, the *Cayley transform*, in (2.2). Specifically, this can be done when $\mathcal{G}$ is a *quadratic* Lie group,

$$\mathcal{G} = \{Y \in \mathrm{GL}(n) : YPY^{\mathrm{T}} = P\}, \tag{4.1}$$

where $P \in \mathrm{GL}(n)$. Examples are $\mathrm{O}(n)$, $\mathrm{O}(n,m)$, $\mathrm{Sp}(n)$ and, replacing ^{T} by the complex transpose ^{H}, the complex unitary group $\mathrm{U}(n)$. The underlying Lie algebra is

$$\mathfrak{g} = \{X \in \mathfrak{gl}(n) : XP + PX^{\mathrm{T}} = O\} \tag{4.2}$$

and the Lie-algebraic equation (2.2) assumes the form

$$\begin{aligned} X' &= \mathrm{dcay}_X^{-1} A(t, \mathrm{cay}(X)Y_0) \\ &= A - \tfrac{1}{2}[X, A] - \tfrac{1}{4}AXA, \quad t \geq t_0, \qquad X(t_0) = O. \end{aligned} \tag{4.3}$$

This equation has been used in [17] to solve Lie–Poisson systems and investigated in the context of Lie-group methods in [8, 9].

Note the term AXA in (4.3). In general, we cannot expect BC to reside in $\mathfrak{g}$ for $B, C \in \mathfrak{g}$, but it is not difficult to prove a 'triple product rule': in a *quadratic Lie algebra* (4.3) $B, C \in \mathfrak{g}$ implies $BCB \in \mathfrak{g}$. As a matter of fact, we note for future reference that in a quadratic Lie algebra

$$B_1, B_2, \ldots, B_r \in \mathfrak{g} \tag{4.4}$$

$$\Rightarrow \quad (\!(B_1, \ldots, B_r)\!)_r := B_1 B_2 \cdots B_r - (-1)^r B_r B_{r-1} \cdots B_1 \in \mathfrak{g}.$$

Let us now focus attention on the linear equation (3.1), except that $\mathcal{G}$ is now assumed to be a quadratic Lie group (4.1). Although this might appear as a fairly specialised special case, it has an important application to the computation of Lyapunov exponents. Moreover, the material can be generalised to nonlinear equations along the lines of the nonlinear Magnus expansions that we have mentioned in Section 3.

Our point of departure is the observation that the solution of (4.3) can be expanded in the form

$$\begin{aligned} X(t) = \int_{t_0}^{t} A(\xi)\mathrm{d}\xi &- \tfrac{1}{2} \int_{t_0}^{t}\!\!\int_{t_0}^{t} \xi_1[A(\xi_2), A(\xi_1)]\mathrm{d}\xi_2\mathrm{d}\xi_1 \\ &+ \tfrac{1}{4} \int_{t_0}^{t}\!\!\int_{t_0}^{\xi_1}\!\!\int_{t_0}^{\xi_2} [[A(\xi_3), A(\xi_2)], A(\xi_1)]\mathrm{d}\xi_3\mathrm{d}\xi_2\mathrm{d}\xi_1 \end{aligned} \tag{4.5}$$

$$-\frac{1}{4}\int_{t_0}^{t}\int_{t_0}^{\xi_1}\int_{t_0}^{\xi_1} A(\xi_2)A(\xi_1)A(\xi_3)\mathrm{d}\xi_3\mathrm{d}\xi_2\mathrm{d}\xi_1 + \cdots.$$

The is reminiscent of a Magnus expansion, except that in place of some iterated commutators we have triple products. It is tempting to extend the graph-theoretic formalism from Section 3, but in the present framework we need to cater for triple products, as well as for iterated commutators. This has been accomplished in [12] by using *bicolour* (i.e., with verices coloured white and black, say) rooted trees with the following composition rules. Here τ, τ_1 and τ_2 are trees which correspond to $[t_0, \infty) \to \mathfrak{g}$ functions $\mathcal{D}_\tau$, $\mathcal{D}_{\tau_1}$ and $\mathcal{D}_{\tau_2}$ respectively. In that case

corresponds to $\int_{t_0}^{t} \mathcal{D}_\tau(\xi)\mathrm{d}\xi$,

corresponds to $[\mathcal{D}_{\tau_1}(t), \mathcal{D}_{\tau_2}(t)]$,

corresponds to $\mathcal{D}_{\tau_1}(t)A(t)\mathcal{D}_{\tau_2}(t)$.

Thus, the graph-theoretic representation of the Magnus-like *Cayley expansion* (4.5) is

$$X(t) \sim \quad - \tfrac{1}{2} \quad + \tfrac{1}{4} \quad - \tfrac{1}{4} \quad + \cdots.$$

With greater generality, we let $\mathbb{S}_m$ stand for the set of all trees that can be obtained in $m \geq 0$ steps of the above procedure,

$\mathbb{S}_0$: ;

$\mathbb{S}_1$: ;

$\mathbb{S}_2$: , ;

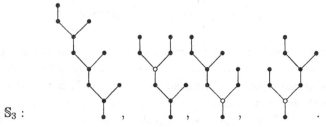

$\mathbb{S}_3$:

The *Cayley expansion* of the solution of (4.3) reads

$$X(t) = \sum_{m=0}^{\infty} \frac{(-1)^m}{2^m} \sum_{\tau \in \mathbb{S}_m} (-1)^{\gamma(\tau)} \mathcal{D}_\tau(t), \qquad (4.6)$$

where $\gamma(\tau)$ is the number of white vertices in a bicolour tree τ [12].

The integrals in (4.6) are of the same form as (3.7) and the quadrature formula (3.8) is equally applicable. As in Section 3, we truncate (4.6) by excluding terms which are $\mathcal{O}(h^{p+1})$. Note that, in variance with the Magnus expansion, this does not imply time symmetry! However, remarkably, time symmetry is restored when integrals are replaced by quadrature with symmetric collocation points (e.g., Gauss or Lobatto quadrature) [12]. In other words, even if the underlying integrals can be evaluated explicitly, much is to be gained by replacing them with time-symmetric quadrature!

Similarly to discretized Magnus expansions, we may attempt to reduce the number of terms in a quadrature formula by expressing function values in a self-adjoint basis. The toolbox of graded algebras, the most important mechanism in economising Magnus expansions, is no longer available. An alternative theory has been developed in [16]. We say that $\mathfrak{h}$ is a *hierarchical algebra* if it is an Abelian group over a field $\mathbb{F}$ of zero characteristic, closed subject to a countable set of r-nary operations,

$$\overbrace{\llbracket \cdot, \ldots, \cdot \rrbracket_r : \mathfrak{h} \times \mathfrak{h} \times \cdots \times \mathfrak{h}}^{r \text{ times}} \to \mathfrak{h}, \qquad r \in \mathbb{N},$$

which obey the following three axioms:

Alternate symmetry: For every $C_1, C_2, \ldots, C_r \in \mathfrak{h}$

$$\llbracket C_1, C_2, \ldots, C_r \rrbracket_r + (-1)^r \llbracket C_r, C_{r-1}, \ldots, C_1 \rrbracket_r = O;$$

Multilinearity:

$$\llbracket C_1, \ldots, C_{k-1}, \alpha_1 B_1 + \alpha_2 B_2, C_{k+1}, \ldots, C_r \rrbracket_r$$
$$= \alpha_1 \llbracket C_1, \ldots, C_{k-1}, B_1, C_{k+1}, \ldots, C_r \rrbracket_r$$

$$+ \alpha_2 [\![C_1, \ldots, C_{k-1}, B_2, C_{k+1}, \ldots, C_r]\!]_r$$

for every $C_1, \ldots, C_{k-1}, C_{k+1}, \ldots, C_r, B_1, B_2 \in \mathfrak{h}$ and $\alpha_1, \alpha_2 \in \mathbb{F}$;

Hierarchy condition:

$$[\![C_1, \ldots, C_{k-1}, [\![B_1, \ldots, B_s]\!]_s, C_{k+1}, \ldots, C_r]\!]_r$$
$$= [\![C_1, \ldots, C_{k-1}, B_1, \ldots, B_s, C_{k+1}, \ldots, C_r]\!]_{r+s-1}$$
$$- (-1)^s [\![C_1, \ldots, C_{k-1}, B_s, \ldots, B_1, C_{k+1}, \ldots, C_r]\!]_{r+s-1}$$

for all $C_1, \ldots, C_r, B_1, \ldots, B_s \in \mathfrak{h}$, $r, s \geq 1$ and $k \in \{1, 2, \ldots, r\}$. Note that, according to (4.4), the symmetric products $(\!(\cdot \ldots, \cdot)\!)_r$ obey all the axioms above and a quadratic Lie group $\mathfrak{g}$ is a hierarchical group. A graded free hierarchical algebra is defined similarly to a graded free Lie algebra.

The dimension of subspaces $\mathcal{K}_r^\nu$ (cf. Section 3) for graded free hierarchical algebras has been determined in [16] for all combinations of grades. In particular, recall that $\omega(B_k) = k$, $k = 1, 2, \ldots, \nu$, in a self-adjoint basis. In that case it follows from [16] that the hierarchical-group equivalent of (3.10) is

$$\dim \mathcal{K}_{2r}^\nu = \frac{1}{2} \sum_{i=1}^\nu \frac{\lambda_i^{-r-1}}{q'(\lambda_i)} \{ 2 - \lambda_i^{-r} - \tfrac{1}{2} [q(\lambda_i^{1/2}) + q(-\lambda_i^{1/2})] \},$$

$$\dim \mathcal{K}_{2r+1}^\nu = \frac{1}{2} \sum_{i=1}^\nu \frac{\lambda_i^{-r-1/2}}{q'(\lambda_i)} \{ -\lambda_i^{-r-1/2} + \tfrac{1}{2} [q(\lambda_i^{1/2}) - q(-\lambda_i^{1/2})] \},$$

where q and $\lambda_1, \lambda_2, \ldots, \lambda_\nu$ have been already defined in Section 3.

Practical implementation of (3.9) in the context of graded free hierarchical algebra requires the representation of quadrature terms as linear combinations of elements from a basis of $\mathcal{K}_r^\nu$ for suitable $r \geq 0$. This must be preceded by the construction of such a basis, a subject that has been explored in [16]. It is similar in spirit, although completely different in detail, to the construction of a *Hall basis* of a graded free Lie algebra. In the latter case it is possible to show that $\mathcal{K}_r^\nu$ is spanned by elements of the form

$$[B_{i_1}, [B_{i_2}, [B_{i_3}, \ldots, [B_{i_{k-1}}, B_{i_k}] \cdots]]], \qquad \sum_{j=1}^k \omega(B_{i_j}) = r,$$

where $B_1, B_2, \ldots, B_\nu$ are the generators and $i_1, i_2, \ldots, i_k \in \{1, 2, \ldots, \nu\}$, and all the constituents of a Hall basis are of this type. Likewise, in a hierarchical algebra we may construct our basis using only elements of the form

$$[\![B_{i_1}, B_{i_2}, \ldots, B_{i_k}]\!]_k, \qquad \sum_{j=1}^{k} \omega(B_{i_j}) = r.$$

This can be accomplished with an algorithm presented in [16] and described, with examples, in [14].

The first three Cayley methods are:

Order 2: $X(t_1) \approx [\![B_1]\!]_1,$

Order 4: $X(t_1) \approx [\![B_1]\!]_1 - \frac{1}{12}[\![B_1, B_2]\!]_2 - \frac{1}{24}[\![B_1, B_1, B_1]\!]_3,$

Order 6: $X(t_1) \approx [\![B_1]\!]_1 - \frac{1}{12}[\![B_1, B_2]\!]_2 - \frac{1}{24}[\![B_1, B_1, B_1]\!]_3$
$$+ \frac{1}{240}[\![B_2, B_3]\!]_2 + \frac{1}{240}[\![B_1, B_2, B_2]\!]_3 - \frac{1}{240}[\![B_1, B_1, B_3]\!]_3$$
$$- \frac{1}{240}[\![B_2, B_1, B_2]\!]_3 - \frac{1}{160}[\![B_1, B_3, B_1]\!]_3$$
$$+ \frac{1}{120}[\![B_1, B_1, B_1, B_2]\!]_4 - \frac{1}{240}[\![B_1, B_1, B_2, B_1]\!]_4$$
$$+ \frac{1}{240}[\![B_1, B_1, B_1, B_1, B_1]\!]_5,$$

where $t_1 = t_0 + h$. Their main advantage, in comparison with Magnus-type methods, is that no matrix exponentials need be computed.

References

[1]   Blanes, S., Casas, F., Oteo, J.A. and Ros, J. (1998). Magnus and Fer expansions for matrix differential equations: The convergence problem, *J. Phys A* **31**, 259–268.

[2]   Blanes, S., Casas, F. and Ros, J. (1999). Improved high order integrators based on Magnus expansion, Tech. Rep. NA1999/08, DAMTP, University of Cambridge.

[3]   Budd, C.J. and Iserles, A. (1999). Geometric integration: Numerical solution of differential equations on manifolds, *Phil. Trans. Roy. Soc. Lond. A* **357**, 945–956.

[4]   Calvo, M.P., Iserles, A. and Zanna, A. (1997). Numerical solution of isospectral flows, *Maths Comput.* **66**, 1461–1486.

[5]   Celledoni, E. and Iserles, A. (2000). Approximating the exponential from a Lie algebra to a Lie group, *Maths Comput.*, to appear.

[6]   Celledoni, E. and Iserles, A. (2000). Methods for the approximation of the matrix exponential in a Lie-algebraic setting, *IMA J. Num. Anal.*, to appear.

[7]   Chu, M.T. (1998). Inverse eigenvalue problems, *SIAM Review* **40**, 1–39.

[8]   Diele, F., Lopez L. and Peluso, R. (1998). The Cayley transform in the numerical solution of unitary differential systems, *Adv. Comput. Maths* **8**, 317–334.

[9]   Engø, K. (1998). On the construction of geometric integrators in the RKMK class, Tech. Rep. 158, Dept Comp. Sc., University of Bergen, Norway.

[10] Hairer, E., Nørsett, S.P. and Wanner, G. (1993). *Solving Ordinary Differential Equations I: Nonstiff Problems* (2nd rev. ed.) (Springer-Verlag, Berlin).

[11] Hausdorff, F. (1906). Die symbolische Exponentialformel in der Gruppentheorie, *Berichte der Sächsischen Akademie der Wißenschaften (Math. Phys. Klasse)* **58**, 19–48.

[12] Iserles, A. (2001). On Cayley-transform methods for the discretization of Lie-group equations, *J. FoCM*, to appear.

[13] Iserles, A. and Nørsett, S.P. (1999). On the solution of linear differential equations in Lie groups, *Phil. Trans. Roy. Soc. Lond. A* **357**, 983–1019.

[14] Iserles, A., Munthe-Kaas, H.Z., Nørsett, S.P. and Zanna, A. (2000). Lie-group methods, *Acta Numerica* **9**, 215–365.

[15] Iserles, A., Nørsett, S.P. and Rasmussen, A.F.(1998). Time-symmetry and high-order Magnus methods, Tech. Rep. 1998/NA06, DAMTP, University of Cambridge.

[16] Iserles, A. and Zanna, A. (2000). On the dimension of certain graded Lie algebras arising in geometric integration of differential equations, *LMS J. Comput. & Maths* **3**, 44–75.

[17] Lewis, D. and Simo, J.C. (1994). Conserving algorithms for the dynamics of Hamiltonian systems of Lie groups, *J. Nonlinear Sci.* **4**, 253–299.

[18] Magnus, W. (1954). On the exponential solution of differential equations for a linear operator, *Comm. Pure and Appl. Maths* **7**, 649–673.

[19] Moan, P.C. (1998). Efficient approximation of Sturm–Liouville problems using Lie-group methods, Tech. Rep. 1998/NA11, DAMTP, University of Cambridge.

[20] Munthe-Kaas H. (1998) Runge–Kutta methods on Lie groups, *BIT* **38**, 92–111.

[21] Munthe-Kaas, H. and Owren, B. (1999). Computations in a free Lie algebra, *Phil. Trans Royal Society A* **357**, 957–982.

[22] Munthe-Kaas, H. and Zanna, A. (1997). Numerical integration of differential equations on homogeneous manifolds, in *Foundations of Computational Mathematics* (F. Cucker and M. Shub, eds) (Springer-Verlag, Berlin), 305–315.

[23] Olver, P.J. (1995). *Equivalence, Invariants, and Symmetry* (Cambridge University Press, Cambridge).

[24] Owren, B. and Marthinsen, A. (1999). Integration methods based on canonical coordinates of the second kind, Tech. Rep. Numerics 5/1999, Norwegian University of Science and Technology, Trondheim.

[25] Varadarajan, V.S. (1984). *Lie Groups, Lie Algebras, and Their Representations* (Springer-Verlag, Berlin).

[26] Zanna, A. (1998). *On the Numerical Solution of Isospectral Flows*, Ph.D. dissertation, DAMTP, University of Cambridge.

[27] Zanna, A. (1999). Collocation and relaxed collocation for the Fer and the Magnus expansions, *SIAM J. Num. Anal.* **36**, 1145–1182.

Feasibility Control in Nonlinear Optimization

Marcelo Marazzi

Department of Industrial Engineering and Management Sciences
Northwestern University, Evanston Il 60208-3118, USA
Email: marazzi@northwestern.edu
Url: www.ece.northwestern.edu/~marazzi

Jorge Nocedal

Department of Electrical and Computer Engineering
Northwestern University, Evanston Il 60208-3118, USA
Email: nocedal@ece.northwestern.edu
Url: www.ece.northwestern.edu/~nocedal

Abstract

We analyze the properties that optimization algorithms must possess in order to prevent convergence to non-stationary points for the merit function. We show that demanding the exact satisfaction of constraint linearizations results in difficulties in a wide range of optimization algorithms. Feasibility control is a mechanism that prevents convergence to spurious solutions by ensuring that sufficient progress towards feasibility is made, even in the presence of certain rank deficiencies. The concept of feasibility control is studied in this paper in the context of Newton methods for nonlinear systems of equations and equality constrained optimization, as well as in interior methods for nonlinear programming.

1 Introduction

We survey some recent developments in nonlinear optimization, paying particular attention to global convergence properties. A common thread in our review is the concept of "feasibility control", which is a name we give to mechanisms that regulate progress toward feasibility.

An example of lack of feasibility control occurs in line search Newton methods for solving systems of nonlinear equations. It has been known since the 1970s (see Powell [24]) that these methods can converge to undesirable points. The difficulties are caused by the requirement that

This work was supported by National Science Foundation grant CDA-9726385 and by Department of Energy grant DE-FG02-87ER25047-A004.

each step satisfy a linearization of the equations, and cannot be overcome simply by performing a line search. The need for more robust algorithms has been one of the main driving forces behind the development of trust region methods. Feasibility control is provided in trust region methods by reformulating the step computation as an optimization problem with a restriction on the length of the step.

This weakness of Newton-type methods manifests itself in a variety of contexts, such as nonlinear systems of equations, equality constrained optimization, active set Sequential Quadratic Programming methods for nonlinear programming, and more surprisingly in interior methods for nonlinear optimization. In this paper we review various techniques for providing feasibility control in these contexts, paying special attention to interior methods.

2 Nonlinear Equations

We begin our study of feasibility control by considering its simplest context, which occurs when solving a system of nonlinear equations

$$c(x) = 0. \tag{2.1}$$

Throughout this section we assume that c is a mapping from $I\!\!R^n$ to $I\!\!R^n$, so that (2.1) represents a system of n equations in n unknowns.

A popular algorithm for solving (2.1) is the line search Newton method

$$A(x)d = -c(x) \tag{2.2a}$$
$$x^+ = x + \alpha d, \tag{2.2b}$$

where $A(x)$ is the Jacobian of c and α is a steplength parameter chosen to decrease a merit function, such as

$$\phi(x) \equiv \|c(x)\|_2^2. \tag{2.3}$$

Note that (2.2a) demands that linear approximations of the functions be exactly satisfied, and as a result, the step d can be exceedingly large, or even be undefined if the Jacobian is rank deficient.

Powell [24] showed that this iteration has a fundamental weakness in that it can converge to a point that is neither a solution of (2.1) nor a stationary point of ϕ. We now discuss this example, which plays an important role throughout this paper.

Example 1 (Powell [24])

Consider the nonlinear system

$$c(x) \equiv \begin{pmatrix} x_1 \\ 10x_1/(x_1 + 0.1) + 2(x_2)^2 \end{pmatrix} = \begin{pmatrix} 0 \\ 0 \end{pmatrix}. \qquad (2.4)$$

This problem has only one solution at $x^* = (0,0)$, which is also the only stationary point of the function (2.3) in the half plane $x_1 > -0.1$. The Jacobian of c is given by

$$A(x) = \begin{bmatrix} 1 & 0 \\ 1/(x_1 + 0.1)^2 & 4x_2 \end{bmatrix},$$

and is singular along the line

$$\mathcal{L} : (x_1, x_2) = (\theta, 0) \qquad \theta \in \mathbb{R}, \qquad (2.5)$$

and in particular, at the solution point $x^* = (0,0)$. The graph of ϕ over the region of interest is plotted in Figure 1.1.

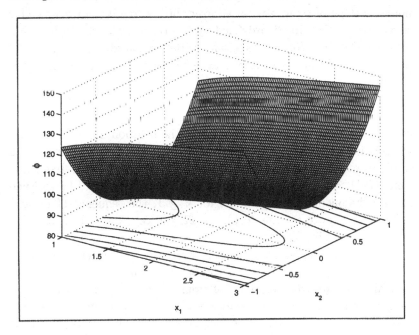

Fig. 1.1. Merit function (2.3) of Example 1 on rectangle $1 \leq x_1 \leq 3$, $-1 \leq x_2 \leq 1$.

The analysis given in [24] studies the Newton iteration (2.2) using a steplength α that minimizes ϕ along the direction d. It shows that starting from $x^1 = (3,1)$, all the iterates remain within the rectangle $1 \leq x_1 \leq 3$, $-1 \leq x_2 \leq 1$, and that the Newton directions d^k quickly tend to become parallel to the x_2-axis, with lengths $\|d^k\|$ that diverge to infinity. As a consequence, the search directions become increasingly perpendicular to $\nabla\phi$ and the steplengths α_k tend to 0. It can be shown that the iterates converge to a point on $\mathcal{L}$ of the form $\tilde{x} = (\beta, 0)$, with $\beta > 1.4$. The value of β depends on the starting point, and a numerical computation reveals that starting from $x^1 = (3,1)$, the iterates converge to $\tilde{x} \approx (1.8016, 0.000)$. It is clear from Figure 1.1 that this point is neither a solution to problem (2.4) nor a stationary point of ϕ, since the slope of ϕ at $\tilde{x}$ in the direction of the solution $(0,0)$ is negative.

What is particularly striking in this example is that the Newton iteration (2.2), which generates directions d^k that are descent directions for ϕ (as is well known and readily verified), can converge to a point with non-zero slope of ϕ, even when an exact line search is performed. This suggests that the direction defined by (2.2a) has an intrinsic flaw, and must be altered in some circumstances. Before considering this, we will pay closer attention to the roles played by the line search and the merit function in this example.

To our surprise we found that, by using a relaxed line search—a backtracking scheme that only requires that ϕ be reduced at each step, trying $\alpha = 1$ first—the iteration converged to the solution $x^* = (0,0)$. Using this inexact line search, the first step was a full Newton step with $\alpha = 1$ that led from $x^0 = (3,1)$ to $x^1 = (0, -1.841)$. The first equation $c_1(x^1) = 0$ is satisfied at x^1, and due to (2.2), it will be satisfied at all subsequent iterates. Thus, for $k \geq 1$, the sequence of iterates (x_1^k, x_2^k) belongs to the line $(0, \theta)$, $\theta \in \mathbb{R}$, and converges to the solution $x^* = (0,0)$ at a linear rate.

We tried another inexact line search strategy and observed that the iteration was again successful. The only way in which we were able to make the Newton iteration fail from the starting point $(3,1)$ was by performing an exact line search, which was unexpected. It is easy, however, to find *other* starting points for which the iteration (2.2) using a relaxed line search fails by converging to a point along the singular line (2.5) and with $x_1 > 0$. This illustrates how difficult it can be to predict when a failure will occur, and suggests that the interaction between the merit function and the search direction in these unfavorable circumstances is worthy of further investigation. We should stress that failures of the line search Newton iteration (2.2) are observed in practical applications, and are not purely of academic interest.

2.1 Feasibility Control for Nonlinear Equations

Since the deficiency of Newton's method just described occurs both with exact and inexact line searches, there is a strong indication that it is necessary to modify the search direction, at least sometimes, to obtain an algorithm with more satisfactory convergence properties. Let us therefore assume that instead of (2.2a) we demand that the step d satisfies

$$c(x) + A(x)d = r,$$

for some vector r.

One way to choose r is indirectly, by using a trust region approach. We can pose the step computation as an optimization problem in which we aim to satisfy (2.2a) as well as possible subject to the restriction that the step is no greater than a given trust region radius Δ. For example, the Levenberg-Marquardt method, which is the classical trust region method for nonlinear equations, makes use of the ℓ_2 norm and poses the step-generation subproblem as

$$\min_d \; \|c(x) + A(x)d\|_2^2 \tag{2.6a}$$

$$\text{subject to} \quad \|d\|_2 \leq \Delta \tag{2.6b}$$

If the linear system (2.2a) admits a solution that is no larger in norm than Δ, it will be taken as the step of the optimization algorithm; otherwise the solution of (2.6) will in general be a vector pointing in a different direction from the pure Newton step (2.2a). We say that this iteration provides "feasibility control" since it automatically determines the degree to which the linear approximation (2.2a) is to be satisfied. The feasibility control mechanism provided by the formulation (2.6) relies on an important algorithmic feature introduced in the 1970s, and used in most† trust region methods: the radius Δ is adjusted adaptively during the course of the optimization according to the success that the linear model has in predicting the behavior of the nonlinear problem during the most recent iteration. The degree of feasibility control is therefore dependent on this trust region adjustment strategy.

It can be shown that trust region methods based on (2.6) cannot

---

† A notable exception is the filter method proposed in [15] where the trust region update strategy depends only on whether the new iterate is acceptable to the filter and on whether the trust region was active during the step generation.

converge to a point where $\nabla\phi$ is nonzero, as is the case in Example 1. Indeed, one can prove (see [21]) that if $A(x)$ is Lipschitz continuous and bounded above in norm in the level set $\mathcal{T} = \{x : \phi(x) \leq \phi(x^0)\}$, then

$$\lim_{k\to\infty} \|\nabla\phi(x^k)\| = 0. \tag{2.7}$$

The iterates must therefore approach a stationary point of ϕ. This is not necessarily a solution of $c(x) = 0$, but since $\nabla\phi(x) = 2A(x)^T c(x)$, this result allows us to identify two possible outcomes. If the smallest singular value of the sequence $\{A(x^k)\}$ is bounded away from zero, then $c(x^k) \to 0$, and the iteration succeeds in finding a solution point. Otherwise, the method is attracted to a point (or a region) where the merit function ϕ cannot be improved to first order.

We can also view the trust region framework as a regularization technique since the solution of (2.6) is given by

$$\left(A(x)^T A(x) + \sigma I\right) d = -A(x)^T c(x), \tag{2.8}$$

for some non-negative parameter σ such that the coefficient matrix in this system is positive semi-definite. The original derivation of the Levenberg-Marquardt method was, in fact, motivated by the need to regularize the system $A(x)^T A(x)d = -A(x)^T c(x)$. We should point out, however, that regularization and feasibility control are fundamentally different. Attempting to improve the robustness of Newton's method by using a regularization approach has not been entirely successful due to the difficulty in choosing an appropriate value of the regularization parameter σ. A lower bound for σ is the value that ensures that (2.8) is numerically nonsingular, but we normally wish to use significantly larger values so that the direction d is not unnecessarily long and potentially unproductive. Our view is that the selection of σ should not be based entirely on linear algebra considerations, but must be tied in with the convergence properties of the optimization process.

As we have mentioned, trust region approaches select σ indirectly through the choice of the trust region radius Δ, which is updated at each iteration according to information about the nonlinear problem gathered during the optimization calculation. In particular, trust region methods provide a strategy for driving the regularizition parameter σ to zero whenever the iterates approach a solution point x^* with a full-rank Jacobian $A(x^*)$.

Variants of the Levenberg-Marquardt approach (2.6) have been proposed to take advantage of quasi-Newton approximations to the Jacobian $A(x)$, or to reduce the cost of computing the step. In the "dogleg method" [24] the subproblem takes the form

$$\min_d \ \|c(x) + A(x)d\|_2^2 \tag{2.9a}$$

$$\text{subject to} \quad x + d \ \in \ \mathcal{D} \tag{2.9b}$$

$$\|d\| \ \leq \ \Delta. \tag{2.9c}$$

The set $\mathcal{D}$ is a piecewise linear path that makes the solution of the problem easy to compute, and yet allows the iteration to be globally convergent in the sense that (2.7) holds. It is defined in terms of the "Cauchy step" p_C and the pure Newton step p_N, which are given by

$$p_C = -\tau A(x)^T c(x), \qquad p_N = -A(x)^{-1} c(x). \tag{2.10}$$

The Cauchy step p_C points in the direction of steepest descent for ϕ (see (2.3)) at the current point x, and τ is the steplength that minimizes the quadratic objective in (2.9a) along this steepest descent direction and subject to the trust region constraint (2.9c). The dogleg path $\mathcal{D}$ is the piecewise linear segment that starts at the current iterate x, passes through $x + p_C$, and terminates at $x + p_N$. As the set $\mathcal{D}$ includes the Cauchy step p_C, the resulting step d will attain a degree of feasibility of the linearized constraints that is no less than that achieved by p_C. This feature gives feasibility control to the method because it ensures—unlike the line search case—that sufficient progress is made at each iteration even when $A(w)$ is rank deficient or nearly deficient. (In the rank-deficient case, as p_N is not defined, we can assume that the dogleg path terminates at $x + p_C$.) The analysis presented in [24], which constitutes one of the founding blocks of trust region convergence theory, shows that any trust region method that decreases the objective (2.9a) at least as well as the Cauchy step p_C will enjoy the same global convergence properties as the Levenberg-Marquardt method. The restriction (2.9b) causes the step to be, in general, of lower quality than the Levenberg-Marquardt step, and this can slow progress toward the solution in some applications. In particular, we will see in section 4.2.2, that an interior method based on the dogleg approach can be inefficient on some simple problems.

The dogleg method was originally developed to take advantage of quasi-Newton approximations B_k to the Jacobian $A(x^k)$. Here computing the Newton-like step $B_k^{-1} c(x^k)$, updating B_k and computing the Cauchy point requires only $O(n^2)$ operations. As a result the dogleg quasi-Newton method allows a reduction in the iteration cost from $O(n^3)$ to $O(n^2)$, in the dense case, compared with the Levenberg-Marquardt method. The increasing availability of derivatives due to automatic differentiation techniques, and the fact that quasi-Newton updates for nonlinear equations are not as successful as those used in optimization, have, however, diminished the benefits of the dogleg method for nonlinear equations. Nevertheless, the dogleg method has seen a revival in the context of nonlinear programming because the generalizations of (2.9) to constrained optimization problems are more amenable to computation than the direct extensions of the Levenberg-Marquardt method (2.6). But as mentioned above, the dogleg method poses some tradeoffs that are currently the subject of investigation in the context of interior methods [6, 32].

3 Equality Constrained Optimization

Before we consider nonlinear programming problems in full generality it is convenient to focus on Newton-type methods for solving the equality constrained optimization problem

$$\min \ f(x) \tag{3.1a}$$

$$\text{subject to} \quad c(x) = 0, \tag{3.1b}$$

where $f : \mathbb{R}^n \to \mathbb{R}$ and $c : \mathbb{R}^n \to \mathbb{R}^t$ are smooth functions, and $t \leq n$.

Classical sequential quadratic programming (SQP) methods define a search direction d from the current iterate x, as the solution of a model of the form

$$\min_{d} \ m(d) \tag{3.2a}$$

$$\text{subject to} \quad c(x) + A(x)d = 0, \tag{3.2b}$$

where $A(x)$ is, as before, the Jacobian of c. Here $m(\cdot)$ is a quadratic Taylor approximation to the Lagrangian of (3.1). A new iterate is computed as

$$x^+ = x + \alpha d, \tag{3.3}$$

where α is a steplength chosen to decrease a merit function.

There is a good reason for choosing the quadratic program (3.2) to generate a search direction, as this amounts (when $m(\cdot)$ is strictly convex on the null space of $A(x)$, and $A(x)$ has full rank) to applying Newton's method to the optimality conditions of (3.1); see for example [14]. The discussion of the previous section suggests, however, that a Newton iteration that uses the linearizations (3.2b) may encounter difficulties that cannot be resolved simply by an appropriate selection of the steplength parameter α. This is indeed the case, as one can show [26, 7] that the lack of feasibility control in (3.2) can lead an SQP method to converge to an infeasible point that is not stationary with respect to a measure of infeasibility, such as $\|c(x)\|_2^2$.

In order to obtain stronger convergence properties, we introduce feasibility control in (3.2). This can be done either by using trust region formulations, or by introducing relaxations of (3.2b) in line search SQP methods.

3.1 Feasibility Control and Trust Regions

We now review some of the most interesting trust region reformulations of (3.2), and discuss the degree to which they succeed in improving the global convergence properties of SQP-type iterations.

In Vardi's method [29] the step computation subproblem takes the form

$$\min_d \ m(d) \tag{3.4a}$$

$$\text{subject to} \quad A(x)d + \theta c(x) = 0 \tag{3.4b}$$

$$\|d\|_2 \le \Delta, \tag{3.4c}$$

where θ is a positive parameter that ensures that the constraints (3.4b) and (3.4c) are compatible. Interestingly, this method does not provide feasibility control, since θ only controls the *length* of the step and not its direction. To be more specific, since θ is given prior to the step computation, we can define $\hat{d} = \theta^{-1}d$, and rewrite (3.4) as

$$\min_{\hat{d}} \ m(\theta\hat{d}) \tag{3.5a}$$

$$\text{subject to} \quad A(x)\hat{d} + c = 0 \tag{3.5b}$$

$$\|\hat{d}\|_2 \le \Delta/\theta. \tag{3.5c}$$

Equations (3.5b) are a standard linearization of the constraints and do not ensure that progress towards feasibility is comparable to that of a steepest descent (Cauchy) step. We speculate that, when applied to the problem [26]

$$\min 0 \quad \text{s.t.} \quad c(x) = 0,$$

where c is given by (2.4), Vardi's method may encounter the same difficulties as the line search Newton method on Example 1.

Due to the trust region constraint, the step cannot be unnecessarily large, but the system (3.5b) can be inconsistent, since it requires that $c(x)$ be in the range of $A(x)$, as in (3.2b). This approach, therefore, does not provide an adequate regularization in the case when $A(x)$ becomes rank-deficient.

Another potential drawback of (3.4) is that there does not appear to be a simple strategy for choosing the parameter θ so as to ensure good practical performance. It is easy to determine a range of values for which the constraints (3.4b)–(3.4c) are consistent, but the particular value chosen can strongly influence performance because it controls whether the step d tends more to satisfy constraint feasibility or to minimize the objective function. On the other hand, Vardi's approach is simpler than the formulations we discuss next.

In order to improve the theoretical deficiencies of Vardi's method, Byrd and Omojokun [4, 23] developed a scheme that emulates the Levenberg-Marquardt method (2.6) in its ability to solve systems of equations—or in the context of problem (3.1), to attain feasibility. They propose to first solve the auxiliary problem

$$\min_p \; q(p) \equiv \|c(x) + A(x)p\|_2^2 \qquad (3.6a)$$

$$\text{subject to} \qquad \|p\|_2 \leq \Delta, \qquad (3.6b)$$

which is the direct extension of (2.6). The step p computed in this manner will make sufficient progress toward feasibility and will prevent the iterates from converging to points that are not stationary for a variety of measures of infeasibility, such as $\phi(x) = \|c(x)\|_2^2$. It also provides regularization since (3.6) is well defined even when $A(x)$ is rank-deficient.

Of course, the step p will, in general, not contribute toward the minimization of f since (3.6) contains no information about the objective function. It will, however, be used to determine the level of linear feasibility that the step d of the optimization algorithm must attain. The step computation subproblem is thus formulated as

$$\min_d \; m(d) \qquad (3.7a)$$

$$\text{subject to} \quad c(x) + A(x)d = c(x) + A(x)p \qquad (3.7b)$$

$$\|d\|_2 \leq 1.2\,\Delta. \qquad (3.7c)$$

The size of the trust region has been expanded to allow room for decreasing the objective function, because when $\|p\|_2 = \Delta$, the only solution of (3.7) could be p.

Methods based on (3.6)–(3.7) have been analyzed in [5, 10, 12]. Under reasonable assumptions† it has been shown that the iterates are not attracted by infeasible points that are not stationary for the measure of infeasibility $\|c(x)\|_2^2$—or for many other measures of infeasibility. In other words, convergence to undesirable points of the type discussed in Example 1, cannot take place.

The feasibility control mechanism is clearly illustrated by (3.7b), where one explicitly determines the degree to which the linear equations $c(x) + A(x)d$ are to be satisfied. As in the case of the Levenberg-Marquardt method (2.6), the degree of linear feasibility is based on the trust region methodology applied to the auxiliary problem (3.6), and not on linear algebra considerations. The subproblems (3.6)–(3.7) can be solved exactly [19] or approximately [20, 23]. In the latter case it is important that the approximate solution provides progress toward feasibility that is comparable to that attained by a Cauchy step p^c for (3.6). This step is defined as the minimizer of the quadratic q in (3.6a) along the steepest descent direction $-\nabla q(0)$, and subject to the trust region constraint (3.6b).

Feasibility control takes a different form in the method proposed by Celis, Dennis and Tapia [8]; see also Powell and Yuan [27]. In this method, the step-computation problem is given by

$$\min_d\ m(d) \tag{3.8a}$$

$$\text{subject to}\quad \|c(x) + A(x)d\|_2 \le \eta \tag{3.8b}$$

$$\|d\|_2 \le \Delta. \tag{3.8c}$$

There are several ways of defining the feasibility control parameter π. Perhaps the most appealing is to let

$$\pi = \|c(x) + A(x)p^c\|_2,$$

where p^c is the Cauchy step for the problem (3.6).

By requiring that the step decreases linear feasibility at least as well as the Cauchy step, this method will not converge to non-stationary points of the type described in Example 1.

In summary, trust region methods based on (3.7) or (3.8) are endowed with feasibility control and regularization, whereas Vardi's approach does not provide either of these mechanisms. The formulation (3.8)

---

† These assumptions made in [5] are, in essence: the problem functions $f(\cdot)$ and $c(\cdot)$ are smooth, the sequence of function values $\{f(x_k)\}$ is bounded below, and the sequences $\{\nabla f(x^k)\}$, $\{c(x^k)\}$, $\{A(x^k)\}$ and the Lagrange Hessian approximations $\{B_k\}$ are all bounded.

proposed by Celis, Dennis and Tapia has not been used much in practice, due to the difficulties of solving a subproblem with two quadratic constraints, but the method of Byrd and Omojokun has been successfully used for large-scale equality constrained problems [20] and within interior methods for nonlinear programming, as we will discuss in the next section.

3.2 Feasibility Control and Line Search SQP

Most line search SQP methods compute a search direction d by solving the quadratic program (3.2). If at some iteration, the linear equations (3.2b) are found to be inconsistent, or if the Jacobian $A(x)$ is believed to be badly conditioned, a relaxation of the quadratic program is introduced.

For example, in the recently developed SNOPT package [18], the iteration enters "elastic mode" if the subproblem (3.2) is infeasible, unbounded, or its Lagrange multipliers become large (which can happen if $A(x)$ is nearly rank deficient). In this case, the nonlinear program is reformulated as

$$\min_{x,v,w} \ f(x) + \gamma e^T(v + w) \tag{3.9a}$$

$$\text{subject to} \quad c(x) - v + w \ = \ 0 \tag{3.9b}$$

$$v \geq 0, w \ \geq \ 0, \tag{3.9c}$$

where γ is a nonnegative penalty parameter. By applying the SQP approach (3.2) to this relaxed problem (3.9), the search direction (d_x, d_v, d_w) is required to satisfy the new linearized constraints

$$c(x) + A(x)\,d_x - (v + d_v) + (w + d_w) \ = \ 0$$
$$v + d_v \geq 0, \ w + d_w \ \geq \ 0,$$

which are always consistent. Once the auxiliary variables v and w are driven to zero, SNOPT leaves elastic mode and returns to the standard SQP subproblem (3.2).

In this approach feasibility control is therefore introduced only when needed. The relaxed problem is an instance of the $S\ell_1 QP$ method advocated by Fletcher [14], which is known to be robust in the presence of Jacobian rank deficiencies. Various other related strategies for relaxing the linear constraints (3.2b) have been proposed; see for example [25] and the references therein.

In a different approach advocated, among others by Biggs [2], the linearized constraints are relaxed at every iteration. This approach can be motivated [22] using the classical penalty function

$$P(x; \nu) = f(x) + \tfrac{1}{2}\nu \sum_{i=1}^{t} c_i(x)^2,$$

where $\nu > 0$ is the penalty parameter. The minimizer x satisfies

$$\nabla P(x; \nu) = \nabla f(x) + \nu \sum_{i=1}^{t} c_i(x)\nabla c_i(x) = 0. \qquad (3.10)$$

Defining the Lagrange multiplier estimates

$$y_i = -c_i(x)\nu, \quad i = 1, \ldots, t,$$

we can rewrite the optimality condition (3.10) as

$$
\begin{aligned}
\nabla f(x) - A^T(x)y &= 0 & (3.11a) \\
c(x) + y/\nu &= 0. & (3.11b)
\end{aligned}
$$

Considering this as a system in the variables x and y, and applying Newton's method to it, we obtain

$$\begin{bmatrix} W(x,y) & -A(x)^T \\ A(x) & \nu^{-1}I \end{bmatrix} \begin{bmatrix} d_x \\ d_y \end{bmatrix} = - \begin{bmatrix} \nabla f(x) - A(x)^T y \\ c(x) + y/\nu \end{bmatrix}, \qquad (3.12)$$

where

$$W(x,y) = \nabla^2 f(x) - \sum_{i=1}^{t} y_i \nabla^2 c_i(x).$$

At every iteration we increase ν in such a way that the sequence $\{\nu_k\}$ diverges. In the limit the term $\nu^{-1}I$ will therefore vanish and the second equation in (3.12) will tend to the standard linearized equation $A(x)d_x + c(x) = 0$.

This type of feasibility control has not been implemented in most SQP codes, but has recently received attention [16] in the context of interior methods for nonlinear programming.

4 Interior Methods

Let us now focus on the general nonlinear programming problem with equality and inequality constraints, which can be written as

$$
\begin{aligned}
\min_{x} \; & f(x) & (4.1a) \\
\text{subject to} \; & c_{\mathrm{E}}(x) = 0 & (4.1b) \\
& c_{\mathrm{I}}(x) \geq 0 & (4.1c)
\end{aligned}
$$

where $c_E : I\!\!R^n \mapsto I\!\!R^t$ and $c_I : I\!\!R^n \mapsto I\!\!R^l$, with $t \leq n$. Interior (or barrier) methods attempt to find a solution to (4.1) by approximately solving a sequence of barrier problems of the form

$$\min_{x,s} \ f(x) - \mu \sum_{i=1}^{l} \log(s_i) \qquad (4.2a)$$

$$\text{subject to} \quad c_E(x) \ = \ 0 \qquad (4.2b)$$

$$c_I(x) - s \ = \ 0, \qquad (4.2c)$$

for decreasing values of μ (see e.g. [13]). Here $s = (s_1, \ldots, s_l)$ is a vector of positive slack variables, and l is the number of inequality constraints. The formulation (4.2) allows one to develop "infeasible" algorithms that can start from an initial guess x^0 that does not satisfy the constraints (4.1b)–(4.1c). Line search and trust region techniques have been developed to solve (4.2), and both must efficiently deal with the implicit constraint that the slacks s must remain positive. The practical performance and theoretical properties of interior methods for nonlinear programming are not yet well understood, and in this section we discuss some recent analytical contributions.

4.1 Line Search Interior Methods

Line search interior methods generate search directions by applying Newton's method to the KKT conditions of the barrier problem (4.2), which can be written in the form

$$\nabla f(x) - A_E(x)^T y - A_I(x)^T z \ = \ 0 \qquad (4.3a)$$

$$Sz - \mu e \ = \ 0 \qquad (4.3b)$$

$$c_E(x) \ = \ 0 \qquad (4.3c)$$

$$c_I(x) - s \ = \ 0. \qquad (4.3d)$$

Here $A_E(x)$ and $A_I(x)$ are the Jacobian matrices of the functions c_E and c_I, respectively, and y and z are their Lagrange multipliers; we also define $S = \text{diag}(s)$ and $Z = \text{diag}(z)$. The Newton equations for this system,

$$
\begin{bmatrix}
W & 0 & A_E{}^T & A_I{}^T \\
0 & Z & 0 & -S \\
A_E & 0 & 0 & 0 \\
A_I & -I & 0 & 0
\end{bmatrix}
\begin{bmatrix}
d_x \\
d_s \\
-d_y \\
-d_z
\end{bmatrix}
= -
\begin{bmatrix}
\nabla f - A_E{}^T y - A_I{}^T z \\
Sz - \mu e \\
c_E \\
c_I - s
\end{bmatrix}
$$

$$(4.4)$$

define the *primal dual* direction $d = (d_x, d_s, d_y, d_z)$. A line search along d determines a steplength $\alpha \in (0, 1]$ such that the new iterate

$$(x^+, s^+, y^+, z^+) = (x, s, y, z) + \alpha (d_x, d_s, d_y, d_z) \qquad (4.5)$$

decreases a merit function and satisfies $s^+ > 0$, $z^+ > 0$. In (4.4), the $n \times n$ matrix W denotes the Hessian, with respect to x, of the Lagrangian function

$$\mathcal{L}(x, y, z) = f(x) - y^T c_E(x) - z^T c_I(x). \qquad (4.6)$$

Many interior methods follow this basic scheme [1, 13, 17, 28, 31]; they differ mainly in the choice of the merit function, in the mechanism for decreasing the barrier parameter μ, and in the way of handling nonconvexities. A careful implementation of a line search interior method is provided in the LOQO software package [28].

This line search approach is appealing due to its simplicity and its close connection to interior methods for linear programming, which are well developed. Numerical results reported, for example, in [11, 28] indicate that these methods represent a very promising approach for solving large scale nonlinear programming problems.

Nevertheless, Wächter and Biegler [30] have recently shown that *all* interior methods based on the scheme (4.4)–(4.5) suffer from convergence difficulties reminiscent of those affecting the Newton iteration in Example 1. We will see that, due to the lack of feasibility control in (4.4), these iterations may not be able to generate a feasible point.

Example 2 (Wächter and Biegler [30]) Consider the problem

$$\min\ f(x) \qquad (4.7a)$$
$$\text{subject to} \quad (x_1)^2 - x_2 - 1 = 0 \qquad (4.7b)$$
$$x_1 - x_3 - 2 = 0 \qquad (4.7c)$$
$$x_2 \geq 0,\ x_3 \geq 0, \qquad (4.7d)$$

where the objective $f(x)$ is any smooth function. (This is a special case of the example presented in [30].) Let us apply an interior method of the form (4.4)–(4.5), starting from the initial point $x^0 = (-2, 1, 1)$. From the third equation in (4.4) we have that the initial search direction d will satisfy the linear system

$$c_E(x^0) + A_E(x^0)d_x = 0, \qquad (4.8)$$

whose solution set can be written in the parametric form

$$d_x = \begin{pmatrix} 0 \\ 2 \\ -5 \end{pmatrix} + \theta \begin{pmatrix} 1 \\ -4 \\ 1 \end{pmatrix} \qquad \theta \in \mathbb{R}.$$

Figure 1.2 illustrates the feasible region (the dotted segment of the parabola) and the set of possible steps d_x, all projected onto the x_1–x_2 plane.

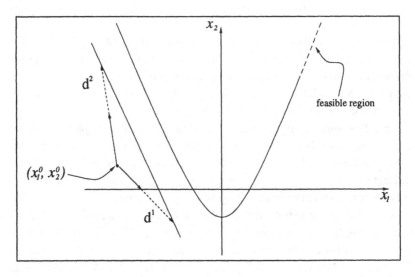

Fig. 1.2. Example 2, projected onto the x_1–x_2 plane.

The steplength $\alpha = 1$ will not be acceptable because, for any value of θ, the point

$$x^0 + d_x = \begin{pmatrix} -2 \\ 1 \\ 1 \end{pmatrix} + \begin{pmatrix} 0 \\ 2 \\ -5 \end{pmatrix} + \theta \begin{pmatrix} 1 \\ -4 \\ 1 \end{pmatrix}$$

violates the positivity bounds (4.7d). The steplength will therefore satisfy $0 < \alpha < 1$. The parabola in Figure 1.2 represents the constraint (4.7b), and the straight line to the left of it is its linearization at x^0. We have drawn two displacements, d^1 and d^2 that satisfy (4.8). Note how the step d^1 is restricted by the bound $x_2 > 0$; the shortened step is drawn as a solid line. The second step, d^2 is restricted by the bound $x_3 > 0$ (which cannot be drawn in this picture), and the shortened step is again drawn as a solid line.

Now, regardless of the value of $\alpha \in (0, 1)$, we see that the new iterate will be confined to the set

$$\{(x_1, x_2) \ : \ x_1 \leq -\sqrt{x_2 + 1}, \ x_2 \geq 0, \ x_3 \geq 0\}. \tag{4.9}$$

All points in this region are infeasible, as the constraint (4.7c) and the bound on x_3 in (4.7d) imply that any feasible point x must satisfy $x_1 \geq 2$.

The argument can now be repeated from the new iterate: one can show that the full step toward the linearized constraints violates at least one of the bounds and that a steplength $\alpha < 1$ must be employed. Using this, Wächter and Biegler showed that the sequence of iterates $\{x^k\}$ never leaves the region (4.9), and cannot generate a feasible iterate (even in the limit). This behavior is obtained for any starting point $x^0 = (x_1^0, x_2^0, x_3^0)$ that belongs to (4.9).

This convergence failure affects *any* method that generates directions that satisfy the linearized constraints (4.8) and that enforces the bounds (4.7d) by means of a backtracking line search. The merit function can only restrict the steplength further, and therefore is incapable of resolving the difficulties. The strategy for reducing μ is also irrelevant, since the proof only makes use of the third equation in the Newton iteration (4.4).

Example 2 is not pathological, in the sense that the Jacobian $A_E(x)$ of the equality constraints given by (4.7b)–(4.7c) has full rank for all $x \in \mathbb{R}^3$. Without specifying the objective function and the merit function, all that can be said is that the algorithm never reaches feasibility, but by choosing f appropriately one can make the sequence of iterates converge to an infeasible point.

More specifically, Wächter and Biegler performed numerical tests defining $f(x) = x_1$ in Example 2, using various starting points, and report convergence to points of the form

$$x = (-\beta, 0, 0), \quad \text{with } \beta > 0. \tag{4.10}$$

In other words, the iteration can converge to a point on the boundary of the set $\{(x_1, x_2) : x_2 \geq 0, \ x_3 \geq 0\}$—a situation that barrier methods are supposed to prevent! It is interesting to note that the points (4.10) are not stationary points for any measure of infeasibility of the form $\|c_E(x)\|_p$ $1 \leq p \leq \infty$ (the bound constraints (4.7d) need not be taken into account in the infeasibility measure, as the limit points (4.10) satisfy both of them) and that the steplengths α_k converge to zero.

We should note that it is not essential that the parabola (4.7b) crosses the horizontal axis for the failure to occur; the analysis given in [30] includes the case in which the entire parabola lies above the x_1-axis.

An intriguing question is whether the failure of the iteration (4.4)–(4.5) is a manifestation of the limitations of Newton's method discussed in Example 1, or whether it is caused by the specific treatment of inequalities in these interior methods. To try to answer this question, we first ask whether the iteration matrix in (4.4) becomes singular in the limit, as is the case in Example 1. By defining $f(x) = x_1$, once more, we can consider x_2 and x_3 as slack variables, and the KKT matrix (4.4) for problem (4.7) is given by

$$
\begin{bmatrix}
W & 0 & 0 & 2x_1 & 1 \\
0 & z_1 & 0 & x_2 & 0 \\
0 & 0 & z_2 & 0 & x_3 \\
2x_1 & 1 & 0 & 0 & 0 \\
1 & 0 & 1 & 0 & 0
\end{bmatrix}.
$$

This matrix is indeed singular on the manifold $\{(x_1, 0, 0) : x_1 \in I\!R\}$. In Example 1, however, the steplengths α_k are forced to converge to zero by the merit function, whereas in Example 2 this is a consequence of the bounds $x_2 \geq 0$, $x_3 \geq 0$. Thus, at least superficially, there appears to be a fundamental difference in these two phenomena, but this topic is worthy of further investigation.

Wächter and Biegler make another interesting observation about their example. At each step the linearized constraints (4.8) and the bounds (4.7d) are inconsistent. In active set SQP methods, this set of conditions comprises the feasible region of the quadratic program, and hence this inconsistency is readily detected. By introducing a relaxation of the quadratic program, as discussed in section 3.2, these methods are able to generate a step. In the context of interior methods, however, all that is known in these unfavorable situations is that the step must be shortened, which is a very common occurrence, and therefore cannot be used as a reliable warning sign. When convergence to non-stationary limit points such as (4.10) is taking place, the steplength will eventually become so small that difficulties will be apparent, but by this time a large amount of computation will have been wasted. This appears to be a theoretical deficiency of the algorithmic class (4.4)–(4.5).

We do not know whether failures of this type are rare in practice, and whether this deficiency of the scheme (4.4)–(4.5) manifests itself more often as inefficient behavior, rather than as outright convergence failures.

An important practical question, which is the subject of current research, is whether a simple modification of the scheme (4.4)–(4.5) can resolve the difficulties. A line search interior method in which the steps do not satisfy the linearization of the constraints was proposed by Forsgren and Gill [16]. We do not know, however, if the convergence properties of this method are superior to those of the basic line search scheme (4.4)–(4.5).

In the next section we will see that a trust region interior method that employs feasibility control will not fail in Example 2, and thus possesses more desirable global convergence properties.

4.2 Trust Region Interior Methods

Let us now discuss interior methods that, in contrast to the schemes just described, generate steps using trust region techniques. For simplicity we will assume that the nonlinear program has only inequality constraints; the extension to the general problem (4.1) is straightforward.

For a given value of the barrier parameter μ, we can apply a trust region method to the barrier problem (4.2), which we now write as

$$\min \ f(x) - \mu \sum_{i=1}^{l} \log(s_i) \qquad (4.11a)$$

$$\text{subject to} \qquad c(x) - s = 0 \qquad (4.11b)$$

(we have omitted the subscript in $c_i(x)$ for simplicity). The interior methods described in [6, 9, 32] differ in the trust region method used to solve (4.11). We will first consider the approach described in [5, 6], which has been implemented in the NITRO software package.

4.2.1 The Algorithm in NITRO

The presence of the implicit bounds $s > 0$ in (4.11) suggests that the trust region should be scaled to discourage the generation of steps that immediately leave the feasible region. We can define, for example,

$$\|(d_x, S^{-1}d_s)\|_2 \leq \Delta, \qquad (4.12)$$

where we have decomposed the step in terms of its x and s components, i.e., $d = (d_x, d_s)$. This trust region becomes more elongated as the iterates approach the boundary of the set $s \geq 0$, and permits the generation of steps that point away from this boundary. In addition, we impose the so-called "fraction to the boundary rule"

$$s + d_s \geq (1 - \tau)s, \qquad \text{with e.g., } \tau = 0.995, \qquad (4.13)$$

to ensure that the slack variables remain strictly positive at every iteration.

The step-generation subproblem is given by

$$\min_{d} \ m(d) \tag{4.14a}$$

$$\text{subject to} \quad c(x) - s + A(x)d_x - Sd_s \ = \ r \tag{4.14b}$$

$$\|d\|_2 \ \leq \ \Delta \tag{4.14c}$$

$$d_s \ \geq \ -\tau e. \tag{4.14d}$$

Here $m(\cdot)$ is a quadratic model of the Lagrangian of the barrier problem (4.11), r is a vector that provides feasibility control, $A(x)$ is the Jacobian of $c(x)$, and e is the vector of all ones. To obtain (4.14) we have introduced the change of variables

$$d_s \leftarrow S^{-1}d_s, \tag{4.15}$$

so that the trust region constraint (4.12) has the familiar spherical form (4.14c).

The vector r is determined by solving a Levenberg-Marquardt type of subproblem, which in this context takes the form

$$\min_{p} \ \|c(x) - s + A(x)p_x - Sp_s\|_2^2 \tag{4.16a}$$

$$\text{subject to} \quad \|p\| \ \leq \ \xi\Delta \tag{4.16b}$$

$$p_s \ \geq \ -\tau\xi e. \tag{4.16c}$$

We call the solution $p = (p_x, p_s)$ of this problem the "normal step" because, as we discuss below, it is often chosen to be normal to the constraints. The trust region radius has been reduced by a factor of ξ (e.g., $\xi = 0.8$) to provide room in the computation of the total step d given by (4.14). Whereas in the equality constrained case (see (3.6)), feasibility control is only provided by the trust region constraint, in this formulation control is also exercised through the bound (4.16c).

After the normal step p has been computed, we define

$$r = c(x) - s + A(x)p_x - Sp_s, \tag{4.17}$$

so that the level of linear feasibility provided by the step d is that obtained by the normal step.

To complete the description of the interior algorithm proposed in [5, 6], we note that the step d is accepted or rejected depending on whether or not it provides sufficient reduction of the merit function

$$\psi(x,s) = f(x) - \mu \sum_{i=1}^{l} \log(s_i) + \nu \|c(x) - s\|_2,$$

where ν is a positive penalty parameter. The trust region radius Δ is updated according to rules that can be viewed as the direct extension of the strategies used in trust region methods for unconstrained and equality constrained optimization.

To obtain a primal dual interior method, as opposed to a primal method, it is necessary to define the model $m(\cdot)$ in (4.14) appropriately. Consider the quadratic function

$$m(d) = (\nabla f, -\mu e)^T (d_x, d_s) + \tfrac{1}{2}(d_x, d_s)^T \begin{bmatrix} W & \\ & S\Sigma S \end{bmatrix} (d_x, d_s), \quad (4.18)$$

where W denotes the Hessian of the Lagrangian function (4.6) of the nonlinear programming program, and Σ is a diagonal matrix. If we choose $\Sigma = \mu S^{-2}$, then $m(\cdot)$ is a quadratic approximation of the Hessian of the barrier function (4.11), and this approach gives rise to a primal method. By defining $\Sigma = S^{-1}Z$, on the other hand, the step generated by (4.14) can be considered as a primal dual method in the sense that, when the problem is locally convex and the trust region is inactive, its solution coincides with the primal dual step defined by (4.4). The choice $\Sigma = S^{-1}Z$ has been observed to provide better performance than the primal method.

Behavior on Example 2. We have seen that the problem described in Example 2 will cause failure of any line search interior method of the form (4.4)–(4.5). Let us now consider the performance of the trust region method just outlined. Since providing analytic expressions for the iterates of this algorithm is laborious when the inequality constraints in the subproblems (4.14) and (4.16) are active, we will simply report the result of running the NITRO package on Example 2.

The iterates generated by the trust region method are plotted in Figure 1.3. During the first three iterations, which start at x^0 and finish at x^3, the trust region is active in both the normal (4.16) and in the step-generation subproblem (4.14). In particular, in these iterations, the normal step is a linear combination of the Cauchy and the Newton steps on the function (4.16a). Hence, the first three displacements did *not* satisfy the linearization of the constraints $c(x) - s + A(x)^T d_x - S d_s = 0$.

We note that x^1 has already left the area in which line search methods get trapped.

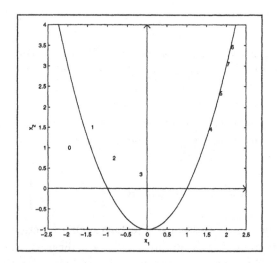

Fig. 1.3. Iterates generated by a trust region interior method with feasibility control on Example 2, projected onto the x_1-x_2 plane. Each iterate x^k is indicated by the integer k.

For all the remaining steps, $x^4, x^5, \ldots$ the trust region was inactive in both subproblems, so that the linearized constraints were satisfied, and the steps were similar to those produced by a line search algorithm of the form (4.4)–(4.5). The trust region algorithm therefore performed very well on this problem, correctly deciding when to include a steepest descent like component on the step.

That this trust region method cannot fail on Example 2 is not surprising in the light of the convergence theory developed in [5]. It shows that the iterates cannot converge to a non-stationary point of the measure of feasibility $\|c(x)\|_2^2$. This result is possible because sufficient progress towards feasibility is required at every iteration, as we now discuss.

The normal step p given by (4.16) plays a similar role to that of the Levenberg-Marquardt step (2.6) for systems of nonlinear equations. In practice we may prefer to solve (4.16) inaccurately, using for example a dogleg method. In this case two conditions are required of the approximate solution p of the normal step. The first is that it must provide a decrease of linear feasibility that is proportional to that attained by the

Cauchy step p_C for (4.16). This is of the form

$$p_C = \tau [A \ - S]^T (c - s),$$

for some scalar τ.

Interestingly, this "Cauchy decrease condition" is not sufficient since it is possible for a very long step to provide the desired degree of linear feasibility but produce a very large increase in the model $m(\cdot)$ defined in (4.18). To control the length of the step we can impose the condition that p lie on the range space of the constraint normals of (4.14b), i.e.

$$p \in \mathcal{R}([A \ - S]^T). \tag{4.19}$$

An alternative to (4.19) that has been studied by El-Alem [12] in the context of equality constrained optimization, is to require that

$$\|p\| = O(\|p_{MN}\|)$$

where p_{MN} is the minimum norm solution of (4.16a).

These are requirements that the step must satisfy to improve feasibility. To make progress toward optimality, the step should also provide a decrease in the model $m(\cdot)$ that is proportional to that attained by the Cauchy step for (4.14). We will not describe this in more detail since it is not central to the main topic of this article.

4.2.2 Tradeoffs in this Trust Region Approach

Even though the trust region interior method implemented in NITRO has more desirable global convergence properties than line search interior methods, we now show through an example that it can be inefficient on some problems when certain cost-saving techniques are used in the solution of the subproblems (4.14) and (4.16).

Example 3. Consider the one variable problem

$$\min x \qquad \text{s.t.} \quad x \geq b,$$

where b is a constant, say $b = 500$. The associated barrier problem is given by

$$\min_{x,s} \ x - \mu \log(s) \quad \text{subject to} \quad x - b - s = 0, \tag{4.20}$$

and its optimal solution is

$$(x, s) = (b + \mu, \mu). \tag{4.21}$$

Let us focus on the first barrier problem, with say $\mu = 0.1$. We choose the initial point $x^0 = 1$, and to simplify the analysis, we set the initial slack to its optimal value $s^0 = \mu$; see Figure 1.4.

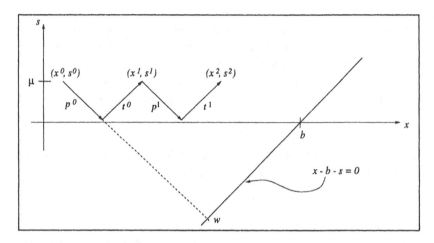

Fig. 1.4. First iterates of a trust region interior method on the barrier problem (4.20).

Suppose that the radius Δ is large enough at x^0 that the trust region is inactive when solving both (4.14) and (4.16). The step generated by the trust region algorithm just described is

$$\left(\frac{\mu - s^0}{z^0} - (x^0 - b - s^0), \frac{\mu - s^0}{s^0 z^0} \right), \qquad (4.22)$$

which can be regarded as a good step: it leads to the point $(b - \mu, 0)$, which is close to the minimizer (4.21).

Since it is expensive, however, to solve the subproblems (4.14) and (4.16) exactly, in the NITRO software package these subproblems are solved inaccurately. In particular, the normal step is required to be in the range of the constraint normals, i.e., p satisfies (4.19). We will now show that for such a normal step, the algorithm can be very inefficient, even if (4.14) is solved exactly.

In this example, there is a unique normal step p^0 from x^0 that lies in the range of the constraint normals; it is illustrated in Figure 1.4. If p^0 were not restricted by the fraction to the boundary rule (4.16c), it would lead to the point w (see Figure 1.4) where the objective in (4.16a) becomes zero. But after being cut back, the ability of the normal step p^0 to reduce infeasibility is greatly impaired. Since the total step d must retain the level of infeasibility provided by the normal step, we can view it as the sum $d = p + t$, for some displacement t perpendicular to p. It is easy to show that, in this example, d_x can be very small compared to b, and that the slack variable will be unchanged after the first step, i.e. $s^1 = s^0$. This behavior can be sustained for a large number of iterations, so that the sequence of iterates x^k moves slowly toward feasibility.

We can show that the number of iterations n_{iter} needed to attain feasibility satisfies

$$n_{\text{iter}} \geq \frac{4}{9\mu}(b - x^0), \tag{4.23}$$

when the primal version of the algorithm is used. This number can be very large; consider e.g. $\mu = 0.1$ and $|b - x^0| = 500$. The bound (4.23) is established under the conditions that (i) $\Delta_0 \geq (9/4)\mu$, (ii) $s^0 = \mu$, and that the merit function is of the form

$$\phi(x, s) = f(x) - \mu \sum_{i=1}^{l} \log(s_i) + \nu \|c(x) - s\|_p$$

for some $p \in [1, \infty]$. It is possible to show that, prior to the generation of a feasible iterate, all the steps are accepted by the merit function, the normal step is always restricted by (4.16c), and the trust region is inactive in both subproblems. The normal and tangential steps then take a simple closed form that enables one to obtain a recurrence relation for the iterates (x^k, s^k), from which (4.23) can be readily established.

We should stress that the interior method will not fail on Example 3, but that it will be inefficient. This is of concern, as we have observed this behavior in several of the problems in the CUTE [3] collection (e.g. in problem GOFFIN). In these problems many iterations are needed to attain feasibility, but convergence to the solution is fast after feasibility is achieved.

In contrast, a line search method of the form (4.4)–(4.5) solves Example 3 in a few iterations. We performed the calculations using an option in the NITRO package that computes steps of the form (4.4)–(4.5). We set $\mu = 0.1$, and report information about the first few iterations in Table 1.1. The first 3 steps where restricted significantly by the fraction to the boundary rule (4.13), (see the "steplength" column) providing small improvement in infeasibility, $\|c(x) - s\|$. But as the value of x became small, the algorithm produced a search direction that was sufficiently horizontal to allow a unit steplength and thus satisfy the linear constraints.

iter	infeasibility	steplength
0	5.000E+02	
1	4.996E+02	8.95E-04
2	4.982E+02	2.75E-03
3	4.853E+02	2.58E-02
4	3.876E+02	2.01E-01
5	6.706E-14	1.00E+00

Table 1.1. *Line search interior method on the barrier problem* (4.20).

We conclude that it is desirable to develop new techniques for approximately solving the subproblems (4.14) and (4.16) that perform well in the situation illustrated in Example 3.

4.3 The Trust Region Method in NUOPT

Yamashita, Yabe and Tanabe [32] have proposed a trust region interior method that promotes global convergence in a different manner from the method implemented in NITRO. They report excellent numerical results on a wide range of problems.

Their algorithm is an infeasible method based on the barrier problem (4.11). Each step d is computed using an extension of the dogleg method described in section 2.1. Specifically, $d = (d_x, d_s)$ is of the form

$$d = \beta d_{\mathrm{SD}} + (1 - \beta)d_{\mathrm{PD}}, \qquad (4.24)$$

where β is a parameter in $[0, 1]$. Here d_{PD} denotes the x-s components of the primal dual step (4.4), and the so-called steepest descent step d_{SD} is also defined by (4.4) but replacing W by a positive definite diagonal matrix D, e.g. $D = I$.

The value of β must be chosen so that the step d satisfies the following three conditions

$$\|d\| \leq \Delta, \qquad s + d_s \geq (1 - \tau)s, \qquad (4.25\mathrm{a})$$
$$M(0) - M(\alpha d) \geq \tfrac{1}{2}\{M(0) - M(\alpha_{\mathrm{SD}}d_{\mathrm{SD}})\}, \qquad (4.25\mathrm{b})$$

where $M(\,\cdot\,)$ is the following model of the barrier problem (4.2),

$$M(d) = (\nabla f, -\mu S^{-1}e)^T(d_x, d_s) + \tfrac{1}{2}(d_x, d_s)^T \begin{bmatrix} W & \\ & \Sigma \end{bmatrix} (d_x, d_s)$$

$$+ \nu \sum_{i=1}^{l} \left(|c_i(x) + \nabla c_i(x)^T d| - |c_i(x)| \right),$$

and $\Sigma = S^{-1} Z$. This model differs from (4.18) in the inclusion of terms measuring changes in the constraints. It is derived from the ℓ_1 merit function

$$f(x) - \mu \sum_{i=1}^{l} \log(s_i) + \nu \sum_{i=1}^{t} \|c(x) - s\|_1, \qquad (4.26)$$

where ν is a positive parameter. The steplengths α and α_{SD} in (4.25b) are the values in $[0,1]$ that minimize $M(\cdot)$ along the directions d and d_{SD} respectively, and subject to the trust region constraint and the fraction to the boundary rule (4.25a).

The definition (4.25b) implies that the step d must provide a decrease in the model that is proportional to that given by the steepest-descent step d_{SD}. Once the step is computed, one follows standard trust region techniques to determine if it is acceptable for the merit function (4.26) and to update the trust region radius.

In the implementation described in [32], the parameter β is computed by a backtracking search, using the initial value $\beta = 0$, which corresponds to the primal dual step d_{PD}, and decreasing it by 0.1 until (4.25) is satisfied.

The similarities of this approach and the one used in the NITRO package are apparent. Both use trust region techniques to generate the step, and employ models of the barrier problem to determine the minimum quality of a step. The algorithms, however, differ in fundamental ways. Since both d_{PD} and d_{SD} are computed by a system of the form (4.4) they satisfy the linearization of the constraints. The total step d therefore does not provide feasibility control, and as a result, this method will fail on the problem described in Example 2.

Let us examine the steepest descent step d_{SD} more closely. Its name derives from an analogy with unconstrained optimization, where a Newton step applied to a quadratic function whose Hessian is the identity, coincides with a steepest descent step. The step d_{SD} is, however, not a steepest descent step for $\|c(x) - s\|_2$, and does not guarantee sufficient progress toward feasibility. Nor is it a steepest descent step for the merit function (4.26). This is in contrast with the model (4.16) which demands that the normal step provides a steepest-descent like reduction in the linearized constraints. Even though the method does not resolve the convergence difficulties of standard line search Newton iterations, the dogleg-like scheme between d_{SD} and d_{PD} constitutes a novel approach to handling nonconvexities, and may provide stability to the iteration in the case when the Hessian W is ill-conditioned.

5 Final Remarks

This article has focused on possible failures of optimization algorithms near non-stationary points of the measure of infeasibility of the problem. At these points the Jacobian is rank-deficient, and they can be avoided by introducing appropriate "feasibility control" in the optimization iteration. We have reviewed a variety of techniques for preventing convergence to non-stationary points, and discussed to what extent they are successful in doing so.

Whereas feasibility control mechanisms have been extensively studied in the context of nonlinear systems of equations and equality constrained optimization, they are only beginning to be investigated within interior methods for nonlinear programming. Designing algorithms with favorable global convergence properties, fast local rate of convergence, and low computational cost, remains an interesting topic for further research.

Acknowledgements. We thank Richard Byrd, Andreas Wächter and Richard Waltz, who provided many useful comments during the preparation of this article, and Mike Powell for an illuminating discussion concerning Example 1.

References

[1] Akrotirianakis. I. and Rustem. B. (1999). A primal-dual interior point algorithm with an exact and differentiable merit function for general nonlinear programming problems. Technical Report 98-09, Department of Computing, Imperial College, London, UK.

[2] M. C. Bartholomew-Biggs (1982). Recursive quadratic programming methods for nonlinear constraints. in *Nonlinear Optimization 1981*, (M.J.D. Powell, ed.) (Academic Press, London), 213–222.

[3] Bongartz, I., Conn, A.R., Gould, N.I.M. and Toint, Ph.L. (1995). CUTE: Constrained and unconstrained testing environment, *ACM Transactions on Mathematical Software* **21**, 123–160.

[4] Byrd, R.H. (1987). Robust trust region methods for constrained optimization, Third SIAM Conference on Optimization, Houston, Texas.

[5] Byrd, R.H., Gilbert, J.Ch. and Nocedal, J. (1996). A trust region method based on interior point techniques for nonlinear programming. Technical Report 96/02, Optimization Technology Center, Northwestern

University, Evanston, IL, USA. Revised August 1998. (Also available as Rapport de Reccherche 2896, INRIA.)

[6]   Byrd, R.H., Hribar, M.E. and Nocedal, J. (1999). An interior point algorithm for large-scale nonlinear programming, *SIAM Journal on Optimization*, **9**, 877–900.

[7]   Byrd, R.H., Marazzi, M. and Nocedal, J. (2000). On the convergence of Newton iterations to non-stationary points. In preparation.

[8]   Celis, M.R., Dennis, J.E. and Tapia, R.A. (1995). A trust region strategy for nonlinear equality constrained optimization. in *Numerical Optimization 1984* (R.B. Schnabel, P.T. Boggs and R.H. Byrd, eds) (Philadelphia, SIAM), 71–82.

[9]   Conn, A.R., Gould, N.I.M., Orban, D. and Toint, Ph.L. (1999). A primal-dual trust-region algorithm for minimizing a non-conve, function subject to general inequality and lineal equality constraints. Tech. Rep. RAL-TR-1999-054, Computational Science and Engineering Department, Atlas Centre, Rutherford Appleton Laboratory, England.

[10] Conn, A.R., Gould, N.I.M. and Toint, Ph. (2000). *Trust region methods* (SIAM, Philadelphia).

[11] Conn, A.R., Gould, N.I.M. and Toint, Ph.L. (1999). A primal-dual algorithm for minimizing a non-convex function soubject to bound and linear equality constraints, in *Nonlinear Optimization and Applications 2* (G. Di Pillo and F. Gianessi, eds) (Kluwer, Dordrecht), to appear.

[12] El-Alem, M. (1999). A global convergence theory for Dennis, El-Alem and Maciel's class of trust region algorithms for constrained optimization without assuming regularity, *SIAM J. Optim.* **9**, 965–990.

[13] El-Bakry, A.S., Tapia, R.A., Tsuchiya, T. and Zhang, Y. (1996). On the formulation and theory of the Newton interior-point method for nonlinear programming. *J. Optim. Th. Al Applic.* **89**, 507–541.

[14] Fletcher, R. (1987). *Practical Methods of Optimization* (2nd ed.) (Wiley, New York).

[15] Fletcher, R. and Leyffer, S. (1997). Nonlinear programming without a penalty function, Tech. Rep. NA/171, Univ. of Dundee, Scotland.

[16] Forsgren, A. and Gill, P.E. (1998). Primal-dual interior methods for nonconvex nonlinear programming, *SIAM J. Optim.* **8**, 1132–1152.

[17] Gay, D.M., Overton, M.L. and Wright, M.H. (1998). A primal-dual interior method for nonconvex nonlinear programming, in *Advances in Nonlinear Programming* (Y. Yuan, ed.) (Kluwer, Dordrecht), 31–56.

[18] Gill, P.E., Murray, W. and Saunders, M.A. (1997). Snopt: An sqp algorithm for large-scale constrained optimization, Tech. Rep. 97-2, Mathematics, University of California, San Diego.

[19] Gomes, F.A.M., Maciel, M.C. and Martinez, J.M. (1999). Nonlinear programming algorithms using trust regions and augmented lagrangians with nonmonotone penalty parameters, *Math. Programming* **84**, 161–200.

[20] Lalee, M., Nocedal, J. and Plantenga, T. (1998). On the implementation of an algorithm for large-scale equality constrained optimization, *SIAM J. Optim.* **8**, 682–706.

[21] Moré, J. (1983). Recent developments in algorithms and software for trust region methods, in *Mathematical Programming: The State of the Art* (A. Bachem, M. Grötschel, and B. Korte, eds) (Springer-Verlag, Berlin), 258–287.

[22] Murray, W. (1969). An algorithm for constrained optimization, in

Optimization (R. Fletcher, ed.) (Academic Press, London).

[23] Omojokun, E.O. (1989). Trust region algorithms for optimization with nonlinear equality and inequality constraints, PhD thesis, University of Colorado, Boulder.

[24] M. J. D. Powell, M.J.D. (1970). A hybrid method for nonlinear equations, in *Numerical Methods for Nonlinear Algebraic Equations* (P. Rabinowitz, ed.) (Gordon & Breach, London), 87–114.

[25] Powell, M.J.D. (1983). Variable metric methods for constrained optimization, in *Mathematical Programming : The State of the Art, Bonn 1982* (A. Bachem, M. Grötschel and B. Korte, eds) (Springer-Verlag, Berlin), 288–311.

[26] Powell, M.J.D. (1987). Methods for nonlinear constraints in optimization calculations, *The State of the Art in Numerical Analysis* (A. Iserles and M.J.D. Powell, eds) (Clarendon Press, Oxford), 325–357.

[27] Powell, M.J.D. and Yuan, Y. (1990). A trust region algorithm for equality constrained optimization, *Math. Programming A* **49**, 189–211.

[28] Vanderbei, R.J. and Shanno, D.F. (1997). An interior-point algorithm for nonconvex nonlinear programming, Tech. Rep. SOR-97-21, Stats & Operations Research, Princeton University.

[29] Vardi, A. (1985). A trust region algorithm for equality constrained minimization: convergence properties and implementation, *SIAM J. Num. Anal.* **22**, 575–591.

[30] Wächter, A. and Biegler, L.T. (1999). Failure of global convergence for a class of interior point methods for nonlinear programming, Tech. Rep. CAPD B-99-07, Dept Chemical Engng, Carnegie Mellon University, Pittsburg.

[31] Yamashita, H. (1994). A globally convergent primal-dual interior-point method for constrained optimization, Tech. Rep., Mathematical System Inst., Tokyo, Japan.

[32] Yamashita, H., Yabe, H. and Tanabe, T. (1997). A globally and superlinearly convergent primal-dual interior point trust region method for large scale constrained optimization, Tech. Rep., Mathematical System Inst., Tokyo, Japan.

Six Lectures on the Geometric Integration of ODEs

Robert McLachlan

IFS, Massey University
Palmerston North, New Zealand
Email: R.McLachlan massey.ac.nz

Reinout Quispel

Department of Mathematics
La Trobe University
Bundoora, Melbourne 3083, Australia
Email: R.Quispel@latrobe.edu.au

Lecture 1: General tools

"A major task of mathematics is to harmonize the continuous and the discrete, to include them in one comprehensive mathematics, and to eliminate obscurity from both."
(E.T. Bell, Men of Mathematics)

1.1 Introduction

Motion is described by differential equations, which are derived from the laws of physics. In the simplest case, they read $m\frac{d^2x}{dt^2} = F(t, x, \frac{dx}{dt})$—Newton's second law. These equations contain within them not just a statement of the current acceleration experienced by the object(s), but *all* the physical laws relevant to the particular situation. Finding these laws and their consequences for the motion has been a major part of physics since the time of Newton. For example, the equations tell us the space in which the system evolves (its *phase space*, which may be ordinary Euclidean space or a curved space such as a sphere); any symmetries of the motion, such as the left–right or forwards–backwards symmetries of a pendulum; and any special quantities such as energy, which for a pendulum is either conserved (if there is no friction) or decreases (if there is friction). Finally and most importantly, the laws describe how all motions starting close to the actual one are constrained in relation to each other. These laws are known as *symplecticity* and *volume preservation*.

155

> *"A gyroscope is an emissary from a six-dimensional symplectic world to our three-dimensional one; in its home world its behavior looks simple and natural."*   *(Yuri Manin)*

Standard methods for simulating motion, called *numerical integrators*, take an initial condition and move the objects in the direction specified by the differential equations. They completely ignore all of the above hidden physical laws contained within the equations. Since about 1990, new methods have been developed, called *geometric integrators*, which obey these extra laws. Since this is physically natural, we can hope that the results will be extremely reliable, especially for long-time simulations.

Before we tell you all the advantages, three caveats:

- The hidden physical law usually has to be known if the integrator is going to obey it. For example, to preserve energy, the energy must be known.
- Because we're asking something more of our method, it may turn out to be computationally more expensive than a standard method. Amazingly (because the laws are so natural?) sometimes it's actually much cheaper.
- Many systems have multiple hidden laws, for which methods are currently known which preserve any one law but not all simultaneously.

Now the advantages:

- Simulations can be run for enormously long times, because there are no spurious non-physical effects, such as dissipation of energy in a conservative system;
- By studying the structure of the equations, very simple, fast, and reliable geometric integrators can often be found;
- In some situations, results can be guaranteed to be *qualitatively* correct, even when the motion is chaotic. This allows one to study systems in a "quick and dirty" mode and explore the system thoroughly, while retaining reliability;
- For some systems, even the actual quantitative errors are much smaller for short, medium, and long times than in standard methods.

Lecture 2 discusses a case where all of these nice features are realized: the solar system.

The first lecture is about general tools which will be useful later on, the second discusses the question "why bother?", and the third to sixth lectures are about how to preserve various specific properties.

These lectures were delivered at ANODE, the Auckland Numerical ODEs workshop, in July 1998. Naturally, they are tailored to our own research interests. They are intended to be suitable for a student's first exposure to the subject, and we have preserved their informality. We are very grateful to John Butcher for inviting us to speak, to all the organizers of ANODE, and especially to Nicolas Robidoux for transcribing the lectures. ANODE and the authors are supported by the Marsden Fund of the Royal Society of New Zealand, the Australian Research Council and the EPSRC. The written form was prepared at the MSRI, Berkeley, supported in part by NSF grant DMS–9701755.

1.2 The exact flow of an ODE, and general properties of integrators

We first define the exact flow (or solution) of an ordinary differential equation (ODE) and discuss what properties one would like an integrator to have. Let $x(t)$ be the exact solution of the system of ordinary equations (ODEs)†

$$\frac{dx}{dt} = f(x), \quad x(0) = x_0, \quad x \in \mathbb{R}^m. \tag{1.1}$$

The exact flow φ_t is defined by

$$x(t + \tau) = \varphi_\tau(x(t)) \quad \forall\, t,\, \tau$$

For each fixed time step τ, φ is a map from phase space to itself, i.e. $\varphi_\tau : \mathbb{R}^m \to \mathbb{R}^m$.

Three properties of exact flows

(i) **(Self-adjointness)** The flow has the continuous group property

$$\varphi_{\tau_1} \circ \varphi_{\tau_2} = \varphi_{\tau_1 + \tau_2} \quad \forall \tau_1, \tau_2 \in \mathbb{R}. \tag{1.2}$$

In particular,

$$\varphi_\tau \circ \varphi_{-\tau} = Id \tag{1.3}$$

Hence the exact flow is *self-adjoint*:

$$\varphi_\tau = \varphi_{-\tau}^{-1}. \tag{1.4}$$

---

† Nonautonomous ODEs $dx/dt = f(x,t)$ can be formulated autonomously as $dx/dt = f(x, x_{m+1})$, $dx_{m+1}/dt = 1$. The geometric integrator is applied to this "extended" system (if possible), and then $t = x_{m+1}$ substituted.

(ii) **(Taylor expansion)**

$$x(\tau) = x(0) + \tau \frac{dx}{dt}(0) + \tau^2 \frac{1}{2}\frac{d^2x}{dt^2}(0) + \dots$$

Substitute

$$\frac{dx}{dt} = f(x)$$
$$\frac{d^2x}{dt^2} = (df)\frac{dx}{dt} = (df)f$$

Hence

$$\varphi_\tau(x_0) = x_0 + \tau f(x_0) + \frac{1}{2}\tau^2(df(x_0))f(x_0) + \dots \qquad (1.5)$$

(iii) **(Formal exact solution)**

$$\varphi_\tau(x) = e^{\tau \sum_{i=1}^n f_i(x)\frac{\partial}{\partial x_i}}(x)$$
$$:= \exp(\tau f)(x) \qquad (1.6)$$

It's impossible to construct integrators with the continuous group property (1.2) for any reasonably general class of ODEs. The closest one can come is to preserve self-adjointness.

Properties of integrators

In general we don't know the flow φ_τ, so we seek maps ψ_τ that approximate φ_τ. We call such ψ_τ integrators. Some properties of integrators:

(i) **(Self-adjointness)** It is useful for ψ_τ to be self-adjoint, i.e.,

$$\psi_\tau = \psi_{-\tau}^{-1}$$

(ii) **(Order of an integrator)** The order of accuracy of ψ_τ is p, if the Taylor series of ψ_τ and the exact flow φ_τ agree to order p:

$$\psi_\tau(x) - \varphi_\tau(x) = \mathcal{O}(\tau^{p+1})$$

(iii) **(Consistency)** A necessary property of ψ_τ is that it be consistent, i.e., first order accurate, i.e.,

$$\psi_\tau(x) = x + \tau f(x) + \mathcal{O}(\tau^2).$$

Note: It is not difficult to show that every self-adjoint integrator is of even order.

There are three types of integrators:

(i) Integrators that form a group
(ii) Integrators that form a symmetric space
(iii) Integrators that form a semigroup

1.3 Integrators that form a group

Suppose we have a set G of integrators which may or may not be consistent. If, for all integrators ψ_τ and χ_τ in G, we have

$$\psi_\tau \circ \chi_\tau \in G$$

and

$$\psi_\tau^{-1} \in G,$$

we say the integrators form a group. That is, they are a group where the group operation is composition of maps.

Examples of integrators that can form a group are

  (i) symplectic integrators (Lecture 3)
 (ii) symmetry-preserving integrators (Lecture 4)
(iii) volume-preserving integrators (Lecture 5)
(iv) integral-preserving integrators (Lecture 6)

For example, for the group of integral-preserving integrators there is a real function $I(x)$ (the *integral*) such that $I(x) = I(\varphi_\tau(x))$ for all x: the value of the integral I is preserved by the integrator. Therefore it is also preserved by $\varphi_\tau \circ \chi_\tau$ and by φ_τ^{-1}: the integrators form a group.†

These groups are infinite-dimensional groups of diffeomorphisms. They share many, but not all of the properties of Lie groups; various extensions of the concept of Lie groups from finite to infinite dimensions have been proposed. One approach is the theory of "Lie pseudogroups" of diffeomorphisms. Cartan discovered in 1913 that in a sense there are just 6 fundamental Lie pseudogroups: the group of all diffeomorphisms; those preserving a symplectic, volume, or contact structure; and those preserving a symplectic or volume structure up to a constant. These correspond to different generic types of dynamics.

How to construct integrators that form a group

The main way to construct integrators that form a group is through *splitting methods*. Splitting methods work for all cases (1)–(4) above, and are discussed further in Lecture 3.

We illustrate splitting for integral-preserving integrators. Assume we don't know an integral-preserving integrator for the vector field f, but f

---

† If φ_τ^{-1} exists, which it does for the methods of Lecture 6, but not necessarily for projection methods.

can be split into two vector fields f_1 and f_2, each with the same integral as f:

$$f(x) = f_1(x) + f_2(x)$$

and assume that we *do* know integral preserving integrators ψ_1 (resp. ψ_2) for f_1 (resp. f_2) separately.

Then we obtain an integral-preserving integrator ψ for f by composition:

$$\psi_\tau = \psi_{2,\tau} \circ \psi_{1,\tau}$$

This is a consistent method for f, because it is the map $\psi_\tau : x \mapsto x''$ given by

$$x' = x + \tau f_1(x) + \mathcal{O}(\tau^2)$$
$$x'' = x' + \tau f_2(x') + \mathcal{O}(\tau^2)$$
$$= x + \tau(f_1(x) + f_2(x)) + \mathcal{O}(\tau^2)$$

Splitting methods are very easy to program—one merely calls routines for ψ_1 and ψ_2 in turn.

Thus the problem becomes:

(i) How to split vector fields while staying in the appropriate class;
(ii) How to construct integrators in the appropriate group;
(iii) How to compose those integrators so as to get an integrator of the original vector field of the desired order.

Each of these will be considered in these lectures.

1.4 Integrators that form a symmetric space

Suppose we have a set G of integrators with the property that, for all integrators ψ_τ and χ_τ in G, we have

$$\psi_\tau \circ \chi_\tau^{-1} \circ \psi_\tau \in G. \tag{1.7}$$

Then G is an example of the algebraic object known as a *symmetric space*, a set G together with a binary operation $*$ obeying the axioms

$$x * x = x$$
$$x * (x * y) = y$$
$$x * (y * z) = (x * y) * (x * z)$$
$$x * y = y \Rightarrow y = x \text{ for all } y \text{ sufficiently close to } x$$

In our case the integrators G form a symmetric space by taking

$$\psi_\tau * \chi_\tau := \psi_\tau \circ \chi_\tau^{-1} \circ \psi_\tau.$$

Notice that every group also forms a symmetric space, but not vice versa: a group may have subsets which are closed under (1.7) but not under simple composition.

The two most important examples of integrators that form a symmetric space are

(i) Self-adjoint integrators.
(ii) Integrators that possess time-reversal symmetry (Lecture 4).

Proof of (1): Let $\theta_\tau = \psi_\tau \circ \chi_\tau^{-1} \circ \psi_\tau$. Then

$$\begin{aligned}
\theta_{-\tau}^{-1} &= \left(\psi_{-\tau} \circ \chi_{-\tau}^{-1} \circ \psi_{-\tau} \right)^{-1} \\
&= \psi_{-\tau}^{-1} \circ \chi_{-\tau} \circ \psi_{-\tau}^{-1} \\
&= \psi_\tau \circ \chi_\tau^{-1} \circ \psi_\tau \\
&= \theta_\tau.
\end{aligned}$$

How to construct integrators that form a symmetric space

There are two main ways:

(i) **(Projection methods)** If ψ_τ is any integrator, then the "projection"

$$\chi_\tau := \psi_{\tau/2} \circ \psi_{-\tau/2}^{-1}$$

is self-adjoint.

(ii) **(Splitting methods)** If f can be split into two vector fields

$$f(x) = f_1(x) + f_2(x)$$

such that we have self-adjoint integrators ψ_1 and ψ_2 for f_1 and f_2 separately, then we obtain a self-adjoint integrator ψ for f from the symmetric composition

$$\psi_\tau := \psi_{1,\tau/2} \circ \psi_{2,\tau} \circ \psi_{1,\tau/2}.$$

These are generalized to other symmetric spaces in Lecture 4. The projection is almost miraculous, because it starts with *any* integrator. There is no analogous projection for groups.

1.5 Integrators that form a semigroup

A set G of integrators forms a semigroup if for all integrators ψ_τ and χ_τ in G, we have $\psi_\tau \circ \chi_\tau \in G$, but not necessarily $\psi_\tau^{-1} \in G$.

These arise from properties that only hold for forwards time:

(i) systems with a Lyapunov function (Lecture 6)

(ii) systems which contract phase space volume

For example, if the Lyapunov function is decreasing as t increases, it is *increasing* as t decreases, and even the flow $\varphi_\tau^{-1} = \varphi_{-\tau}$ does not have the Lyapunov property. This means one cannot use backwards time steps when composing these integrators, which can be proved to limit the order of composition methods to 2.

1.6 Creating higher order integrators: composition methods

Having obtained a geometric integrator ψ_τ, a higher order method can be obtained from the composition

$$\chi_\tau = \psi_{\alpha_n \tau} \psi_{-\alpha_{n-1}\tau}^{-1} \psi_{\alpha_{n-2}\tau} \psi_{-\alpha_{n-3}\tau}^{-1} \psi_{\alpha_{n-3}\tau} \psi_{-\alpha_{n-2}\tau}^{-1} \psi_{\alpha_{n-1}\tau} \psi_{-\alpha_n\tau}^{-1}$$

which has been chosen to be self-adjoint, i.e. $\chi_\tau = \chi_{-\tau}^{-1}$. Here the number of integrators n and the coefficients α_n can be adjusted to obtain the desired order. High order methods can also be designed in the context of splitting methods, using the two flows $\varphi_{1,\tau}$ and $\varphi_{2,\tau}$ of f_1 and f_2 respectively. One uses the composition

$$\varphi_{1,\beta_1\tau}\varphi_{2,\gamma_1\tau}\cdots\varphi_{1,\beta_n\tau}.$$

However, these two approaches turn out to be equivalent [7].

Example 1 If ψ is self-adjoint, then a fourth-order integrator is obtained as follows:

$$\chi_\tau = \psi_{\gamma\tau} \circ \psi_{(1-2\gamma)\tau} \circ \psi_{\gamma\tau}$$

where $\gamma := (2 - 2^{1/3})^{-1}$.

This example is generalized in Theorem 1 below. Note that $1 - 2\gamma < 0$.

An example of composition methods: the generalized Yoshida method

Theorem 1 (Yoshida, Qin, and Zhu) *Let ψ be a self-adjoint integrator of order $2n$. Then*

$$\chi_\tau = \psi_{\gamma\tau} \circ \psi_{(1-2\gamma)\tau} \circ \psi_{\gamma\tau}, \quad \gamma = (2 - 2^{\frac{1}{2n+1}})^{-1}$$

is a self-adjoint integrator of order $2n + 2$.

Proof Let

$$\psi_\tau(f) = \varphi_\tau(f) + \delta\tau^{2n+1} + \dots.$$

Then, using the flow property of the exact flow φ_τ,

$$\psi_{\gamma\tau} \circ \psi_{(1-2\gamma)\tau} \circ \psi_{\gamma\tau} = \varphi_\tau(f) + \left(\gamma^{2n+1} + (1-2\gamma)^{2n+1} + \gamma^{2n+1}\right)\delta\tau^{2n+1} + \dots$$

which has order $2n + 1$ if γ is as given in the theorem. However, it is self-adjoint by construction, so it has even order, hence the order is $2n + 2$. $\square$

References

Books

[1] Stuart, A.M. and Humphries, A.R. (1996). *Dynamical Systems and Numerical Analysis* (Cambridge University Press, Cambridge).

[2] Sanz-Serna, J.M. and Calvo, M.-P. (1994). *Numerical Hamiltonian Problems* (Chapman & Hall, London).

Survey articles

[3] Sanz-Serna, J.M. (1997). Geometric integration, in *The State of the Art in Numerical Analysis* (I.S. Duff and G.A. Watson, eds) (Clarendon Press, Oxford), 121–143.

[4] Quispel, G.R.W. and Dyt, C. (1997). Solving ODE's numerically while preserving symmetries, Hamiltonian structure, phase space volume, or first integrals, in *Proc. 15th IMACS World Congress, vol. 2* (A. Sydow, ed.) (Wissenschaft & Technik, Berlin), 601–607.

[5] Budd, C.J. and Iserles, A., eds (1999). Geometric integration: Numerical solution of differential equations on manifolds, *Phil. Trans. Roy. Soc. A* **357**, 943–1133.

Composition methods

The best place to start is

[6] Yoshida, H. (1990). Construction of higher order symplectic integrators, *Phys. Lett.* **150A**, 262–268.

which was generalized in

[7] Qin, M. and Zhu, W.J. (1992). Construction of higher order symplectic schemes by composition, *Computing* **27**, 309–321.

and further in

[8.] McLachlan, R.I. (1995). On the numerical integration of ODE's by symmetric composition methods, *SIAM J. Numer. Anal.* **16**, 151–168.
[9.] Murua, A. and Sanz-Serna, J.M. (1999). Order conditions for numerical integrators obtained by composing simpler integrators, *Phil. Trans. Roy. Soc. A* **357**, 1079–1100.

Lecture 2: Why preserve structure?

2.1 Introduction

Let's start with an example of a simulation of the outer solar system by Jack Wisdom and coworkers. Part of its appeal is the long history of modelling the solar system. The people who do this are not from the numerical analysis community, but they have their own history of methods which they have developed and tweaked.

In the 1980's, a special-purpose supercomputer, the "Digital Orrery", simulated the outer planets for 845 million years. With a lot of tweaking, an energy error of about 10^{-9} was achieved with a time step of 45 days (a six month calculation!). A calculation with a very high order symmetric multistep method achieved an energy error of about 10^{-10} in a 3 million year simulation, with a time step of 0.75 days. In a completely different approach, Laskar (1990) used classical perturbation theory (expanding in mass ratios and eccentricities about circular orbits) to eliminate the fast (annual) frequencies. This required 250,000 terms, but a time step of 500 years could be taken.

All of these attempts were roundly routed by the calculation of Jack Wisdom et al., using a very simple, elegant symplectic integrator. Their billion year simulation with a time step of 7.5 days gave an energy error of only 2×10^{-11}. Moreover, only one force evaluation was used per time step, making the method very fast.

Roughly speaking, they wrote the ODE as a sum of uncoupled Kepler 2-body problems and the potential which couples the planets: $f = f_1 + f_2 = f_{\text{Kepler}} + f_{\text{coupling}}$. Each f_i is a Hamiltonian system, and the flow $\varphi_{\tau,i}$ of each can be found exactly and quickly (the 2-body problems using an elegant method of Gauss). The time stepping is simply the simplest composition $\psi_\tau = \varphi_{\tau,2} \circ \varphi_{\tau,1}$—a form of the "leapfrog" method. Since

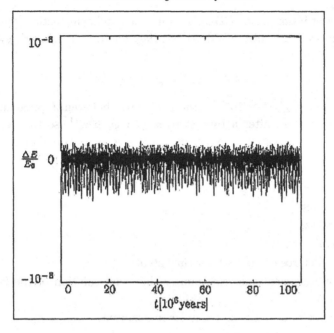

Fig. 2.1. Energy error of leapfrog applied to the whole solar system over 10^8 years (Wisdom et al.)

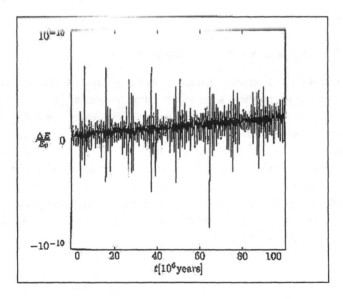

Fig. 2.2. Energy error after application of corrector χ_τ.

the flow of Hamiltonian ODEs is symplectic, and symplectic maps form a group, ψ_τ is symplectic. Moreover, they found a "corrector" χ_τ such that

$$\chi_\tau \circ \psi_\tau \circ \chi_\tau^{-1} = \varphi_\tau + \mathcal{O}(m^2 \tau^3)$$

where $m \approx |f_2/f_1| \approx 10^{-3}$ is the mass ratio between Jupiter and the sun. (The result after n time steps is $\chi_\tau \circ \psi_\tau^n \circ \chi_\tau^{-1}$, so that χ_τ only needs to be evaluated once, no matter how long the simulation.) This method:

- is symplectic;
- is one-step;
- is explicit;
- is second order;
- uses one force evaluation per time step;
- exploits classical analysis, namely the exact solution of the 2-body problem;
- preserves total linear and angular momentum;
- is self-adjoint and reversible;
- has an extra factor of $m^2 = 10^{-6}$ in its local truncation error, compared to classical methods;
- for moderate times ($\approx 2 \times 10^7$ years), has linear growth of global errors, compared to quadratic growth for classical methods;
- has bounded energy errors for long times.

This is almost a dream situation, where geometric integration has lead to a simple method with vastly improved local (time τ), global (time T), and structural (time ∞) errors. This calculation discovered chaos in the outer solar system with a Lyapunov time, the time for the separation between nearby orbits to grow by a factor e, of 20 million years. Over the billion year calculation, they would separate by $e^{50} \approx 10^{22}$, and integration errors would be magnified by this amount also. Thus, the final angular positions of the planets are not expected to be accurate. However, we can be confident that the qualitative or statistical properties of the solution are correct.

2.2 Phase space and phase flow

"In phase space, no one can hear you scream."

(Caltech T-shirt)

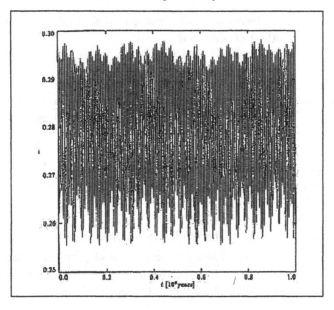

Fig. 2.3. Inclination of Pluto over 10^9 years, showing chaos. Even after 10^9 years the inclination has reached a new maximum.

The fundamental idea to keep in mind is to think in phase space. It's a simple idea but one which you have to keep reminding yourself of: a simple definition in a dynamical systems class just isn't enough. Considering that differential equations were studied for 200 years before Poincaré adopted this point of view, this may not be too surprising.

> *"Consider the fluid molecules which initially form a certain figure F_0; when these molecules are displaced, their ensemble forms a new figure which will be deformed in a continuous manner, and at the instant t the envisaged ensemble of molecules will form a new figure F."*
>
> *(Poincaré,* Celestial Mechanics, *1899)*

In a *trajectory* $\varphi_t(x_0)$, one thinks of the initial condition x_0 as fixed, and the time t increasing; in the *flow map* $\varphi_\tau(x)$, one thinks of all initial conditions x flowing forward for some fixed time τ. We'll only consider one-step methods, so that the numerical approximation for one time-step τ is a map

$$\psi_\tau : \mathbb{R}^m \to \mathbb{R}^m .$$

Now classical approximation theory, e.g. for Runge–Kutta methods, shows that chaos always wins: the best bound that can be obtained in general for a method of order p is

$$\left| \psi_\tau^{T/\tau}(x) - \varphi_T(x) \right| \le (\Delta t)^p C \frac{e^{\Lambda T} - 1}{\Lambda}$$

The precise value of Λ depends on the Lipschitz constant of the vector field and on the method, but $\Lambda > 0$ and consequent exponential growth of error cannot be avoided in general. But dynamical systems theory teaches that ψ can be "close" to φ in other ways: their phase portraits may be qualitatively or even quantitatively similar; the stability of their orbits may be the same; for strange attractors, their Lyapunov exponents or fractal dimensions may be close.

The pendulum: theory

Systems can have many geometric or structural properties. Before we get into definitions, let's look at the planar pendulum. It is a two-dimensional system with phase space $\mathbb{R}^2$, and dynamics

$$\dot{q} = p, \quad \dot{p} = -\sin q \qquad (2.1)$$

where q is the angle of the pendulum, and p its angular momentum. (Here we are taking $q \in \mathbb{R}$, the covering space of the actual angle.) Here are some of the properties of the pendulum:

- It conserves the total energy $\dot{H} = \frac{1}{2}p^2 - \cos q$. That is, its flow stays on the level sets of this function. Because this is a two-dimensional system, these level sets are curves in the plane.
- Being a Hamiltonian system, its flow is symplectic. For two-dimensional systems, this is equivalent to being area-preserving.
- It has one discrete symmetry and one discrete reversing symmetry (see Lecture 4). The symmetry, $(q,p) \mapsto (-q,-p)$, maps the vector field into itself; the reversing symmetry, $(q,p) \mapsto (q,-p)$, maps the vector field into minus itself. Imagining flowing along one of the solution curves, you can see that the motion of the reflected points is constrained.

Because this is such a simple system, preserving any of these three properties gives a geometric integrator with good long-time behavior for almost all initial conditions. A picture of its phase portrait will look very similar to the true phase portrait; we'll see examples of this in Section 2.7. By contrast, standard methods (e.g. Euler's method) destroy the qualitative phase portrait completely.

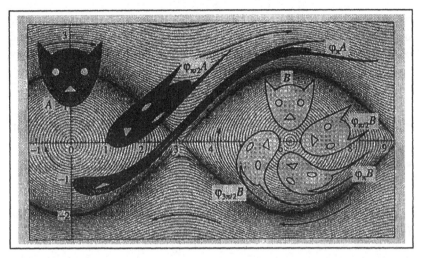

Fig. 2.4. Phase portrait and flow of the pendulum (from Hairer and Wanner). The area of each cat is preserved in time, the manifestation of symplecticity. Energy, whose levels sets are the curves shown, is preserved. Rotation by 180° $((q, p) \mapsto (-q, -p))$ is a symmetry, while flipping up-down $(p \mapsto -p)$ is a reversing symmetry.

2.3 Philosophy of geometric integration

In any numerical study, one should

- examine any geometric or structural properties of the ODE or its flow;
- design numerical methods which also have these structural properties; and
- examine the consequences, hopefully over and above the immediate ones.

This encourages us to

- confront questions of phase space and degrees of freedom;
- think about the significance of local, global, and qualitative errors; and
- think about the kinds of tools and functions allowed in numerical analysis.

For example, multistep methods do not define a map on phase space, because more than one initial condition is required. They can have geometric properties, but in a different (product) phase space, which can alter the effects of the properties. (See Fig. 2.12.) This puts geometric integration firmly into the "single step" camp. If a system is defined on a sphere, one should stay on that sphere: anything else introduces spurious, non-physical degrees of freedom.

The *direct* consequences of geometric integration are that we are

- studying a dynamical system which is close to the true one, and in the right class; and
- this class may have restricted orbit types, stability, and long-time behavior.

In addition, because the structural properties are so natural, some *indirect* consequences have been observed. For example,

- symplectic integrators have good energy behavior;
- symplectic integrators can conserve angular momentum and other conserved quantities;
- geometric integrators can have smaller local truncation errors for special problems, and smaller global truncation errors for special problems/initial conditions (even though they're larger in the "generic" case);
- some problems (particle scattering, isospectral problems) can have errors tending to zero at long times.

Here's a pictorial survey showing what you can expect from geometric integration. Fig. 2.5 appears in Channell and Scovel [3], one of the first symplectic integration papers. Orbits starting on the smooth curves ("invariant circles") stay on them forever. Of course, the orbit may be going around the circle at the wrong speed, but the "orbital error" does not grow in time. Compare this to the traditional approach to numerical integration, with its overwhelming emphasis on the estimation and control of local errors. The idea that errors grow in time and, once committed, cannot be undone, was deeply ingrained. Pictures like Fig. 2.5 did a lot to revise this traditional point of view.

Other orbits in Fig. 2.5 are chaotic, and their position errors grow exponentially. But, they can never jump across the invariant circles, and because it's the *right kind* of chaos (namely, the solution of some nearby Hamiltonian system), statistical observations of this chaos will have small errors.

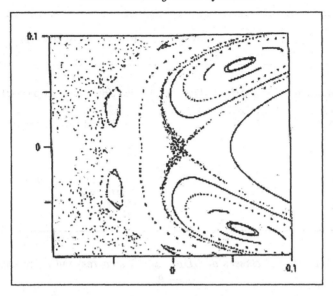

Fig. 2.5. Phase portrait of a symplectic integration, from Channell and Scovel. 10^5 time steps for 10 different initial conditions are shown. Smooth curves ("KAM tori") correspond to regular, quasiperiodic motion; clouds correspond to chaotic motion.

2.4 Types of geometric properties

Study the list in the Table. The left hand column gives properties of vector fields, and the right hand column gives the corresponding properties of their flow. It's the right hand property that must be preserved by the integrator. Usually the flow properties are named the same as the ODE property.

(The standard example of a *Lie group* G is the set of orthogonal 3×3 matrices, $A^T A = I$, which represent rotations. Its Lie algebra $\mathfrak{g}$ is the set of antisymmetric 3×3 matrices. G is a manifold whereas $\mathfrak{g}$ is a linear space, a much simpler object to work with.)

To bring some order to this table, consider the following features.

- Is the structure linear in some sense?
  All of the ODE properties are linear in f, but all of the flow properties are nonlinear in φ, except for linear symmetries. Symplecticity, Poisson, and reversibility are quadratic; volume preservation and isospectrality are degree m when $x \in \mathbb{R}^m$.
- Does the structure appear explicitly or implicitly in the ODE?

ODE $\dot{x} = f(x)$		flow φ_t, derivative $d\varphi_t$

Hamiltonian	$f = J\nabla H(x)$, $J = \begin{pmatrix} 0 & I \\ -I & 0 \end{pmatrix}$	$d\varphi^T J d\varphi = J$ (symplectic)
Poisson	$f = J(x)\nabla H(x)$	$d\varphi^T J d\varphi = J \circ \varphi$
source-free	$\nabla \cdot f = 0$	$\det d\varphi = 1$ (volume preserving)
symmetric	$dS.f = f \circ S$	$S \circ \varphi = \varphi \circ S$
reversible	$-dR.f = f \circ R$	$R \circ \varphi^{-1} = \varphi \circ R$
Lie group	$f = a(x)x$, $x \in G$, $a \in \mathfrak{g}$	$\varphi \in G$
isospectral	$f = [b(x), x]$, $x, b \in \mathfrak{g}$	eigenvalues $\lambda(x)$ constant
integral	$f \cdot \nabla I = 0$	$I(x(t)) = I(x(0))$
dissipative	$f \cdot \nabla V \leq 0$	$V(x(t)) \leq V(x(0))$

Table 1.1. *Special classes of ODEs, and the corresponding properties of their flows.*

Hamiltonian, Poisson, Lie group, and isospectral ODEs are explicit (e.g. $f = J\nabla H$ generates *all* Hamiltonian ODEs); the rest are implicit —there are side conditions which f has to satisfy.

• Does the flow property depend on φ or $d\varphi$?
Symplecticity, Poisson, and volume preservation depend on the Jacobian $d\varphi$. This makes them harder to preserve.

These will be explored further in the other lectures. Briefly, it is easier to work on *linear* and *explicit* properties, so we concentrate on bringing all flow properties into this form. (See §3.1 on splitting.) This has been achieved for all the properties in the Table, but not for some of their nonlinear generalisations and combinations.

A major justification for geometric integration comes from *backward error analysis*. This theoretical tool writes the integrator ψ_τ as the time-τ flow of *some* vector field $\tilde{f}$, i.e. $\psi_\tau(f) = \varphi_\tau(\tilde{f})$. If the method is of order p, we have $\tilde{f} = f + \mathcal{O}(\tau^p)$. Then, in many cases one can argue that since ψ_τ is in some class (e.g. symplectic), the perturbed vector field must be in the appropriate class too (e.g. Hamiltonian). So we know that by studying the dynamics of the method, we are at least studying dynamics in the right class. The reliability of the results then depends on the "structural stability" of the original system: a difficult problem, but a standard one in dynamical systems.

In the Hamiltonian case, $\tilde{f} = J\nabla\tilde{H}$ for some Hamiltonian $\tilde{H}$, which is conserved by the method. Since we don't know $\tilde{H}$ and can only measure the original energy H, it (H) will be seen to oscillate, but (if the levels sets of H and $\tilde{H}$ are bounded) will not drift away from its original level.

Technically, one suspends the map ψ_τ to a time-dependent flow $\varphi_\tau(\tilde{g}(x,t))$, from which, when ψ_τ is analytic, nearly all the time dependence can be removed by a change of variables, giving $\tilde{f}(x) + \mathcal{O}(e^{-1/\tau}, t)$. This introduction of an exponentially small nonautonomous term is inevitable, because most maps, even those close to the identity, are not actually flows. If the time step is too large these exponentially small terms can actually pollute the calculation, and one observes, for example, the energy drifting.

2.5 Miscellaneous topics

Some other branches of geometric integration are

- *ODEs on manifolds*, such as homogeneous spaces. Although ultimately one can only compute in a linear space, it's best to formulate the method on the manifold and transfer to coordinates as late as possible. A special case is when the manifold is a Lie group [4]; Lie group methods are one of the major themes in geometric integration which we don't have space to discuss here.

- *Mapping methods* approximate the equations in x as well as in t, for example, by Taylor series. Maps defined by series can then be manipulated analytically.

- When evaluating *Lyapunov exponents* one should try to preserve their structure, e.g., that the Jacobians used are symplectic or volume-preserving.

- For *partial differential equations* one can either discretize in space first, seeking a finite-dimensional version of, e.g., the Hamiltonian structure, or discretize space-time directly.

- One can discretize phase space itself and study *lattice maps*, a form of cellular automata. This has been used in studies of the effect of roundoff error.

- Instead of trying to construct special methods that preserve particular properties, one can see how well standard methods do. Usually the property has to be fairly robust, e.g., dissipation of the type $d|x|^2/dt < 0$ for $|x| > R$ is studied, instead of $dV/dt \leq 0$ for all x. This approach

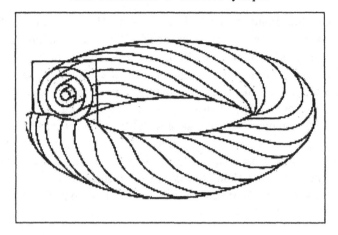

Fig. 2.6. Flow on a family of invariant tori. From V.I. Arnol'd, Small denominators and problems of stability of motion in classical and celestial mechanics, *Uspehi Mat. Nauk (Russ. Math. Surv.)* **18** (1963) no. 6 (114) 91–192.

is thoroughly treated in Stuart and Humphries, *Dynamical Systems and Numerical Analysis*.

2.6 Growth of global errors

The global error is $\psi_\tau^{T/\tau}(x) - \varphi_\tau(x)$ where T is a large, but fixed, time. Geometric integrators are not expressly designed to control the global error. Nevertheless, sometimes it grows linearly in a symplectic integrator and quadratically in a standard integrator. This will make the symplectic integrator superior if T is large enough.

This property has been observed in many systems of different types. It is associated with preservation of *invariant tori* by the method. An invariant torus is a subset of initial conditions, topologically a torus, which orbits starting on stay on for all forwards and backwards time. A torus is *preserved* if the integrator has an invariant torus of its own, which tends to the torus of the ODE as $\tau \to 0$.

Invariant tori

Invariant tori are ubiquitous in dynamics. They're found in:

- Hamiltonian systems (tori have dimension $n/2$);
- reversible systems (when orbits intersect the symmetry plane; tori often have dimension $n/2$);

- volume-preserving systems (tori have any dimension $< n$).

They are important because they

- form positive-measure families of neutrally stable orbits, which
- mostly persist under small perturbations of the system;
- form "sticky sets," dominating behavior of nearby orbits on intermediate time scales

Nearby orbits diverge like

- $\mathcal{O}(1)$ on same torus
- $\mathcal{O}(T)$ on a nearby or perturbed torus
- $\mathcal{O}(T^2)$ if $\mathcal{O}(T)$ drift across tori
- $\mathcal{O}(T, e^{\lambda T})$ on nearby chaotic orbits; λ depends on the order of resonance, but can be very small.

Therefore, in an integrator we should try to preserve tori *of the correct dimension*. In a standard method, they are not preserved, and orbits drift transversely, leading to $\mathcal{O}(T^2)$ growth of global errors. If the torus is preserved, orbits only move around the torus at a slightly wrong angle or speed, leading to $\mathcal{O}(T)$ errors.

It turns out to be an extraordinarily subtle question to determine when which tori persist under which perturbations. Finally, in the 1960's, conditions were found by Kolmogorov, Arnol'd, and Moser under which most tori do persist under appropriate perturbations, although some are destroyed. This forms the subject of KAM theory.

For Hamiltonian systems, an appropriate perturbation is Hamiltonian, so the results apply to symplectic integrators.

In between invariant tori, or if tori were destroyed by taking too large a time step, orbits can be chaotic. But, because of the nearby tori, exponential separation can be very slow, and the linear error growth can dominate for long times.

2.7 The pendulum: numerical experiments

"Theories come and go, but examples stay forever." (Gelfand)

We illustrate the above points on the simplest meaningful example, the pendulum (Eq. (2.1)). The simplest symmetric, reversible, self-adjoint

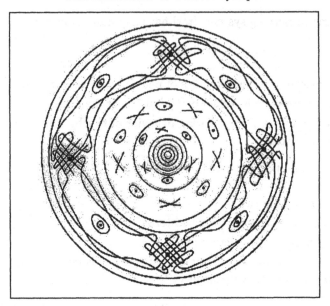

Fig. 2.7. Cross-section of the tori in Fig 2.6 after perturbation (Arnol'd). Some are destroyed and replaced by chaos, some persist.

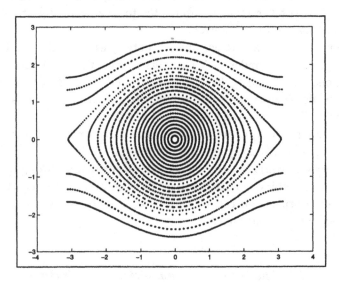

Fig. 2.8. 1000 times steps of symplectic leapfrog applied to the pendulum, time step $\tau = 0.1$.

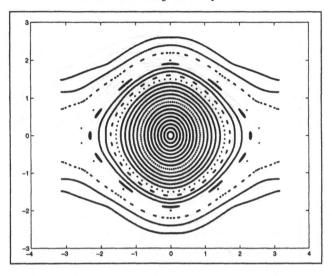

Fig. 2.9. As in Fig. 2.8, but $\tau = 1$.

symplectic method is leapfrog:

$$q' = q + \frac{1}{2}\tau p$$

$$p' = p - \tau \sin q'$$

$$q'' = q' + \frac{1}{2}\tau p'$$

The results of this method are shown in Fig. 2.8 for a small time step ($\tau = 0.1$) and in Fig. 2.9 for a much larger time step ($\tau = 1$). Even for the larger time step, the left-right and up-down symmetries are preserved, as are most of the invariant circles, as promised by KAM theory for symplectic integrators. Chaos is significant only in a small neighbourhood of the homoclinic orbit connecting $(\pi, 0)$ and $(-\pi, 0)$.

A symplectic method which is *not* symmetric or self-adjoint is shown in Fig. 2.10; the lack of symmetry is plain to see. In this case, invariant circles are still preserved. In higher-dimensional systems, there is a more complicated interaction between symplecticity and reversibility.

What is the effect of the chaos created by the numerical integrator? Fig. 2.11 shows one chaotic orbit of leapfrog at the large time step $\tau = 1$, obtained with initial condition $(q, p) = (0, 1.8)$. It was found to have a large Lyapunov exponent of 10^{-2}. By $T \sim 100$, the chaos would dominate the numerical errors. by contrast, with initial condition

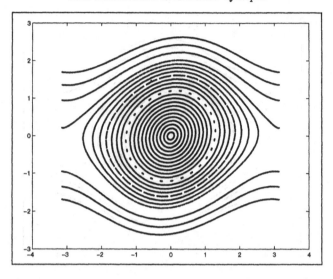

Fig. 2.10. A nonsymmetric symplectic integration of the pendulum, $\tau = 0.1$.

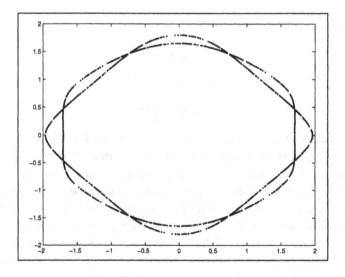

Fig. 2.11. 10^6 time steps of leapfrog at $\tau = 1$, showing a chaotic orbit.

$(q, p) = (0, 1.6)$, the Lyapunov exponent is already reduced to 10^{-7}, and phase errors (moving around the circle at the wrong speed) would dominate until $T \sim 10^7$. Thus, even when the numerical orbit does not

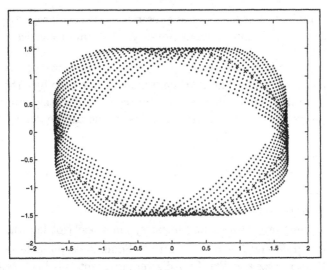

Fig. 2.12. A symplectic multistep method: the torus has dimension 2 instead of 1 as in Figs. 2.8–2.10. A single orbit is shown, with the first 50 time steps marked by ×.

lie on an invariant torus, the preservation of *some* invariant tori nearby helps a great deal.

In Section 2.3, we talked about the importance of staying in the right phase space. The multistep method $x_{n+1} = x_{n-1} + 2\tau f(x_n)$ is a map on the *product* phase space $\mathbb{R}^2 \times \mathbb{R}^2$. It can be shown to be symplectic in this larger space, but its KAM tori have dimension 2, instead of 1 as in the real system. When projected to the original phase space, they fill out a solid region, instead of a curve—a disaster for long-time simulations. This effect is illustrated in Fig. 2.12.

2.8 Summary

Systems may have many geometric or structural features. Integrators must balance costs, local, global, and long-time errors, stability, and structural preservation. You can't expect to do well at all of these simultaneously! Also, numerical studies can have different goals. Demanding very small local errors for a large class of ODEs tilts the balance in favour of highly-developed standard methods; seeking reliability over long times with simple, fast methods tilts in favour of geometric integrators.

180   R.I. McLachlan & G.R.W. Quispel

The remaining lectures look at preserving different properties. Here we sum up what is known about preserving several properties at once.

(i) Symplecticity and energy: If, by "integrator", we mean that the method is defined for *all* Hamiltonian ODEs, then by a theorem of Ge, this is impossible. For exceptional ("completely integrable") problems, such as the free rigid body, this can be done.

(ii) Symplecticity and integrals apart from energy: Not known, although doable in principle.

(iii) Symplecticity and linear symmetries: Achieved by, e.g., the implicit midpoint rule.

(iv) Poisson and linear symmetries: Not known.

(v) Volume preservation and linear symmetries: Not known.

(vi) Integrals and linear symmetries: Sometimes possible using the Harten, Lax and Van Leer discrete gradient (see Lecture 6).

(vii) Volume and an integral: Can be done by splitting for all systems with some integrals and for some systems with any integrals. Not known in general.

Extending the concept of geometric integration to PDEs is much less developed. Work has been done, e.g., on integral preservation [7], symmetry preservation [8], and Lagrangian (variational) structure [9].

References

Background on numerical ODEs

[1] Hairer, E. Nørsett, S.P. and Wanner, G. (1993). *Solving Ordinary Differential Equations I: Nonstiff Problems*, 2nd ed. (Springer, Berlin).

Historical illustrations

[2] Wisdom, J. and Holman, M. (1991). Symplectic maps for the N-body problem, *Astron. J.* **102**, 1528–1538; Wisdom, J., Holman, M. and Touma, J. (1996). Symplectic correctors, in *Integration Algorithms for Classical Mechanics*, Fields Institute Communications **10**, 217–244.
[3] Channell, P.J. and Scovel, J.C. (1990). Symplectic integration of Hamiltonian systems, *Nonlinearity* **3**, 231–259.

Equations on manifolds and Lie groups

[4] Iserles, A., Munthe-Kaas, H.Z., Nørsett, S.P. and Zanna, A. (2000). Lie-group methods, *Acta Numerica* **9**, 215–365.

Backward error analysis, invariant tori, and error growth

[5] Hairer, H. and Lubich, Ch. (1997). The life-span of backward error analysis for numerical integrators, *Numer. Math.* **76**, 441–462.

[6] Cano, B.and Sanz-Serna, J.M. (1997). Error growth in the numerical integration of periodic orbits, with application to Hamiltonian and reversible systems, *SIAM J. Numer. Anal.* **34**, 1391–1417.

PDEs

[7] McLachlan, R.I. and Robidoux, N. (2000). Antisymmetry, pseudospectral methods, and conservative PDEs, in *Proc. Int. Conf. EQUADIFF '99*, to appear.

[8] Dorodnitsyn, V. (1993). Finite difference methods entirely inheriting the symmetry of the original equations, in *Modern Group Analysis: Advanced Analytical and Computational Methods in Mathematical Physics* (N. Ibragimov, ed.) (Kluwer, Dordrecht), 191–201; Budd, C.J.,Collins, G.J., Huang, W.Z. and Russell, R.D. (1999). Self-similar numerical solutions of the porous medium equation using moving mesh methods, *Phil. Trans. Roy. Soc. A* **357**, 1047–1077.

[9] Marsden, J.E., Patrick, G.W. and Shkoller, S. (1998). Multisymplectic geometry, variational integrators, and nonlinear PDEs, *Comm. Math. Phys.* **199**, 351–395.

Lecture 3: Symplectic integrators: A case study of the molecular dynamics of water

"Chemistry is a science, but not Science; for the criterion of true science lies in its relation to mathematics"　　　　　*(Kant)*

"Chemistry will only reach the rank of science when it shall be found possible to explain chemical reactions in the light of their causal relations to the velocities, tensions and conditions of equilibrium of the constituent molecules; that the chemistry of the future must deal with molecular mechanics by the methods and in the strict language of mathematics, as the astronomy of Newton and Laplace deals with the stars in their courses"

(Du Bois Reymond)

This quote (from D'Arcy Thompson's *On Growth and Form*) could not be more apt: symplectic integrators, developed to deal with the stars in their courses, are now applied to the velocities of molecules.

There are many fine surveys of symplectic integration, so here we'll discuss *Poisson systems*, or noncanonical Hamiltonian systems, and how they arose in a study of water. Water, the "king of polar fluids," has many strange phases and anomalous properties, which statistical mechanics has a hard time explaining. Therefore people turn to numerical simulations.

3.1 Splitting

Recall the problem of splitting—how can we write $f = f_1 + f_2$ so that the f_i retain some properties of f? The idea is to represent all f in the given class explicitly by a "generating function." Then we split the generating function. This can be done for Hamiltonian systems by splitting the Hamiltonian. Look at Table 2.1: Hamiltonian systems are expressed explicitly.

Example 2 *Hamiltonian systems.* The generating function is the Hamiltonian H.

$$f = J\nabla H = J\nabla\left(\sum_i H_i\right) = J\nabla H_1 + \ldots + J\nabla H_n.$$

Properties due to J, which is not split, are retained—symplecticity. Properties due to H, which is split, are lost—conservation of H.

Example 3 *Systems with an integral.* The generating function is the skew-symmetric matrix function J.

$$f = J\nabla H = \left(\sum_i J_i\right)\nabla H = J_1\nabla H + \ldots + J_n\nabla H$$

Properties due to J, which is split, are lost—symplecticity. Properties due to H, which is not split, are retained—conservation of H.

We'll return to systems with an integral in Lecture 6, and see how to apply splitting to volume-preserving systems in Lecture 5.

3.2 Poisson systems

Consider a standard, canonical Hamiltonian system.

$$\dot{x} = J\nabla_x H(x), \quad J = \begin{pmatrix} 0 & I \\ -I & 0 \end{pmatrix}.$$

It only has this special form when written in special variables. If we apply an arbitrary change of variables, writing the system in terms of $y = g(x)$, it becomes

$$\begin{aligned}
\dot{y} &= dg \cdot \dot{x} \\
&= dg \cdot J\nabla_x H(x) \\
&= dg \cdot J \cdot dg^T \nabla_y H(x) \\
&= \tilde{J}(y)\nabla_y \tilde{H}(y),
\end{aligned}$$

where

$$\tilde{J} = dg \cdot J \cdot (dg)^T$$
$$\tilde{H}(y) = H(x).$$

This is an example of a "Poisson system," the most obvious change being that the matrix J now depends on y. However, the class of Poisson systems is invariant under changes of variables. Since the history of mathematics and of physics is a history of requiring invariance under more operations, it seems we should study Poisson systems in their own right.

(There are many other motivations for the introduction of Poisson systems, from PDEs, systems on Lie groups and other manifolds, and symmetry reduction.)

An important special case are the "Lie"-Poisson systems. Let $x \in \mathbb{R}^m$ be an element of a Lie algebra. Let $[x_i, x_j] = \sum_{k=1}^{m} c_{ij}^k x_k$ be the Lie bracket. Let $J_{ij} = [x_i, x_j]$, so that the entries of J are linear functions of x. Then

$$\dot{x} = J(x) \nabla H(x)$$

or

$$\dot{x}_i = \sum_{j,k} c_{ij}^k x_k \frac{\partial H}{\partial x_j}$$

is called a Lie-Poisson system.

Example 4 *The free rigid body in $\mathbb{R}^3$.* The variables are π_1, π_2, π_3, the angular momenta of the body in body-fitted coordinates.

$$J = \begin{pmatrix} 0 & \pi_3 & -\pi_2 \\ -\pi_3 & 0 & \pi_1 \\ \pi_2 & -\pi_1 & 0 \end{pmatrix}$$

$$H = \frac{1}{2} \left(\frac{\pi_1^2}{I_1} + \frac{\pi_2^2}{I_2} + \frac{\pi_3^2}{I_3} \right)$$

Here the Lie algebra is $so(3)$, the antisymmetric 3×3 matrices.

3.3 Splitting into solvable pieces

Earlier we showed how to split a vector field into appropriate pieces, and how to compose their flows. But, it is still important to be able to apply a geometric integrator to each piece. Here we achieve this by requiring the pieces to be (easily) integrable.

Observation I If $J_{ij} = 0$ for $1 \leq i, j \leq k < n$ and $H = H(x_1, \ldots, x_k)$, then the ODEs are

$$\dot{x} = \begin{pmatrix} 0 & * \\ * & * \end{pmatrix} \begin{pmatrix} * \\ 0 \end{pmatrix}$$

or

$$\dot{x}_i = \begin{cases} 0 & i = 1, 2, \ldots, k \\ \sum_{j,l} c_{ij}^l x_l f_j(x_1, \ldots, x_k) & i = k+1, \ldots, m \end{cases}$$

which are linear with constant coefficients, hence easily solved (although the coefficients depend parametrically on the other variables $x_1, \ldots, x_k$).

Observation II Systems with $H = \sum_i H_i(x_i)$ can be split into easily solved parts. The rigid body has this form.

Observation III Quadratic Hamiltonians can be diagonalized, i.e., put in the form of Observation II, and hence split.

3.4 Molecular dynamics

The basic steps in a molecular dynamics simulation are the following.

   (i) Take a large sea of particles.
  (ii) Impose boundary conditions (e.g. 3D periodic) and impose constancy of any three out of the four quantities pressure, volume, temperature, and number of particles; the fourth is determined.
 (iii) Find a classical model of the interparticle forces.
  (iv) Move the particles for a long time.
   (v) Collect statistics of the motion.

Applications are to exploring states of matter (phase transitions, liquid, liquid crystal, colloidal), protein folding, the design of large organic molecules and drugs, nanotechnology. It's a big field.

Current limits are about 10^8 simple atoms on a supercomputer, 10^5 simple atoms on a workstation, and 10^2–10^3 water molecules on a workstation. For water, even $27 = 3 \times 3 \times 3$ molecules with periodic boundary conditions are enough to see solid, liquid, and gas phases.

What's the best way to move the particles? The method should

- obey the physical laws;
- exhibit the correct statistical equilibrium in the face of chaos; and

- be fast and cheap, since forces are expensive to evaluate.

In fact, the forces are *so* expensive that users don't want to evaluate them more than once per time step. For decades they've been using the Verlet method for point masses:

$$H = \text{kinetic } + \text{potential} = \frac{1}{2}p^2 + V(q),$$

$$q_{n+1} = q_n + \tau p_n$$
$$p_{n+1} = p_n - \tau \nabla V(q_{n+1})$$

We now know that it's so good because it's the simplest symplectic integrator, and comes from splitting the Hamiltonian.

How can we extend the Verlet algorithm to non-symmetric molecules like water? Many approaches have been considered.

- Move each atom separately. This involves modelling the interatomic forces, which means simulating the many modes of vibration within each molecule. Their time scale is very short and they are not believed to affect the macroscopic properties of water, which rules out this approach.

- Model as a rigid body. This is the preferred option. It can be done in various ways:

- Consider the molecule as a set of particles subjected to constraints on interatomic distances and angles. This is possible, but constraints lead to expensive, implicit methods. They are needed for problems involving flexible chains such as proteins.

- Model as a rigid body, with orientation represented by Euler angles or unit quaternions. The Hamiltonian in these variables is complicated and nonseparable; this makes symplectic methods expensive. In the most popular variant, unit quaternions are used and the ODEs are passed to a black-box solver.

- Represent orientation by a 3 × 3 orthogonal matrix, and update this by rotations only.

This last is what we now do. First, let's look at the free rigid body.

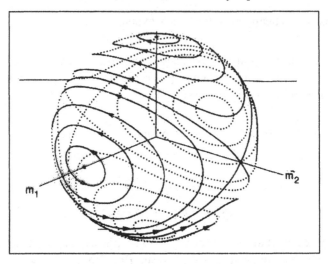

Fig. 3.1. The phase portrait of the free rigid body (from Bender & Orszag, *Advanced Mathematical Methods for Scientists and Engineers.*) Orbits on a sphere of constant angular momentum $|\pi|$ are shown. There are three pairs of fixed points, corresponding to rotation about each of the three principal axes; two are stable and one is unstable. To observe passage along the "homoclinic orbit" joining the unstable pair, hold a hammer with the head horizontal and toss so it rotates once about a left-right axis.

3.5 The free rigid body

Let the angular momentum vector be π, the orientation be $Q \in SO(3)$, i.e. $Q^T Q = Id$; the Hamiltonian for the free rigid body is

$$H = \frac{1}{2}\left(\frac{\pi_1^2}{I_1} + \frac{\pi_2^2}{I_2} + \frac{\pi_3^2}{I_3}\right).$$

The phase portrait for this system is shown in the famous Figure 3.1.

As noted above, splitting methods work excellently for Lie-Poisson systems with Hamiltonians of this type. The flow of $H = \frac{\pi_1^2}{2I_1}$ is

$$\pi(t) = R\pi(0)$$
$$Q(t) = Q(0)R^T$$

where

$$R = \begin{pmatrix} 1 & 0 & 0 \\ 0 & \cos\theta & -\sin\theta \\ 0 & \sin\theta & \cos\theta \end{pmatrix}, \quad \theta = t\frac{\pi_1}{I_1}.$$

This decomposes the motion into three elementary rotations.   The method is fast, accurate, reversible and symplectic. Q is always or-thogonal up to round-off error, because it is updated only by rotations.

How does this fit into a full simulation of water? For each molecule the variables are the q, the position of the centre of mass; p, the linear momentum; Q; and π. The total energy has the form

$$H = T^{\text{rotation}}(\pi) + T^{\text{translation}}(p) + V(q, Q)$$

together with the constraints $Q^T Q = Id$. We apply a Verlet-like splitting into kinetic and potential parts. For each molecule, we have

Potential part:

$$\dot{q} = 0$$
$$\dot{Q} = 0$$
$$\dot{p} = -\frac{\partial V}{\partial q} =: f \quad \text{(force)}$$
$$\dot{\pi} = (Q^T f) \times x \quad \text{(torque)}$$

where x is the point at which the force acts. Since the positions and orientations are here held constant, these equations are easy to solve.

Kinetic part:

$$\dot{q} = p$$
$$\dot{Q} = Q \left(\text{skew}(I^{-1}\pi) \right)$$
$$\dot{p} = 0$$
$$\dot{\pi} = \pi \times I^{-1}\pi$$

where $I = \text{diag}(I_1, I_2, I_3)$ is the inertia tensor. The centres of mass undergo free, straight-line motion, while the orientations move as a free rigid body. The latter could be solved explicitly, although this has never been implemented in a production code; in practice, we approximate its flow by the previously-given splitting method.

Composing these pieces gives an analogue of the Verlet method for this non-canonical Hamiltonian system. The final method uses only one force evaluation per time step, but is still explicit, symplectic, reversible, and conserves total linear and angular momentum (because each piece does). As expected for such a method, energy errors are bounded in time. When implemented in the existing research code ORIENT using existing error criteria, this method was about ten times faster than the old method (2-level leapfrog with Bulirsch-Stoer extrapolation).

References

Hamiltonian systems

For an introduction to Hamiltonian and Poisson systems, see

[1]   Olver, P.J. (1986). *Applications of Lie groups to differential equations* (Springer-Verlag, New York).

and to their dynamics,

[2]   Arrowsmith, D.K. and Place, C.M. (1992). *Dynamical systems: differential equations, maps, and chaotic behavior* (Chapman & Hall, New York).

Splitting

[3]   McLachlan, R.I. and Quispel, G.R.W. (1997). Generating functions for dynamical systems with symmetries, integrals, and differential invariants, *Physica D* **112**, 298–309.

Molecular dynamics

[4]   Allen, M.P. and Tildesley, D.J. (1987). *Computer Simulation of Liquids* (Oxford Science, Oxford).
[5]   Dullweber, A., Leimkuhler, B. and McLachlan, R. (1997). Split-Hamiltonian methods for rigid body molecular dynamics, *J. Chem. Phys.* **107**, 5840.

Lecture 4: Symmetries and reversing symmetries

4.1 Symmetries of ODEs

A symmetry is a map $h : \mathbb{R}^m \to \mathbb{R}^m$ from phase space to itself, such as $x \mapsto -x$. In a system with symmetries, the vector field at the two points x and $h(x)$ are related to each other. This is shown for the pendulum in Fig. 2.4. Under the 180° rotation $(q, p) \mapsto (-q, -p)$, arrows (the vector field) map to arrows: a symmetry. Under the reflection $p \to -p$, arrows map to arrows if we also reverse their direction: a reversing symmetry. The analogous properties for flows can be seen by tracing along the flow lines.

Symmetries and reversing symmetries both reduce the possible complexity of the phase portrait, and should be preserved.

In reversible Hamiltonian systems, reversing symmetries are a bit easier to preserve than symplecticity (although one can have both, if desired). For example, for simple mechanical systems there are explicit, variable-step-size reversible methods.

Consider the ODE $\dot{x} = f(x)$ under the change of variables $y = h(x)$. The new system is

$$\dot{y} = \widetilde{f}(y) := ((dh \cdot f)h^{-1})(y).$$

Definition 1 *The vector field f has symmetry h if $f = \widetilde{f}$, i.e., if $dh \cdot f = fh$. The vector field f has reversing symmetry h if $f = -\widetilde{f}$, i.e., $dh \cdot f = -fh$, and f is called reversible.*

The notation fh indicates composition, i.e., $(fh)(x) = f(h(x))$.

Example 5 *The pendulum.* For the vector field

$$f : \quad \dot{q} = p, \quad \dot{p} = -\sin q$$

we have

$$h_1 : \quad \widetilde{q} = q, \quad \widetilde{p} = -p \Rightarrow \dot{\widetilde{q}} = -\widetilde{p}, \quad \dot{\widetilde{p}} = \sin \widetilde{q},$$

—a reversing symmetry;

$$h_2 : \quad \widetilde{q} = -q, \quad \widetilde{p} = p \Rightarrow \dot{\widetilde{q}} = -\widetilde{p}, \quad \dot{\widetilde{p}} = \sin \widetilde{q},$$

—a reversing symmetry; and so

$$h_1 \circ h_2 : \quad \widetilde{q} = -q, \quad \widetilde{p} = -p \Rightarrow \dot{\widetilde{q}} = \widetilde{p}, \quad \dot{\widetilde{p}} = -\sin \widetilde{q}$$

is a symmetry. So the pendulum has "reversing symmetry group" (the group of all symmetries and reversing symmetries)

$$\Gamma = \{\mathrm{id}, h_1, h_2, h_1 h_2\}.$$

In general, half of the elements of Γ are symmetries, and the composition of two reversing symmetries is a symmetry.

We will use S for a symmetry and R for a reversing symmetry.

Definition 2 *The fixed set of S is*

$$\mathrm{fix}(S) := \{x : x = S(x)\}.$$

The fixed set is invariant under the flow of f. So preserving symmetries is one way of staying on a submanifold.

Example 6 *A nonlinear symmetry.* For the pendulum, the elements of Γ were all linear maps. Here is an example of a matrix ODE with a nonlinear symmetry. It is related to the famous Toda lattice. Let

$$X, L_0 \in \mathbb{R}^{n \times n},$$
$$\dot{X} = B(XL_0X^{-1})X,$$
$$B(L) = L_+ - L_-,$$

where L_+ (L_-) is the upper (lower) triangular part of L. This system has $h(X) = X^{-T}$ as a symmetry. The fixed set is $X = h(X) = X^{-T}$ or $XX^T = I$, i.e., $X \in O(n)$, the orthogonal group. A symmetry-preserving integrator for this system would also have $O(n)$ as an invariant set.

4.2 Symmetries of maps

Definition 3 A map ψ has h as a symmetry if $h\psi = \psi h$, i.e.,

$$\psi = \mathcal{N}_h\psi := h^{-1}\psi h.$$

A map ψ has h as a reversing symmetry if $h\psi = \psi^{-1}h$, i.e.

$$\psi = \mathcal{N}_h\mathcal{I}\psi := h^{-1}\psi^{-1}h.$$

The important property of the operators $\mathcal{N}_h$, $\mathcal{I}$ is how they act on compositions of maps. $\mathcal{N}_h$ acts as an *automorphism*, i.e.

$$\mathcal{N}_h(\psi_1\psi_2) = (\mathcal{N}_h\psi_1)(\mathcal{N}_h\psi_2),$$

while $\mathcal{I}$ acts as an *antiautomorphism*, i.e.

$$\mathcal{I}(\psi_1\psi_2) = (\psi_1\psi_2)^{-1} = \psi_2^{-1}\psi_1^{-1} = (\mathcal{I}\psi_2)(\mathcal{I}\psi_1)$$

For a map, having a (reversing) symmetry is equivalent to being in the fixed set of an (anti)automorphism. Therefore, we study how to construct maps in such fixed sets. We shall see that for antiautomorphisms this is relatively simple, while for automorphisms it is unsolved.

Thus, paradoxically, we know how to construct reversible integrators, (which is good, because reversibility brings good long-time behavior, e.g., through invariant tori), but not symmetric integrators, which looks at first sight simpler.

4.3 Covariance

Why are Runge–Kutta methods called linear methods? One explanation is that they are linearly *covariant*. Consider methods ψ which associate to each ODE f a map $\psi_\tau(f)$, where τ is the time step.

Definition 4 *A method ψ is h-covariant if the following diagram commutes.*

$$
\begin{array}{ccc}
\dot{x} = f(x) & \overset{x=h(y)}{\longrightarrow} & \dot{y} = \tilde{f}(y) \\
\downarrow & & \downarrow \\
\tilde{x} = \psi_\tau(f)(x) & \overset{x=h(y)}{\longrightarrow} & \tilde{y} = \psi_\tau(\tilde{f})(y)
\end{array}
$$

That is, if

$$
\psi = \mathcal{K}_h \psi := h^{-1}\psi((dh \cdot f)h^{-1})h
$$

where $\mathcal{K}_h$ is an automorphism.

In words, we get the "same" integrator whether we take the ODE in variables x or y. Notice that if h is a symmetry of f, then $f = \tilde{f}$, and hence h is a symmetry of ψ. So an h-covariant method is automatically h-symmetric, even if we don't know what the symmetry is!

So, we should classify methods by their covariance group.

Example 7 *Euler's method*

$$
\psi_\tau(f) : x \mapsto x + \tau f(x)
$$

(or, more generally, any Runge–Kutta method), is covariant under any affine map $x = Ay + b$.

Example 8 *The exact solution $\varphi_\tau(f)$ is covariant under all maps $x = h(y)$.*

Example 9 *The splitting method for $\dot{q} = f(p)$, $\dot{p} = g(q)$,*

$$
\psi_\tau(f) : q' = q + \tau f(p), \ p' = p + \tau g(q')
$$

is covariant under all maps $\tilde{q} = h_1(q)$, $\tilde{p} = h_2(p)$. It is not even covariant under linear maps which couple the q, p variables.

When composing symmetric or reversible methods, we can use the properties

(i) The fixed sets of automorphisms form a group.

(ii) The fixed sets of antiautomorphisms form a symmetric space.

Property (1) is immediate, while property (2) follows from the following

Lemma 2 (The Generalized Scovel Projection.)   *Let $\mathcal{A}_-$ be an antiautomorphism with order 2, i.e. $\mathcal{A}_-^2 = \mathrm{id}$. Let $\chi = \mathcal{A}_-\chi$, i.e. $\chi \in \mathrm{fix}\,\mathcal{A}_-$. Then*

$$\psi\chi\mathcal{A}_-\psi \in \mathrm{fix}\,\mathcal{A}_- \quad \forall\psi.$$

Proof $\mathcal{A}_-(\psi\chi\mathcal{A}_-\psi) = \mathcal{A}_-^2\psi\mathcal{A}_-\chi\mathcal{A}_-\psi = \psi\chi\mathcal{A}_-\psi.$ □

Example 10 *For any (anti)automorphism $\mathcal{A}$, we have $\mathcal{A}(\mathrm{id}) = \mathrm{id}$ and hence $\mathcal{A}(\chi^{-1}) = \mathcal{A}(\chi)^{-1}$. Therefore, $\chi \in \mathrm{fix}\,\mathcal{A}_- \Rightarrow \chi^{-1} \in \mathrm{fix}\,\mathcal{A}_-$. Taking $\psi = \mathcal{A}_-\psi$ gives the symmetric space property $\psi\chi^{-1}\psi \in \mathrm{fix}\,\mathcal{A}_-$.*

Example 11 $\chi = \mathrm{id} \Rightarrow \psi\mathcal{A}_-\psi \in \mathrm{fix}\,\mathcal{A}_-$. *This gives a way of constructing elements fixed under any antiautomorphism, starting from any element.*

Example 12 *With $\mathcal{A}_-\psi_\tau := \psi_{-\tau}^{-1}$ and $\chi = \mathrm{id}$, this builds self-adjoint methods of the form $\psi_\tau\psi_{-\tau}^{-1}$.*

Example 13 *With $\mathcal{A}_-\psi := h^{-1}\psi^{-1}h$, $h^2 = \mathrm{id}$, and $\chi = \mathrm{id}$, this builds reversible methods of the form $\psi h^{-1}\psi^{-1}h$.*

4.4 Building symmetric methods

This is unsolved except in two simple cases.

   (i) If the method is h-covariant and h is a symmetry, then the method is h-symmetric.
   (ii) If the symmetry group H is linear and the map ψ belongs to a linear space, then we can average over H:

$$\overline{\psi} := \frac{1}{|H|}\sum_{h\in H}\psi h$$

   is H-symmetric.

Since preserving symmetries is difficult, we should try not to destroy symmetries in the first place, by doing non-symmetric splittings, for example.

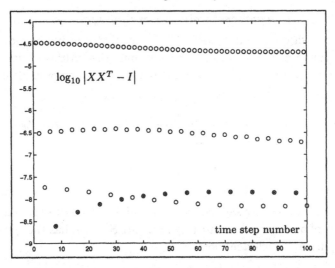

Fig. 4.1. The discrepancy for $\psi_3 =$ '01101001' applied to Example 6, where the base method ψ_0 is Euler. Notice how the symmetry error is drastically reduced every 2, then every 4, then every 8 time steps.

Hamiltonian systems

A symplectic integrator preserves an integral I if it preserves the symmetry associated with that integral, namely, the flow of $J\nabla I$.

Example 14 *Quadratic integrals* are associated with linear symmetries, so are preserved by any linearly covariant symplectic integrator, such as the midpoint rule (by Noether's theorem).

4.5 Building reversible methods

Here the situation is much nicer.

Theorem 3 *Let Γ be a group of automorphisms and antiautomorphisms. Let φ be fixed under the automorphisms. Then*

$$\psi = \varphi \mathcal{A}_- \varphi$$

is fixed under $\mathcal{A}_g$ for all $g \in \Gamma$, where $\mathcal{A}_-$ is any antiautomorphism in Γ.

For example, if all symmetries are linear, then we can use this theorem to construct integrators having the full reversing symmetry group.

4.6 Approximately preserving symmetries

The composition used in Lemma 2 is so nice that it would be nice to use it for symmetries as well as reversing symmetries. Although it doesn't eliminate the symmetry error, it does reduce it by one power of the time step.

Theorem 4 *Let $\mathcal{A}_+$ be an automorphism of order 2. Let ψ_τ be a method with $\psi = \mathcal{A}_+\psi + \mathcal{O}(\delta)$. Let $\psi_1 := \psi\mathcal{A}_+\psi$. Then $\psi_1 = \mathcal{A}_+\psi_1 + \mathcal{O}(\tau\delta)$, where $\delta = \mathcal{O}(\tau)$.*

Proof The proof is an illustration of backward error analysis and manipulation of flows considered as exponentials. We write the map ψ as the flow of a vector field consisting of a part S which has the symmetry and a part M which does not:

$$\psi = \exp(\tau S + \delta M).$$

Therefore,

$$\mathcal{A}_+\psi = \exp(\tau S + \delta N)$$

for some vector field N. Now

$$
\begin{aligned}
\psi_1(\mathcal{A}_+\psi_1)^{-1} &= (\psi\,\mathcal{A}_+\psi)(\mathcal{A}_+\psi\,\psi)^{-1} \\
&= \psi\,\mathcal{A}_+\psi\,\psi^{-1}\,(\mathcal{A}_+\psi)^{-1} \\
&= \exp([\tau S + \delta M, \tau S + \delta N] + \ldots) \\
&= \exp(\tau\delta[S, N - M] + \ldots)
\end{aligned}
$$

□

Usually, the initial symmetry error δ will be $\mathcal{O}(\tau^{p+1})$ for a method of order p, and this composition reduces it to $\mathcal{O}(\tau^{p+2})$. The idea can be applied iteratively: if

$$\psi_{n+1} = \psi_n\mathcal{A}_+\psi_n,$$

then ψ_n has symmetry error $\mathcal{O}(\tau^n\delta)$. This gives methods of the form

$$\psi_2 = \psi\,\mathcal{A}_+\psi\,\mathcal{A}_+\psi\,\psi = \text{`0110'},$$

$$\psi_3 = \text{`01101001'},$$

and so on, given by the initial elements of the famous 'Thue-Morse' sequence.

In the matrix example given previously, it is desired to leave the fixed set $XX^T = I$ invariant. This could be done by, e.g., the midpoint rule, but this is implicit and, given the form of the ODE, very expensive. Instead, one can use a simple explicit method for ψ, and reduce the symmetry error to any desired order using ψ_n. This leaves X orthogonal to any desired order.

References

Background on symmetries

[1] Roberts, J.A.G. and Quispel, G.R.W. (1992). Chaos and time-reversal symmetry: order and chaos in reversible dynamical systems, *Phys. Rep.* **216**, 63–177.

Symmetric integration

[2] McLachlan, R.I., Quispel, G.R.W. and Turner, G.S. (1998). Numerical integrators that preserve symmetries and reversing symmetries, *SIAM J. Numer. Anal.* **35**, 586–599.
[3] Iserles, A., McLachlan, R. and Zanna, A. (1999). Approximately preserving symmetries in the numerical integration of ordinary differential equations, *Eur. J. Appl. Math.* **10**, 419–445.
[4] Huang, W. and Leimkuhler, B. (1997). The adaptive Verlet method, *SIAM J. Sci. Comput.* **18**, 239–256.
[5] Munthe-Kaas, H., Quispel, G.R.W. and Zanna, A. (2000). Applications of symmetric spaces and Lie triple systems in numerical analysis. To appear.

Lecture 5: Volume-preserving integrators

Remember that the ODE

$$\frac{dx}{dt} = f(x)$$

is source-free (or divergence-free) if

$$\nabla \cdot f = \sum_{i=1}^{m} \frac{\partial f_i}{\partial x_i} = 0$$

for all x. Let $df = (\partial f_i / \partial x_j)$ be the derivative of f and $A = \partial \varphi_\tau / \partial x$ be the Jacobian of its flow. A evolves according to

$$\frac{dA}{dt} = df A, \quad A(0) = Id$$

and one can show that

$$\frac{d}{dt} \det A = \text{tr}(df) \det(A).$$

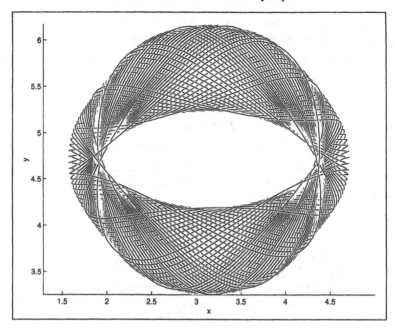

Fig. 5.1. Orbit of the 3D "ABC" flow computed with a second-order volume-preserving integrator. The system is $\dot{x} = A\sin z + C\sin y$, $\dot{y} = B\sin x + A\cos z$, $\dot{z} = C\sin y + B\cos x$, with parameters $A = B = 1$, $C = 2$ and initial conditions $(2, 5, 0)$. The phase space $\mathbb{T}^3$ is here viewed along the z-axis, the long axis of the torus. The integration time is 750, equivalent to 240 circuits of the z-axis, and the time step is $\tau = 0.1$. The orbit lies on a torus and its regular, quasiperiodic behaviour is apparent.

Consequently, if $\nabla \cdot f = \text{tr}(df) = 0$, then $\det A = 1$ for all time; the flow is volume preserving.

Volume preserving systems may be seen as one of the very few fundamental types of dynamics; their flows belong to one of the "Lie pseudogroups" of diffeomorphisms. They arise in tracking particles in incompressible fluid flow, in perturbations of Hamiltonian systems, and in discretizations of the wave equations of mathematical physics; volume preservation (and not symplecticity, for example) is the key conservation law underlying statistical mechanics. An example comparing volume- and non-volume-preserving integration is shown in Figs. 5.1–5.3.

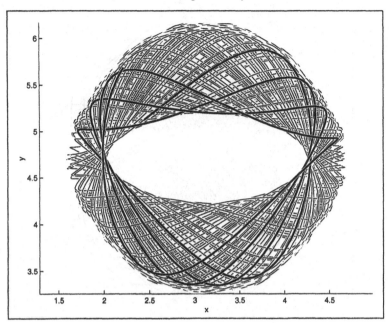

Fig. 5.2. As in Fig. 5.1, but computed with a non-volume preserving second order Runge–Kutta method with the same time step. The computed motion gradually decreases in amplitude (the last 3% of the orbit is shown in bold), tending to a periodic orbit on a (Lyapunov) time scale of 1.5×10^4. The dissipation and lack of quasiperiodicity can be seen even on the short time scale shown here. However, this method does preserve the 16 linear symmetries of the ODE (Lecture 4, [2]), which may explain why the results are better than in Fig. (5.3)

The integrator ψ_τ is volume preserving (VP) if

$$\det \left(\frac{\partial \psi_{\tau,i}}{\partial x_j} \right) = 1$$

for all x. There are two general ways to construct VP integrators:

(i) the splitting method, and

(ii) the correction method.

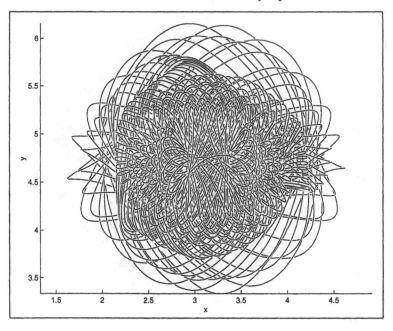

Fig. 5.3. As in Fig. 5.1, but computed with MATLAB's ODE45 routine. The use of time-adaptivity has broken the spatial symmetries, with drastic consequences.

5.1 Volume-preserving splitting method

Starting with the system of ODEs

$$\frac{dx_1}{dt} = f_1(x)$$

$$\vdots$$

$$\frac{dx_m}{dt} = f_m(x)$$

we substitute

$$f_m(x) = \int \frac{\partial f_m}{\partial x_m}\, dx_m$$

$$= -\int \sum_{i=1}^{m-1} \frac{\partial f_i}{\partial x_i}\, dx_m$$

with appropriately chosen constants of integration, to get the equivalent form

$$\frac{dx_1}{dt} = f_1(x)$$

$$\vdots$$

$$\frac{dx_{m-1}}{dt} = f_{m-1}(x)$$

$$\frac{dx_m}{dt} = -\sum_{i=1}^{m-1} \int \frac{\partial f_i}{\partial x_i}\, dx_m.$$

Now we split f, writing f as the sum of the $m-1$ vector fields

$$\frac{dx_i}{dt} = 0 \quad i \neq j, m$$

$$\frac{dx_j}{dt} = f_j(x)$$

$$\frac{dx_m}{dt} = -\int \frac{\partial f_j}{\partial x_j}\, dx_m$$

for $j = 1, \ldots, m-1$. Note that:

(i) Each of these $m-1$ vector fields is source-free.

(ii) We have split one big problem into $m-1$ small problems. But we know the solution to each small problem! They each correspond to a two-dimensional Hamiltonian system

$$\frac{dx_j}{dt} = \frac{\partial H_j}{\partial x_m}$$

$$\frac{dx_m}{dt} = -\frac{\partial H_j}{\partial x_j}$$

with Hamiltonian $H_j := \int f_j(x)dx_m$, treating x_i for $i \neq j, m$ as fixed parameters. Each of these 2D problems can either be solved exactly (if possible), or approximated with any symplectic integrator ψ_j. Even though ψ_j is not symplectic in the whole space $\mathbb{R}^m$, it *is* volume-preserving.

A volume-preserving integrator for f is then given by

$$\psi = \psi_1 \circ \psi_2 \circ \ldots \psi_{m-1}.$$

Example 15 *(for illustration only)* The 3D Volterra system

$$\frac{dx_1}{dt} = x_1(x_2 - x_3)$$

$$\frac{dx_2}{dt} = x_2(x_3 - x_1)$$

$$\frac{dx_3}{dt} = x_3(x_1 - x_2)$$

is source-free, and splits as

$$\left.\begin{array}{l} \dfrac{dx_1}{dt} = x_1 x_2 - x_1 x_3 \\[2mm] \dfrac{dx_2}{dt} = 0 \\[2mm] \dfrac{dx_3}{dt} = -x_2 x_3 + \dfrac{1}{2}x_3^2 \end{array}\right\} \tilde{f}_1,$$

$$\left.\begin{array}{l} \dfrac{dx_1}{dt} = 0 \\[2mm] \dfrac{dx_2}{dt} = x_2 x_3 - x_2 x_1 \\[2mm] \dfrac{dx_3}{dt} = x_1 x_3 - \dfrac{1}{2}x_3^2 \end{array}\right\} \tilde{f}_2,$$

where volume-preserving integrators for $\tilde{f}_1$ and $\tilde{f}_2$ are given by the implicit midpoint rule,

$$x' = x + \tau \tilde{f}_1\left(\frac{x + x'}{2}\right), \quad x'' = x' + \tau \tilde{f}_2\left(\frac{x' + x''}{2}\right).$$

Note that the x_3^2 terms were not in the original system, but on combining the two steps they cancel.

The splitting is an example of a generating function method: we construct source-free f's without any side conditions.

5.2 *The volume-preserving correction method*

The simplest case is the semi-implicit method

$$x_1' = x_1 + \tau f_1(\tilde{x})$$

$$\vdots$$

$$x_{m-1}' = x_{m-1} + \tau f_{m-1}(\tilde{x})$$

$$x_m = \int^{x_m'} J(\tilde{x}) \, dx_m'$$

where

$$J := \det \left(\frac{\partial x_i'}{\partial x_j} \right)_{i,j=1,\ldots,m-1}$$

and

$$\tilde{x} = (x_1, \ldots, x_{m-1}, x_m').$$

For a proof of consistency and volume-preservation, see [3].

Example 16 *(for illustration only)* For the 3D Volterra system

$$\frac{dx_1}{dt} = x_1(x_2 - x_3)$$

$$\frac{dx_2}{dt} = x_2(x_3 - x_1)$$

$$\frac{dx_3}{dt} = x_3(x_1 - x_2)$$

we get

$$x_1' = x_1 + \tau x_1(x_2 - x_3')$$
$$x_2' = x_2 + \tau x_2(x_3' - x_1)$$

and

$$J = \begin{vmatrix} \frac{\partial x_1'}{\partial x_1} & \frac{\partial x_1'}{\partial x_2} \\ \frac{\partial x_2'}{\partial x_1} & \frac{\partial x_2'}{\partial x_2} \end{vmatrix}$$

$$= \begin{vmatrix} 1 + \tau(x_2 - x_3') & \tau x_1 \\ -\tau x_2 & 1 + \tau(x_3' - x_1) \end{vmatrix}$$

$$= 1 + \tau(x_2 - x_1) + \tau^2(x_2 x_3' + x_1 x_3' - x_3'^2)$$

and the last component of the method is $x_3 = \int^{x_3'} J dx_3'$ or

$$x_3 = x_3' + \tau x_3'(x_2 - x_1) + \frac{\tau^2}{2}\left(x_2 x_3'^2 + x_1 x_3'^2 - \frac{2}{3}x_3'^3\right).$$

References

The splitting method

[1] Feng K. and Shang Z.-J. (1995). Volume-preserving algorithms for source-free dynamical systems, *Numer. Math.* **71**, 451.
[2] McLachlan, R.I. and Quispel, G.R.W. (1997). Generating functions for dynamical systems with symmetries, integrals, and differential invariants, *Physica D* **112**, 298–309.

The correction method

[3] Quispel, G.R.W. (1995). Volume-preserving integrators, *Phys. Lett.* **206A**, 26–30.

[4] Shang, Z.-J. (1994). Construction of volume-preserving difference schemes for source-free systems via generating functions, *J. Comput. Math.* **12**, 265–272.

Error growth

[5] Quispel, G.R.W. and Dyt, C.P. (1998). Volume-preserving integrators have linear error growth, *Phys. Lett.* **242A**, 25–30.

Lecture 6: Integrators that preserve integrals and/or Lyapunov functions

Definition 5 *$I(x)$ is a (first) integral or a conserved quantity of an ODE if*

$$\frac{d}{dt}I(x(t)) = 0$$

for solutions $x(t)$ of the ODE $\frac{dx}{dt} = f(x)$, $x \in \mathbb{R}^m$.

By the chain rule, this requires $\sum \frac{dI}{dx_i}\frac{dx_i}{dt} = 0$ for all solutions $x(t)$, or equivalently

$$\sum \frac{dI}{dx_i}f_i(x) = f \cdot \nabla I = 0$$

for all x.

Definition 6 *$V(x)$ is a Lyapunov function if*

$$\frac{d}{dt}V(x) \leq 0.$$

Equivalently,

$$f \cdot \nabla V \leq 0 \text{ for all } x \in \mathbb{R}^m.$$

ODEs with one or more first integrals occur frequently in physics. Many examples come from two main classes:

(i) **Hamiltonian systems.** For example, the pendulum

$$\frac{dx_1}{dt} = x_2$$

$$\frac{dx_2}{dt} = -\sin(x_1)$$

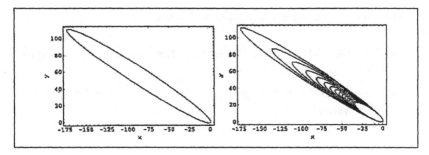

Fig. 6.1. An orbit of the Kepler 2-body problem with the eccentricity of the Hale–Bopp comet, computed with an integral-preserving method (left) and Runge–Kutta (right).

where x_1 is the angular position of the pendulum and x_2 its angular momentum, has the first integral

$$I(x_1, x_2) = \frac{1}{2}x_2^2 - \cos(x_1).$$

(ii) **Poisson systems.** For example, the free rigid body with moments of inertia I_1, I_2, and I_3, and angular momentum π_1, π_2, π_3 in body-fixed coordinates,

$$\frac{d\pi_1}{dt} = \left(\frac{1}{I_2} \quad \frac{1}{I_3} \right) \pi_2 \pi_3$$

$$\frac{d\pi_2}{dt} = \left(\frac{1}{I_3} - \frac{1}{I_1} \right) \pi_3 \pi_1$$

$$\frac{d\pi_3}{dt} = \left(\frac{1}{I_1} - \frac{1}{I_2} \right) \pi_1 \pi_2$$

has the first integral

$$I(\pi_1, \pi_2, \pi_3) = \pi_1^2 + \pi_2^2 + \pi_3^2,$$

which is the body's total angular momentum.

6.1 Preserving a first integral

Before presenting the theory, here is a picture. Fig. 6.1 shows an orbit in the Kepler problem; it's an ellipse. This remains true if you use an integral preserving method. With a standard method such as Runge–Kutta, the orbit spirals down to the origin.

The general method we present is as follows:

(i) For every ODE with a first integral $I(x)$, we construct an equivalent "skew-gradient system";

(ii) we discretize the skew-gradient system to a "skew discrete-gradient system";

(iii) we show that the skew discrete-gradient system has the same integral $I(x)$.

More specifically,

(i) Given the system $\frac{dx}{dt} = f(x)$ and first integral $I(x)$ such that $\frac{d}{dt} I(x(t)) = 0$, we construct the equivalent skew gradient system

$$\frac{dx}{dt} = S\nabla I, \quad S^T = -S;$$

(ii) we discretize this to the skew discrete gradient system

$$\frac{x' - x}{\tau} = S\overline{\nabla} I(x, x'),$$

where $\overline{\nabla}$ is a "discrete gradient;"

(iii) we show that $I(x') = I(x)$.

Constructing an equivalent skew gradient system

We want to solve $S\nabla I = f$ for the antisymmetric matrix S, where f and the integral I are given. Because I is an integral,

$$\frac{d}{dt} I = \frac{dx}{dt} \cdot \nabla I = f^T \nabla I = 0.$$

One solution for S is

$$S = \frac{f(\nabla I)^T - (\nabla I) f^T}{|\nabla I|^2}$$

but S is not unique. In particular, if the critical points of I (points where $\nabla I(x) = 0$) are nondegenerate, then there is an S which is as smooth as f and ∇I. Sometimes, as in Poisson systems, S is already known and does not need to be constructed.

Discretizing the skew-gradient system to a skew discrete-gradient system

A discrete gradient $\overline{\nabla} I$ is defined by the two axioms

$$I(x') - I(x) = (\overline{\nabla} I) \cdot (x' - x)$$
$$\overline{\nabla} I(x, x') = \nabla I(x) + \mathcal{O}(x' - x).$$

For any such discrete gradient we can construct the skew discrete-gradient system

$$\frac{x' - x}{\tau} = \widetilde{S}\overline{\nabla} I$$

where $\widetilde{S}$ is any consistent antisymmetric matrix, such as $\widetilde{S}(x, x') = S((x + x')/2)$.

This discretization has the same integral I:

$$I(x') - I(x) = (\overline{\nabla} I) \cdot (x' - x)$$
$$= \tau(\overline{\nabla} I)^T S(\overline{\nabla} T)$$
$$= 0$$

Examples of discrete gradients

The problem is reduced to finding discrete gradients $\overline{\nabla}$ satisfying the axioms. The general solution is known. Here are two particular solutions:

(i) **(Itoh and Abe)**

$$\overline{\nabla} I(x, x')_i := \left(I(x^{(i)}) - I(x^{(i-1)}) \right) / \left(x_i' - x_i \right),$$

where

$$x^{(i)} := (x_1', \ldots, x_i', x_{i+1}, \ldots, x_m).$$

(ii) **(Harten, Lax and Van Leer)**

$$\overline{\nabla} I(x, x') := \int_0^1 \nabla I \left(x + \xi(x' - x) \right) d\xi$$

Numerical example preserving one integral

Consider the system

$$\frac{dx}{dt} = yz + x(1 + 0.1y^2 + y^3 + 0.1y^5)$$
$$\frac{dy}{dt} = -x^2 + z^2 - 0.1x^2y^2$$
$$\frac{dz}{dt} = -z - xy - y^3z$$

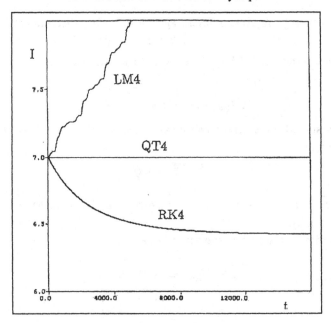

Fig. 6.2. Evolution of an integral $I(x)$ in a 3D system with three fourth-order methods. QT4: Integral preserving; RK4: Classical Runge–Kutta; LM4: linear multistep.

which has first integral

$$I = \frac{x^2}{2} + \frac{y^4}{4} + y + \frac{z^2}{2},$$

but is *not* Hamiltonian or Poisson. $I(x)$ has compact level surfaces, so staying on them means that the numerical integration is stable. This system is written as a skew-gradient system as

$$\frac{d}{dt}\begin{pmatrix} x \\ y \\ z \end{pmatrix} = \begin{pmatrix} 0 & x(1+0.1y^2) & y \\ -x(1+0.1y^2) & 0 & z \\ -y & -z & 0 \end{pmatrix} \nabla I$$

and a comparison between a skew-discrete gradient integrator and two classical methods is shown in Fig. 6.2.

6.2 Preserving a Lyapunov function

An integral is a function that is preserved in time, $dI/dt = 0$. A Lyapunov function decreases in time, $dV/dt \leq 0$. These also arise frequently:

(i) **Gradient systems** (these arise in dynamical systems theory)

$$\frac{dx}{dt} = -\nabla V(x)$$

Here $\frac{dV}{dt} = -(\nabla V)^T \nabla V = -|\nabla V|^2 \le 0$, so V is a Lyapunov function.

(ii) **Systems with dissipation** For example, the damped pendulum with friction $\alpha \ge 0$,

$$\frac{dx_1}{dt} = x_2$$

$$\frac{dx_2}{dt} = -\sin(x_1) - \alpha x_2$$

has Lyapunov function $V(x_1, x_2) = \frac{1}{2}x_2^2 - \cos(x_1)$, because

$$\frac{dV}{dt} = -\alpha x_2^2 \le 0.$$

(iii) **Systems with an asymptotically stable fixed point** Here the construction of the Lyapunov function is a standard (although difficult) problem in dynamical systems.

These systems can be discretized very similarly to systems with an integral. Namely,

(i) Given the system $\frac{dx}{dt} = f(x)$ with Lyapunov function V, we construct the equivalent "linear-gradient system"

$$\frac{dx}{dt} = L\nabla V$$

where L is negative semidefinite, i.e. $v^T L v \le 0$ for all vectors v;

(ii) we take the linear-discrete gradient system

$$\frac{x' - x}{\tau} = L\overline{\nabla}V(x, x'),$$

(iii) we show that $V(x') \le V(x)$.

Note that L is *not* necessarily symmetric. This is important, because as the dissipation tends to zero, we want L to smoothly tend to an antisymmetric matrix, to recover the integral-preserving case.

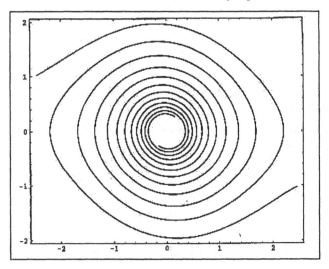

Fig. 6.3. The damped pendulum, computed with a linear-discrete gradient method. Orbits spiral in correctly even if the dissipation rate tends to zero.

Constructing an equivalent linear-gradient system

We want to solve $L\nabla V = f$ where f and V are given, $\frac{dV}{dt} = f \cdot \nabla V \leq 0$, and L is a negative semidefinite matrix. One solution is

$$L = \frac{f(\nabla V)^T - (\nabla V)f^T + (f \cdot \nabla V)I}{|\nabla V|^2}.$$

One can check that $v^T L v = |v|^2 (f \cdot \nabla V)/|\nabla V|^2 \leq 0$, so that L is negative semi-definite, and that $L\nabla V = f$. As with skew-gradient systems, the matrix L is not unique, and special care is required near critical points of V.

The linear-discrete gradient system

By analogy with the integral-preserving case, we check that the linear-discrete gradient system has the same Lyapunov function as the original system:

$$V(x') - V(x) = (\overline{\nabla} V) \cdot (x' - x)$$
$$= \tau (\overline{\nabla} V)^T L (\overline{\nabla} V)$$
$$\leq 0.$$

Numerical example

The damped pendulum

$$\frac{dx_1}{dt} = x_2$$

$$\frac{dx_2}{dt} = -\sin(x_1) - \alpha x_2$$

with Lyapunov function $V(x_1, x_2) = \frac{1}{2}x_2^2 - \cos(x_1)$, can be written in the linear-gradient form

$$\frac{d}{dt}\begin{pmatrix} x_1 \\ x_2 \end{pmatrix} = \begin{pmatrix} 0 & 1 \\ -1 & -\alpha \end{pmatrix}\begin{pmatrix} \sin(x_1) \\ x_2 \end{pmatrix} = L\nabla V.$$

(Note that L is negative semi-definite, because the eigenvalues of its symmetric part are 0 and $-\alpha$.) Using the Itoh–Abe discrete gradient we get the discretization

$$\frac{1}{\tau}\begin{pmatrix} x_1' - x_1 \\ x_2' - x_2 \end{pmatrix} = L\overline{\nabla}V$$

$$= \begin{pmatrix} 0 & 1 \\ -1 & -\alpha \end{pmatrix}\begin{pmatrix} \frac{-\cos(x_1')+\cos(x_1)}{x_1'-x_1} \\ \frac{x_2'+x_2}{2} \end{pmatrix},$$

whose phase portrait is sketched in Fig. 6.3. The behaviour of the non-dissipative Euler's method is quite different. It increases energy near $p = 0$ for all time steps τ. Globally, for $\tau > \sim 2\alpha$ orbits move out across the separatrix; and for $\alpha < \sim \tau < \sim 2\alpha$ there are spurious asymptotically stable periodic orbits inside the separatrix.

Extensions and generalizations

(i) The above methods can be generalized to ODEs with any number of integrals and/or Lyapunov functions.

(ii) There are discrete gradient methods of order 2, but higher order is desirable. For systems with an integral, this can be done by composition. For Lyapunov functions, the maps form only a semigroup, so the order cannot be increased beyond 2 by composition.

(iii) Given a system in skew- or linear-gradient form, the matrix can be split, leading to 2D systems with the same integral or Lyapunov function.

(iv) Satisfactory treatment of *nonautonomous* ODEs is still an open problem.

References

Discrete gradients were introduced for Hamiltonian ODEs in

[1] Gonzalez, O. (1996).Time integration and discrete Hamiltonian systems, *J. Nonlinear Sci.* **6**, 449–467.

and generalized as presented here in

[2] McLachlan, R.I., Quispel, G.R.W.and Robidoux, N. (1999). Geometric integration using discrete gradients, *Phil. Trans. Roy. Soc.* **357**, 1021–1045.

of which a shorter version appears in

[3] McLachlan, R.I., Quispel, G.R.W.and Robidoux, N. (1998). A unified approach to Hamiltonian systems, Poisson systems, gradient systems, and systems with Lyapunov functions and/or first integrals, *Phys. Rev. Lett.* **81**, 2399–2403.

When are Integration and Discrepancy tractable?

Erich Novak

Mathematisches Institut
Universität Erlangen-Nürnberg
Bismarckstraße 1 1/2
91054 Erlangen, Germany, and
Department of Computer Science
KU Leuven
Celestijnenlaan 200A, B-3001 Heverlee, Belgium
Email: novak@mi.uni-erlangen.de

Henryk Woźniakowski

Department of Computer Science
Columbia University
New York, NY 10027, USA, and
Institute of Applied Mathematics
University of Warsaw
ul. Banacha 2, 02-097 Warsaw, Poland
Email: henryk@cs.columbia.edu

1 Introduction

There has been an increasing interest in studying computational aspects of high dimensional problems. Such problems are defined on spaces of functions of d variables and occur in many applications, with d that can be hundreds or even thousands. Examples include:

- *High dimensional integrals* or *path integrals* with respect to the *Wiener measure*. These are important for many applications, in particular, in physics and in finance. High dimensional integrals also occur when we want to compute certain parameters of stochastic processes. Moreover, path integrals arise as solutions of partial differential equations given, for example, by the *Feynman–Kac formula*. See [25, 40, 66, 82, 85, 91].

- *Global optimization* where we need to compute the (global) minimum of a function of d variables. This occurs in many applications, for example, in *pattern recognition* and in *image processing*, see [97], or in the modelling and prediction of the geometry of proteins, see [45]. Simulated annealing strategies and genetic algorithms are often used, as well as smoothing techniques and other stochastic algorithms, see

[10] and [74]. Some error bounds for deterministic and stochastic algorithms can be found in [42, 43, 44, 48, 53].

• The *Schrödinger equation* for $m > 1$ particles in $\mathbb{R}^3$ is a $d = 3m$-dimensional problem.

The solutions of many high dimensional problems are difficult to obtain or even to approximate. Typically, the cost of solving these problems increases exponentially with the dimension d. This is true for numerical integration, optimal recovery, global optimization, and for differential and integral equations over most classical function spaces. However, this is *not* true for systems of ordinary differential equations and for convex optimization, see [85] for a survey. For example, the minimal error for integration of C^r-functions in one variable is of order n^{-r}, where n is the number of function samples used by the algorithm. Using product rules, it is easy to obtain an error of order $n^{-r/d}$ in dimension $d > 1$. It is known that the exponent r/d is best possible. Hence, if we want to guarantee an error ε we must take n of order $\varepsilon^{-d/r}$. For fixed regularity r, this is an exponential function of d. For large d, we cannot afford to sample the function so many times. This exponential dependence is called *intractability* or the *curse of dimensionality*; the latter notion goes back to Bellman, see [7].

It is natural to ask which high dimensional problems are *tractable* and which are *intractable*. Tractability means that there is an algorithm that approximates the solution with error ε using $n = n(\varepsilon, d)$ samples of the function, where $n(\varepsilon, d)$ is polynomially bounded in ε^{-1} and d. If $n(\varepsilon, d)$ has a bound independent of d and polynomially dependent on ε^{-1} then we have *strong tractability*. The number $n(\varepsilon, d)$ of samples is usually directly related to the cost of an algorithm, and therefore tractability means that we can solve the problem to within ε in cost polynomially dependent on ε^{-1} and d. The error ε of an algorithm can be defined in various settings including the worst case, average case, randomised or probabilistic settings.

This definition is similar† to the notion of tractability for discrete problems using the Turing machine as a model of computation. To prove the lack of the existence of polynomial-time algorithms for a number of discrete problems is at the core of theoretical computer science. This is the essence of the famous conjecture P$\neq$NP. According to Steve Smale,

---

† For discrete problems, d is replaced by the number of input data, and often we want to solve the problem exactly, i.e., with $\varepsilon = 0$. For many discrete problems, intractability is conjectured (often modulo the famous open problem that P$\neq$NP), whereas for high dimensional continuous problems intractability is often *proven*.

see [75], the question of whether P$\neq$NP holds is one of the three most important problems in mathematics‡.

However, we use the *real number model with oracles*† in our study, see [8, 9, 39, 49, 84, 86]. In this model we assume that we can exactly compute arithmetic operations over reals, comparisons of real numbers and function values as oracles. This model is typically used for numerical and scientific computations since it is an abstraction of floating point arithmetic in fixed precision. More about this model of computation can be found in [60, 61, 63, 94, 101].

The subject of tractability and strong tractability for high dimensional problems over the real number model with oracles was introduced in [99] and [100]. Today this topic is being thoroughly studied in information-based complexity and a number of characterisation results have been recently obtained, see [11, 29, 32, 51, 52, 59, 62, 72, 73, 90, 91, 93, 99, 100, 102].

The notion of strong tractability seems very demanding and one may suspect that only trivial problems are strongly tractable. Indeed, many problems defined over classical function spaces are not strongly tractable or even not tractable. Sometimes, the same problems become strongly tractable over function spaces with suitable weights that moderate the behaviour with respect to successive components. This subject will be explained in detail later.

To simplify the exposition we restrict ourselves primarily only to the very important problem of high dimensional integration in the worst case setting. For some classes of functions integration is related to various notions of *discrepancy*, which is widely used and studied in many areas such as number theory, approximation, stochastic analysis, combinatorics, ergodic theory and numerical analysis. The notion of discrepancy is related to Sobolev spaces, Wiener measure, VC dimension and Ramsey theory, see [5, 6, 19, 20, 35, 38, 46, 47, 83, 98].

---

‡ Smale writes: "In fact, included are what I believe to be the three greatest open problems of mathematics: the Riemann Hypothesis, the Poincaré Conjecture, and 'Does P = NP?' "

† Some problems can be analysed in the Turing machine model as well as in the real number model. Which model is more appropriate depends on the application. A problem can be tractable in one model and intractable in the other, because the assumptions concerning cost and size of a problem are quite different in both models. It is well known that the problem of linear optimization is tractable with respect to the Turing model. It is still unknown, and an important open problem, whether the same is true with respect to the real number model, see [87]. We stress that a conjecture of the type P$\neq$NP has been defined also over the real (or the complex) numbers, see [8] and [9].

This paper is a survey and proofs are only sketched or completely skipped. We will present a few classical results from the theory of numerical quadrature. Most results, however, are quite recent since most classical error estimates do not address the dependence on d. They contain unknown factors that depend on dimension and often increase exponentially with d. If we want to establish tractability, we must know how the error depends on d. Therefore we need to revisit all the classical problems and seek sharp estimates also in terms of d.

Tractability of multivariate integration obviously depends on the class of functions. We will consider a number of classes and present three types of results:

- Classes of functions for which multivariate integration is intractable and suffers the curse of dimensionality. This will be done by establishing exponential lower bounds in d on the number of function samples.
- Classes of functions for which multivariate integration is tractable or strongly tractable. This will be done by proving the *existence* of polynomial-time algorithms by non-constructive arguments.
- Classes of functions for which multivariate integration is tractable or strongly tractable. This will be done by *constructing* polynomial-time algorithms.

For problems belonging to the first case we *prove* intractability. This is in contrast to discrete complexity theory, where it is only *believed*, not proved, that certain problems are difficult (in the sense of exponential lower bounds).

An important area for future research is to move problems from the second to the third case. That is, to provide *constructive* algorithms for problems where we currently have only existence. Obviously, we are especially interested in the case when the construction is simple and does not involve extensive precomputations. Classical algorithms are often based on grids, and the computing time is exponential in the dimension for such algorithms. Only algorithms based on non-grids may have the computing time bounded by a polynomial in the dimension.

2 Multivariate integration and tractability

In this paper we survey results for the following multivariate integration problem. Let F_d be a normed space of integrable functions $f : D_d \to \mathbb{R}$ where $D_d \subset \mathbb{R}^d$. Let $\varrho_d : D_d \to \mathbb{R}_+$ be a weight function such that

$\int_{D_d} \varrho_d(t)\,dt = 1$. For $f \in F_d$ we want to approximate the multivariate integral

$$I_d(f) = \int_{D_d} f(t)\,\varrho_d(t)\,dt. \qquad (2.1)$$

Typical examples of such problems are *uniform* integration with $D_d = [0,1]^d$ and $\varrho_d(t) = 1$, and *Gaussian* integration with $D_d = \mathbb{R}^d$ and

$$\varrho_d(t) = (2\pi)^{-d/2}(\sigma_1 \cdots \sigma_d)^{-1} \exp(-t_1^2/(2\sigma_1^2) - \cdots - t_d^2/(2\sigma_d^2))$$

for given positive σ_i.

2.1 Algorithms and Errors

We approximate $I_d(f)$ by algorithms that use finitely many function values. We consider a number of classes of algorithms:

- The class QMC of *quasi-Monte Carlo* algorithms, which is widely used for financial problems. The QMC algorithms are of the form

$$Q_{n,d}(f) = \frac{1}{n} \sum_{i=1}^{n} f(t_i). \qquad (2.2)$$

  The sample points t_i are deterministic, belong to D_d, and may depend on n and d as well as on the weight ϱ_d and the space F_d. The sample points t_i are *nonadaptive*. That is, they are given a priori and do not depend on the integrand f. This is a desirable property for parallel or distributed computations, since the function values $f(t_i)$ can be computed simultaneously by different processors or computers. The weights of QMC algorithms are by definition all equal to $1/n$. This means that QMC algorithms integrate constant functions exactly.

- The class POS of algorithms for which the weights $1/n$ of QMC algorithms are replaced by *positive or non-negative* weights a_i. The POS algorithms are of the form

$$Q_{n,d}(f) = \sum_{i=1}^{n} a_i f(t_i), \qquad a_i \geq 0. \qquad (2.3)$$

  The coefficients a_i are, as the sample points t_i, deterministic and may depend on n, d, ϱ and F_d. Algorithms in the class POS having small error for constant functions are numerically stable, since $\sum_{i=1}^{n} a_i$ is close to 1.

- The class LIN is the class of *linear* algorithms of the form (2.3) with arbitrary real weights a_i. In this case, some of a_i's may be negative. The sample points t_i for this class satisfy the same conditions as for the classes QMC and LIN, i.e., they are nonadaptive. As we shall see the class LIN can sometimes be much more powerful than the class POS.

- The class ARB is the class of *nonlinear* algorithms that use *adaptive* sample points. That is, the sample point t_1 is selected based on our knowledge of the space F_d, and each sample point t_i for $i \geq 2$ may be now dependent on the previously computed function values $f(t_1), \ldots, f(t_{i-1})$. Thus, ARB algorithms are of the form

$$Q_{n,d}(f) = \varphi(f(t_1), f(t_2), \ldots, f(t_n)), \qquad (2.4)$$

where φ is an arbitrary (possible non-linear) mapping from $\mathbb{R}^n$ to $\mathbb{R}$, see e.g., [84] for more details.

The class ARB is the most general of the four. Since we do not restrict in any way how the adaptive choice of the sample points t_i is performed and how we combine the computed function values, one may be afraid that good error properties of the class ARB cannot be computationally realized. As we shall see, the class ARB, despite its generality, is *not* more powerful than the class LIN in many cases.

We are ready to define the error of an algorithm $Q_{n,d}$ for any of these classes. In this paper we restrict ourselves to the worst case error, which is defined by the worst case performance of $Q_{n,d}$ over the unit ball of F_d,

$$e(Q_{n,d}) = \sup_{f \in F_d, \|f\|_d \leq 1} |I_d(f) - Q_{n,d}(f)|. \qquad (2.5)$$

For $n = 0$ we do not sample the function and we† set $Q_{0,d} = 0$. Then

$$e(Q_{0,d}) = \sup_{f \in F_d, \|f\|_d \leq 1} |I_d(f)| = \|I_d\|$$

is the *initial error*. This is the a priori error in multivariate integration without sampling the function. We call $e(Q_{n,d})$ the *(absolute) error* of $Q_{n,d}$ and $e(Q_{n,d})/e(Q_{0,d})$ its *normalised error*.

For each class of algorithms we want to find $Q_{n,d}$ having minimal error. Let A be one of the class QMC, POS, LIN or ARB, and let

$$e(n, F_d, A) = \inf\{e(Q_{n,d}) : Q_{n,d} \in A\} \qquad (2.6)$$

---

† For $n = 0$ we could take $Q_{0,d} = c$ for some number c. It is easy to see that $c = 0$ minimises the error of $Q_{0,d}$.

be the minimal error of algorithms from the class A when we use n function values. Clearly,

$$e(0, F_d, A) = e(Q_{0,d}) = \|I_d\|, \quad A \in \{\text{QMC,POS,LIN,ARB}\}.$$

Since the classes are nested we also have

$$e(n, F_d, \text{ARB}) \leq e(n, F_d, \text{LIN}) \leq e(n, F_d, \text{POS}) \leq e(n, F_d, \text{QMC}).$$

Surprisingly enough, we have

$$e(n, F_d, \text{LIN}) = e(n, F_d, \text{ARB}). \tag{2.7}$$

That is, the worst case error is minimised by linear algorithms using non-adaptive sample points. This important result is due to Bakhvalov and Smolyak, see [3].

We first discuss the result of Bakhvalov, who proved that adaption does not help when we approximate linear functionals over balanced and convex sets†. Note that these assumptions obviously hold for multivariate integration (which is a linear functional) and for the unit ball of F_d (which is balanced and convex). It is also known that adaption does not help very much when we approximate linear operators over balanced and convex sets. In this case, adaption may help at most by a factor of 2, as proven in [23] and [86]; the factor 2 cannot be replaced by 1, see [34]. All the assumptions in the theorem about adaption are essential. That is, if the operator is not linear, the set is not balanced, or the set is not convex, then adaption may help; moreover adaption may be significantly better than non-adaption. The reader interested in this issue is referred to the survey [50].

The result of Smolyak is on the power of linear algorithms. He proved that if we approximate a linear functional over a balanced and convex set, then an arbitrary non-linear algorithm must have a worst case error at least as large as that of an appropriately chosen linear algorithm. His theorem has many different generalisations. This subject is beyond the scope of this paper and the reader is referred to [84, 85].

2.2 Tractability

In view of (2.7), we may restrict ourselves to the three classes QMC, POS and LIN. We now formally define tractability and strong tractability for the class $A \in \{\text{QMC,POS,LIN}\}$. We would like to reduce the initial

---

† A set B is balanced if $f \in B$ implies that $-f \in B$, and convex if $f_1, f_2 \in B$ implies that $tf_1 + (1-t)f_2 \in B$ for all $t \in [0,1]$.

error by a factor ε, where $\varepsilon \in (0,1)$. We are looking for the smallest $n = n(\varepsilon, F_d, A)$ for which there exists an algorithm from the class A such that $e(Q_{n,d}) \leq \varepsilon\, e(Q_{0,d})$. That is,

$$n(\varepsilon, F_d, A) = \min\{\, n : e(n, F_d, A) \leq \varepsilon\, e(Q_{0,d}) \,\}. \qquad (2.8)$$

We say that integration for a sequence $\{F_d\}$ of spaces is *tractable* (with respect to the normalised error) in the class A iff there exist nonnegative C, q and p such that

$$n(\varepsilon, F_d, A) \leq C\, d^q\, \varepsilon^{-p} \qquad \forall d = 1, 2, \dots, \forall \varepsilon \in (0,1). \qquad (2.9)$$

Tractability means that we can reduce the initial error† by a factor ε by using a number of function values which is polynomial in d and ε^{-1}. The smallest q and p (or the infima of q and p) in (2.9) are called the *d-exponent* and the *ε-exponent* of tractability in the class A.

If $q = 0$ in (2.9) then we say that integration for a sequence $\{F_d\}$ of spaces is *strongly tractable* in the class A. In this case, the number of samples is independent of d and depends polynomially on ε^{-1}. The smallest (or the infimum of) p in (2.9) is now called the *strong exponent* of tractability in the class A.

We stress that the minimal number $n(\varepsilon, F_d, A)$ of function samples is directly related to the complexity, which is the minimal cost of computing an approximation with error ε. Indeed, this is obvious for the class QMC. In this case, we need only to precompute the sample points $t_1, t_2, \dots, t_n$ with $n = n(\varepsilon, F_d, A)$. For any $f \in F_d$, we compute n function values at t_i, and then perform $n - 1$ additions and one division. For the classes POS and LIN we also need to precompute the weights a_i; for each $f \in F_d$, we will need to perform additionally at most n multiplications, instead of one division. The cost c of one function value is usually much greater than the cost of one arithmetic operation. Therefore, the cost of algorithms from the classes QMC, POS and LIN is roughly equal to $c\, n$. This means that the complexity is roughly equal to $c\, n(\varepsilon, F_d, A)$.

2.3 Hilbert Spaces

In general, it is hard to find a formula for the worst case error $e(Q_{n,d})$. One important exception is the case when the space F_d is a reproducing

---

† It is also of interest to study tractability for the absolute error instead of the normalised error. Then in (2.8) we replace $e(n, F_d, A) \leq \varepsilon\, e(Q_{0,d})$ by $e(n, F_d, A) \leq \varepsilon$. In general, tractability results for the absolute and the normalised error are different. We shall see such examples later.

kernel Hilbert space. The theory of such spaces can be found in, e.g., [1] and [88]. Here, we recall that a Hilbert space F_d is a reproducing kernel Hilbert space iff the linear functional $L(f) = f(t)$, $\forall f \in F_d$, is continuous for any $t \in D_d$. If so, then there exists a *reproducing kernel* $K_d : D_d \times D_d \to \mathbb{R}$ such that $K_d(\cdot, t) \in F_d$ for any $t \in D_d$ and

$$f(t) = \langle f, K_d(\cdot, t) \rangle \qquad \forall f \in F_d, \; \forall t \in D_d.$$

Here, $\langle \cdot, \cdot \rangle$ denotes the inner product of F_d, and $\|f\| = \langle f, f \rangle^{1/2}$. Let

$$h_d(x) = \int_{D_d} K_d(x, t) \varrho_d(t) \, dt \qquad \forall x \in D_d. \tag{2.10}$$

We assume that $h_d \in F_d$. Then the linear functional I_d given by (2.1) is continuous and

$$I_d(f) = \langle f, h_d \rangle.$$

The initial error is now given by

$$e(Q_{0,d}) = \|I_d\| = \|h_d\| = \left(\int_{D_d \times D_d} K_d(x, t) \varrho_d(x) \varrho_d(t) \, dt \, dx \right)^{1/2}. \tag{2.11}$$

It is also easy to check the well-known fact that the worst case error of $Q_{n,d}(f) = \sum_{i=1}^{n} a_i f(t_i)$ is of the form

$$e^2(Q_{n,d}) = e^2(Q_{0,d}) - 2 \sum_{i=1}^{n} a_i \int_{D_d} K_d(t, t_i) \varrho_d(t) \, dt + \sum_{i,j=1}^{n} a_i a_j K_d(t_i, t_j). \tag{2.12}$$

2.4 The Spaces $C^r([0,1]^d)$

We end this section by an example of a sequence $\{F_d\}$ of spaces for which uniform integration is intractable. This example is due Bakhvalov, see [2]. Let $F_d = C^r([0,1]^d)$ be the space of r times continuously differentiable functions with the norm

$$\|f\| = \max_{|\alpha| \le r} \|D^\alpha f\|_\infty.$$

Here $\alpha = [\alpha_1, \ldots, \alpha_d]$ is a multi-index of non-negative integers and $|\alpha| = \sum_{i=1}^{d} \alpha_i$. The operator D^α stands for partial differentiation,

$$D^\alpha f(x) = \frac{\partial^{|\alpha|} f}{\partial x_1^{\alpha_1} \cdots \partial x_d^{\alpha_d}}(x) \quad \text{with } x = [x_1, \ldots, x_d].$$

It is relatively easy to show that uniform integration

$$I_d(f) = \int_{[0,1]^d} f(x)\, dx, \quad f \in F_d,$$

is intractable. It suffices to show that there are positive numbers $c_{r,d}$ and $C_{r,d}$ such that

$$c_{r,d} \cdot \varepsilon^{-d/r} \le n(\varepsilon, F_d, \mathrm{LIN}) \le C_{r,d} \cdot \varepsilon^{-d/r} \quad \forall \varepsilon \in (0,1). \tag{2.13}$$

We sketch the idea of the proof. We consider algorithms of the form

$$Q_{n,d}(f) = \sum_{i=1}^{n} a_i\, f(t_i).$$

The upper bound on $n(\varepsilon, F_d, \mathrm{LIN})$ can be proved with product formulas, for example, we may take products of Gaussian rules or products of the Clenshaw–Curtis formulas. Then for $n = O(\varepsilon^{-d/r})$, the error is at most ε.

For the lower bound, we consider an arbitrary $Q_{n,d}$ with $2n = \ell^d$, where $\ell \in \mathbb{N}$ and n is of order $\varepsilon^{-d/r}$. We take the partition of $[0,1]^d$ into ℓ^d cubes having edge size ℓ^{-1}. Then there are at least n small cubes without a sample point, hence $Q_{n,d}(f)$ cannot depend on the behaviour of $f \in F_d$ over these unsampled cubes. We then can construct two functions $f_1, f_2 \in F_d$ such that they only differ on the unsampled cubes and for which

$$I_d(f_1) - I_d(f_2) \ge 2\varepsilon$$

Since $Q_{n,d}(f_1) = Q_{n,d}(f_2)$ we obtain

$$e(Q_{n,d}) \ge \varepsilon$$

and (2.13) is proven, see e.g. [48] for the details.

We comment on this typical result and its proof. For any d, we obtain the rate

$$n(\varepsilon, F_d, \mathrm{LIN}) \asymp \varepsilon^{-d/r}$$

constructively. In fact, it is easy to write down corresponding product rules with error at most ε in an explicit way. Nontrivial product rules need, however, at least 2^d sample points. As long as d is small this is not a problem. On the other hand, if $d = 100$ this can hardly be called a meaningful construction. This is where the curse of dimensionality strikes. Since r is fixed, the exponent d/r in (2.13) goes to infinity with d. Hence, $n(\varepsilon, d, \mathrm{LIN})$ does *not* depend polynomially on ε^{-1}; that is, the problem is intractable.

Assume now that the smoothness parameter r depends on d, say $r = d$. Then the power of ε^{-1} in (2.13) is just 1 and the dependence on ε^{-1} is polynomial. How about the dependence on d? To claim tractability it would be enough to prove that $C_{r,d}$ depends polynomially on d, and to claim intractability it would be enough to prove that $c_{r,d}$ is exponentially large in d. If one checks the proof of (2.13) more carefully then one will notice that, unfortunately, $C_{r,d}$ is exponentially large in d, and $c_{r,d}$ is exponentially small in d. In fact, the gap between $C_{r,d}$ and $c_{r,d}$ is something like

$$\frac{C_{r,d}}{c_{r,d}} \approx 5^{d-1}.$$

This is a typical situation, in which the classical analysis is not enough to claim either tractability or intractability. In fact, tractability of uniform integration for the class $C^r([0,1]^d)$ when $r = d$, or more generally, when $r/d \leq c$ for some positive c, is still open.

3 Discrepancy

The notion of *discrepancy* goes back to the work of Weyl in 1916, see [96], and van der Corput in the thirties, see [14] and [15].

Discrepancy is a quantitative measure of the uniformity of the distribution of points in the d-dimensional unit cube. Today we have various notions of discrepancy, and there are literally thousands of papers studying different aspects of discrepancy. Research on discrepancy is very intensive, and the reader is referred to the recent books [5, 6, 19, 38, 46, 71] and [81]. In this paper we limit ourselves to a few basic notions of discrepancy, and show their relation to uniform integration

$$I_d(f) = \int_{[0,1]^d} f(t)\,dt.$$

In Sections 3.1 to 3.3 we study the classical L_2-discrepancy, which is sometimes also called L_2-star discrepancy or L_2-star discrepancy with boundary condition. We first recall the definition of the L_2-discrepancy and then present its absolute and normalised bounds. In Section 3.4 we discuss the weighted L_2-discrepancy (without boundary conditions) and in Section 3.5 we explain its relation to errors for integration in weighted Sobolev classes. Then we discuss the L_p-star discrepancy for finite p, see Section 3.6, and the *-discrepancy, where $p = \infty$, see Section 3.7.

For all these discrepancies, boxes of the form $[0, x)$ are used as test-sets, see, for example, the classical L_2-discrepancy (3.2). One can avoid this unsymmetric situation and can study notions of symmetric discrepancy, see [30]. We discuss two such cases: the extreme or unanchored discrepancy in Section 3.8, and the centred discrepancy in Section 3.9.

3.1 L_2-Discrepancy

Let $x = [x_1, \ldots, x_d] \in [0,1]^d$. By the box $[0,x)$ we mean the set $[0,x_1) \times \cdots \times [0,x_d)$ whose (Lebesgue) volume is clearly $x_1 \cdots x_d$. For given points $t_1, \ldots, t_n \in [0,1]^d$, we approximate the volume of $[0,x)$ by the fraction of the points t_i which are in the box $[0,x)$. The error of such an approximation is

$$x_1 \cdots x_d - \frac{1}{n} \sum_{i=1}^{n} 1_{[0,x)}(t_i),$$

where $1_{[0,x)}(t_i)$ is the indicator (characteristic) function, which is equal to 1 if $t_i \in [0,x)$, and to 0 otherwise.

Observe that we use equal coefficients n^{-1} in the previous approximation scheme. As we shall see, this corresponds to QMC algorithms for integration. We generalise this approach by allowing arbitrary coefficients a_i instead of n^{-1}. That is, we approximate the volume of $[0,x)$ by the weighted sum

$$\mathrm{disc}(x) := x_1 \cdots x_d - \sum_{i=1}^{n} a_i 1_{[0,x)}(t_i). \tag{3.1}$$

The L_2-discrepancy of points $t_1, \ldots, t_n$ and coefficients $a_1, \ldots, a_n$ is just the L_2-norm of the error function (3.1),

$$\mathrm{disc}_2(\{t_i\}, \{a_i\}) = \left(\int_{[0,1]^d} \left(x_1 \cdots x_d - \sum_{i=1}^{n} a_i 1_{[0,x)}(t_i) \right)^2 dx \right)^{1/2}. \tag{3.2}$$

By direct integration, we have the explicit formula

$$\mathrm{disc}_2^2(\{t_i\}, \{a_i\}) = \tag{3.3}$$

$$\frac{1}{3^d} - \frac{1}{2^{d-1}} \sum_{i=1}^{n} a_i \prod_{k=1}^{d} (1 - t_{i,k}^2) + \sum_{i,j=1}^{n} a_i a_j \prod_{k=1}^{d} (1 - \max(t_{i,k}, t_{j,k})),$$

for the L_2-discrepancy, where $t_i = [t_{i,1}, \ldots, t_{i,d}]$.

Hence, $\mathrm{disc}_2(\{t_i\}, \{a_i\})$ can be computed with a cost of $O(dn^2)$ arithmetic operations. Faster algorithms for computing L_2-discrepancy for relatively small d have been found by Heinrich, see [28] and also [20].

The major problem of L_2-discrepancy is to find points $t_1, \ldots, t_n$ and coefficients $a_1, \ldots, a_n$ for which $\mathrm{disc}_2(\{t_i\}, \{a_i\})$ is minimised. Let

$$\overline{\mathrm{disc}}_2(n, d) = \inf_{t_1, \ldots, t_n} \mathrm{disc}_2(\{t_i\}, \{n^{-1}\})$$

and

$$\mathrm{disc}_2(n, d) = \inf_{t_1, \ldots, t_n, a_1, \ldots, a_n} \mathrm{disc}_2(\{t_i\}, \{a_i\})$$

denote the minimal L_2-discrepancy when we use n points in dimension d. For the minimal L_2-discrepancy $\overline{\mathrm{disc}}_2(n, d)$ we choose optimal t_i for coefficients $a_i = n^{-1}$ whereas for $\mathrm{disc}_2(n, d)$ we also choose optimal a_i.

Observe that for $n = 0$ we do not use the points t_i or the coefficients a_i, and obtain the initial L_2-discrepancy

$$\overline{\mathrm{disc}}_2(0, d) = \mathrm{disc}_2(0, d) = \left(\int_{[0,1]^d} x_1^2 \cdots x_d^2 \, dx \right)^{1/2} = 3^{-d/2}. \quad (3.4)$$

Hence, the initial L_2-discrepancy is exponentially small in d.

3.2 Bounds for the (absolute) L_2-Discrepancy

We now discuss bounds on the minimal (absolute) L_2-discrepancy, which are related to tractability of integration for the absolute error. For a fixed d, the asymptotic behaviour of the minimal L_2-discrepancy as a function of n is known. There exist positive numbers c_d and C_d such that

$$c_d \frac{\log^{(d-1)/2} n}{n} \le \mathrm{disc}_2(n, d) \le \overline{\mathrm{disc}}_2(n, d) \le C_d \frac{\log^{(d-1)/2} n}{n} \quad (3.5)$$

for all $n \ge 2$. The lower bound was proven by Roth in 1954 for $a_i = n^{-1}$, see [69], but essentially the same proof covers arbitrary a_i, see [12] and [13]. The upper bound was proven by Roth and Frolov in 1980, see [70] and [22].

The bound (3.5) justifies the definition of *low-discrepancy* sequences (and points which we do not cover here), for the coefficients a_i equal n^{-1}. Namely, the sequence $\{t_i\}$ is a low-discrepancy sequence if there is a positive number C_d such that

$$\mathrm{disc}_2(\{t_i\}, \{n^{-1}\}) \le C_d \frac{\log^d n}{n} \quad \forall n \ge 2. \quad (3.6)$$

That is, the L_2-discrepancy of low-discrepancy sequences enjoys almost the same asymptotic as the minimal L_2-discrepancy with the only difference being in the power of a logarithmic factor. The search for low-discrepancy sequences has been a very active research area, and many beautiful and deep constructions have been obtained. Such sequences usually bear the name of their authors. Today we know low-discrepancy sequences (and points) of Faure, Halton, Hammersley, Niederreiter, Sobol, and Tezuka, as well as (t, m, s) points and (t, m) nets, and lattice points as their counterparts for the periodic case, see [5, 6, 19, 38, 46, 71, 81]. There are also points and sequences satisfying (3.6) with more general coefficients a_i than n^{-1} in (3.1). An example is provided by hyperbolic points, see [80, 90], although in this case we have $\log^{1.5\,d} n$ instead of $\log^d n$ in (3.6). Explicit bounds on the L_2-discrepancy for hyperbolic points can be found in [90]. In particular, hyperbolic points $t_1, \ldots, t_n$ and coefficients $a_1, \ldots, a_n$ were constructed such that $\mathrm{disc}_2(\{t_i\}, \{a_i\}) \leq \varepsilon$ with

$$
\begin{aligned}
n \;\; &\leq \;\; 3.304 \left(1.77959 + 2.714 \frac{-1.12167 + \log \varepsilon^{-1}}{d-1} \right)^{1.5(d-1)} \frac{1}{\varepsilon} \\
&\leq \;\; 7.26 \left(\frac{1}{\varepsilon} \right)^{2.454}.
\end{aligned}
$$

The essence of (3.5) is that modulo a logarithmic factor, the L_2-discrepancy behaves asymptotically in n like n^{-1} independently of d. The power of the logarithmic factor is $(d-1)/2$ and as long as d is not too large this factor is negligible. On the other hand, if d is large, say $d = 360$ as in some financial applications, the factor $\log^{(d-1)/2} n$ is essential. Indeed, the function $n^{-1} \log^{(d-1)/2} n$ is increasing for $n \leq \exp\left((d-1)/2\right)$. The latter number for $d = 360$ is $\exp(179.5) \approx 9 \cdot 10^{77}$.

Obviously, it is impossible to use n so large, and therefore for large d, the good asymptotic behaviour of $\mathrm{disc}_2(n, d)$ is dubious for practical purposes.

Obviously if d is large, then the numbers c_d and C_d from (3.5) are also very important. We do not know much about them. However, we know that the asymptotic constant

$$
A_d \;=\; \limsup_{n \to \infty} \overline{\mathrm{disc}_2}(n, d) \, \frac{n}{\log^{(d-1)/2} n}
$$

is super-exponentially small in d.

For large d and a relatively small n, we need other estimates on $\mathrm{disc}_2(n, d)$. By a simple averaging argument with respect to t_i and for

$a_i = n^{-1}$, we have

$$\overline{\text{disc}}_2(n, d) \leq \left(\int_{[0,1]^{nd}} \text{disc}_2^2(\{t_i\}, \{n^{-1}\}) \, dt_1 \cdots dt_n \right)^{1/2} \leq \frac{2^{-d/2}}{n^{1/2}}.$$
(3.7)

The last estimate looks very promising since we have an exponentially small dependence on d through $2^{-d/2}$. However, we should keep in mind that even the initial L_2-discrepancy is $3^{-d/2}$ which is much smaller than $2^{-d/2}$ for large d. From the last estimate we can easily conclude by applying Chebyshev's inequality that for any number $c > 1$, the set of sample points

$$A_c = \left\{ (t_1, \ldots, t_n) \; : \; \text{disc}_2(\{t_i\}, \{n^{-1}\}) \leq c \, 2^{-d/2} n^{-1/2} \right\}$$

has Lebesgue measure at least $1 - c^{-2}$. Hence, for $c = 10$ we have a set of points $t_1, \ldots, t_n$ of measure at least 0.99 for which the L_2-discrepancy is at most $10 \cdot 2^{-d/2} n^{-1/2}$. Surprisingly enough, we still do *not* know how to construct such points. Of course, such points can be found computationally. Indeed, it is enough to take points $t_1, \ldots, t_n$ randomly, and compute their discrepancy with $a_i = n^{-1}$. If their discrepancy is at most $10 \cdot 2^{-d/2} n^{-1/2}$ we are done. If not, we repeat random selection of points $t_1, \ldots, t_n$. If random samples are chosen with uniform distribution, then after a few such selections we will get the desired points.

3.2.1 The Exponent of L_2-Discrepancy

We now briefly discuss the exponent p^* of the absolute L_2-discrepancy, see [92]. This is defined as the smallest number (or the infimum) of positive p for which there exists a positive C such that

$$\text{disc}_2(n, d) \leq C \, n^{-1/p} \qquad \forall n, d = 1, 2, \ldots. \qquad (3.8)$$

We stress that the last estimate holds for all n and d; hence both C and p do not depend on n and d. The bound $p^* \geq 1$ is obvious, since for $d = 1$ we have

$$\text{disc}_2(n, 1) = \Theta \left(\overline{\text{disc}}_2(n, 1) \right) = \Theta(n^{-1}).$$

For $a_i = n^{-1}$, it is proven in [37] that p in (3.8) must be at least 1.0669. This means that the case of arbitrary d is harder and the presence of the logarithmic factors in (3.5) cannot be entirely neglected.

The upper bound

$$p^* \leq 1.4779$$

is proven in [92]. The proof of this upper bound is *non-constructive*. The best constructive bound currently known is given in [90], and states that

$$\text{disc}_2(n, d) \leq \frac{2.244}{n^{0.408}} \qquad \forall\, n, d,$$

which corresponds to $p = 2.454$. This bound is obtained by hyperbolic points.

Recently, it was proven in [65] that using hyperbolic points or more generally nested sparse grids points leads to $p \geq 2.1933$. Hence, to obtain $p < 2.1933$ we must use other than sparse grids points.

There are two challenging problems concerning the exponent of L_2-discrepancy. The first is to find p^*, and the second is to construct points t_i for which (3.8) holds with $p < 2$.

3.3 Normalised L_2-Discrepancy

We now discuss bounds on the normalised L_2-discrepancy. By the normalised L_2-discrepancy we mean $\text{disc}_2(\{t_i\}, \{a_i\})/\text{disc}_2(0, d)$. That is, we normalise by the initial value of the L_2-discrepancy, which is $3^{-d/2}$. This case is directly related to tractability of integration when we reduce the initial error by a factor of ε. Similarly to (2.8), we define

$$\bar{n}(\varepsilon, d) \;=\; \min\{\, n \;:\; \overline{\text{disc}_2}(n, d) \leq \varepsilon\, \text{disc}_2(0, d)\,\}, \qquad (3.9)$$

$$n(\varepsilon, d) \;=\; \min\{\, n \;:\; \text{disc}_2(n, d) \leq \varepsilon\, \text{disc}_2(0, d)\,\} \qquad (3.10)$$

and ask whether $\bar{n}(\varepsilon, d)$ and $n(\varepsilon, d)$ behave polynomially in ε^{-1} and d. We wish to stress that polynomial bounds on the absolute value of the L_2-discrepancy, which we presented so far, are useless for the normalised case since we now have to compare the minimal L_2-discrepancy to $\varepsilon\, 3^{-d/2}$ instead of ε.

The problem how $\bar{n}(\varepsilon, d)$ and $n(\varepsilon, d)$ depend on d has been recently solved. First of all, notice that it directly follows from (3.4) and (3.7) that

$$\bar{n}(\varepsilon, d) \leq 1.5^d \, \varepsilon^{-2}. \qquad (3.11)$$

It was proven in [102], see also [73] and Section 4.1, that

$$\bar{n}(\varepsilon, d) \geq (9/8)^d\, (1 - \varepsilon^2). \qquad (3.12)$$

The bound (3.12) is also valid if $a_i \geq 0$, which corresponds to the assumption defining the class POS. Hence, we have exponential dependence on d.

For arbitrary a_i, it was proven in [62] that

$$1.0628^d \left(1 + o(1)\right) \leq n(\varepsilon, d) \leq 1.5^d \, \varepsilon^{-2} \qquad \text{as } d \to \infty, \qquad (3.13)$$

the lower bound holding for any fixed $\varepsilon < 1$. Hence in the general case, $n(\varepsilon, d)$ goes to infinity faster than any power of d and the normalised L_2-discrepancy is intractable. The upper bounds in (3.11) and in (3.13) coincide. We do not know whether arbitrary weights are better than positive or equal weights.

3.4 Weighted L_2-Discrepancy

We now discuss a slightly different L_2-discrepancy for functions without a boundary condition, see Section 3.5 and [30]. The *weighted* version of L_2-discrepancy was introduced in [73], and plays a major role for the study of tractability in weighted Sobolev spaces, as will be explained in the next subsection.

We are given a sequence $\gamma = \{\gamma_j\}$ of weights such that

$$\gamma_1 \geq \cdots \geq \gamma_j \geq \cdots \geq 0.$$

For any subset u of $U_d = \{1, 2, \ldots, d\}$, we define $\gamma_\emptyset = 1$ and $\gamma_u = \prod_{j \in u} \gamma_j$ for non-empty u. For a vector $x \in [0, 1]^d$, we denote x_u as the vector from $[0, 1]^{|u|}$, where $|u|$ is the cardinality of u, with the components of x whose indices are in u. For example, for $d = 5$ and $u = \{2, 4, 5\}$ we have $x_u = [x_2, x_4, x_5]$. Then $dx_u = \prod_{j \in u} dx_j$. By $(x_u, 1)$ we mean the vector from $[0, 1]^d$ with the same components as x for indices in u and with the rest of components being replaced by 1. For our example, we have $(x_u, 1) = [1, x_2, 1, x_4, x_5]$. Recall that for given points $t_1, \ldots, t_n$ and coefficients $a_1, \ldots, a_n$, the function $\mathrm{disc}_2(x_u, 1)$ is given by (3.1) and takes the form

$$\mathrm{disc}_2(x_u, 1) = \prod_{k \in u} x_k - \sum_{i=1}^n a_i 1_{[0, x_u)} \left((t_i)_u \right).$$

The *weighted* L_2-discrepancy is then defined as

$$\mathrm{disc}_{2,\gamma}(\{t_i\}, \{a_i\}) = \left(\sum_{u \subset U_d} \gamma_u \int_{[0,1]^{|u|}} \mathrm{disc}_2^2(x_u, 1) \, dx_u \right)^{1/2}. \qquad (3.14)$$

It is easy to obtain the explicit formula

$$\mathrm{disc}_{2,\gamma}^2(\{t_i\}, \{a_i\}) =$$

$$\sum_{u \subset U_d} \gamma_u \left(\frac{1}{3^{|u|}} - \frac{1}{2^{|u|-1}} \sum_{i=1}^{n} a_i \prod_{k \in u} (1 - t_{i,k}^2) + \right.$$

$$\left. \sum_{i,j=1}^{n} a_i a_j \prod_{k \in u} (1 - \max(t_{i,k}, t_{j,k})) \right)$$

for the weighted L_2-discrepancy. The classical (unweighted) case corresponds to $\gamma = 1$, i.e., $\gamma_j = 1$. Since

$$\sum_{u \subset U_d} 3^{-|u|} = \sum_{k=0}^{d} \binom{d}{k} 3^{-k} = (1 + 3^{-1})^d,$$

$$\sum_{u \subset U_d} \prod_{k \in u} \frac{1 - t_{i,k}^2}{2} = \prod_{k=1}^{d} \left(1 + \frac{1 - t_{i,k}^2}{2} \right),$$

we have

$$\mathrm{disc}_{2,1}^2(\{t_i\}, \{a_i\}) =$$

$$\left(\frac{4}{3} \right)^d - 2 \sum_{i=1}^{n} a_i \prod_{k=1}^{d} \frac{3 - t_{i,k}^2}{2} + \sum_{i,j=1}^{n} a_i a_j \prod_{k=1}^{d} (2 - \max(t_{i,k}, t_{j,k})).$$

The square of the weighted L_2-discrepancy is defined as the sum of 2^d terms. One might expect that the cost of computing $\mathrm{disc}_{2,\gamma}(\{t_i\}, \{a_i\})$ is exponential in d. Surprisingly, this is *not* the case. It is shown in [33] that the weighted discrepancy can be computed with cost of order $d n^2$.

As before, let

$$\overline{\mathrm{disc}}_{2,\gamma}(n, d) = \inf_{t_1, \ldots, t_n} \mathrm{disc}_{2,\gamma}(\{t_i\}, \{n^{-1}\}), \qquad (3.15)$$

$$\mathrm{disc}_{2,\gamma}(n, d) = \inf_{t_1, \ldots, t_n \; a_1, \ldots, a_n} \mathrm{disc}_{2,\gamma}(\{t_i\}, \{a_i\}) \qquad (3.16)$$

be the minimal weighted L_2-discrepancies. For $n = 0$ we obtain

$$\overline{\mathrm{disc}}_{2,\gamma}^2(n, d) = \mathrm{disc}_{2,\gamma}^2(n, d) = \sum_{u \subset U_d} \gamma_u 3^{-|u|} = \prod_{j=1}^{d} \left(1 + \frac{1}{3} \gamma_j \right).$$

Observe that for the unweighted case, $\gamma_j = 1$, we have $\mathrm{disc}_2(0, d) = (4/3)^d$, which is exponentially large in d. On the other hand, the initial weighted L_2-discrepancy is uniformly bounded in d iff $\sum_{j=1}^{\infty} \gamma_j < \infty$.

Let us define

$$\overline{n}_\gamma(\varepsilon, d) = \min\{ n : \overline{\mathrm{disc}}_{2,\gamma}(n, d) \leq \varepsilon \, \mathrm{disc}_{2,\gamma}(n, d) \}, \quad (3.17)$$

$$n_\gamma(\varepsilon, d) = \min\{ n : \mathrm{disc}_{2,\gamma}(n, d) \leq \varepsilon \, \mathrm{disc}_{2,\gamma}(n, d) \} \quad (3.18)$$

as the minimal n for which we reduce the initial weighted L_2-discrepancy by a factor of ε.

The following bounds on $\bar{n}_\gamma(\varepsilon, d)$ were proven in [73],

$$1.055^{s_d}\left(1 - \varepsilon^2\right) \le \bar{n}_\gamma(\varepsilon, d) \le 1.1836^{s_d}\varepsilon^{-2} - \varepsilon^{-2} \qquad \forall \varepsilon \in (0,1),\, d \ge 1. \tag{3.19}$$

Here, $s_d = \sum_{j=1}^{d} \gamma_j$ is the partial sum of the weights. From these estimates it is easy to determine when $\bar{n}_\gamma(\varepsilon, d)$ depends polynomially on d. It was proven in [62] that the case of arbitrary a_i leads to the same conclusion and we have:

- $n_\gamma(\varepsilon, d)$ is uniformly bounded in d iff $\sum_{j=1}^{\infty} \gamma_j < \infty$,

- $n_\gamma(\varepsilon, d)$ is polynomially bounded in d iff $\displaystyle\limsup_{d \to \infty} \frac{\sum_{j=1}^{d} \gamma_j}{\log d} < \infty$.

Hence, for the unweighted case, $\gamma_j = 1$, we do not have polynomial dependence on d. We obtain

$$1.0563^d\left(1 - \varepsilon^2\right) \le \bar{n}_\gamma(\varepsilon, d) \le 1.125^d\,\varepsilon^{-2}, \tag{3.20}$$

for the case of equal (or positive) weights and

$$1.0463^d\left(1 + o(1)\right) \le n_\gamma(\varepsilon, d) \le 1.125^d\,\varepsilon^{-2} \qquad \text{as } d \to \infty, \tag{3.21}$$

for arbitrary weights. The upper bound is proved by an average case argument, as in Section 3.3. The lower bound in (3.20) follows from the technique in [102], see Section 4.1. The lower bound in (3.21) is proved in [62], see Section 4.2.

On the other hand, for weights γ_j for which the series $s_\gamma = \sum_{j=1}^{\infty} \gamma_j$ is finite, there is no dependence on d. For example, take $\gamma_j = j^{-2}$. Then $s_\gamma = \pi^2/6$ and

$$\bar{n}_\gamma(\varepsilon, d) \le 0.3195\,\varepsilon^{-2}.$$

If $a := \limsup_{d \to \infty} \sum_{j=1}^{d} \gamma_j / \log d < \infty$ then there exists a positive C such that

$$\bar{n}_\gamma(\varepsilon, d) \le C\,d^{a/6}\varepsilon^{-2}.$$

The exponent p_γ^* of the weighted normalised L_2-discrepancy can be defined analogously to the exponent of the L_2-discrepancy. Namely, p_γ^* is defined as the smallest number (or infimum) of positive p for which there exists a positive C such that

$$n_\gamma(\varepsilon, d) \le C\,\varepsilon^{-p} \qquad \forall \varepsilon \in (0,1),\, d \ge 1.$$

Then

$$p_\gamma^* \text{ exists iff } \sum_{j=1}^\infty \gamma_j < \infty.$$

Furthermore, if $\sum_{j=1}^\infty \gamma_j < \infty$ then $p_\gamma^* \in [1,2]$. It is also known that $\sum_{j=1}^\infty \gamma_j^{1/2} < \infty$ implies that $p_\gamma^* = 1$. The proof is non-constructive and is given in [32]. It is not known whether the condition $\sum_{j=1}^\infty \gamma_j^{1/2} < \infty$ is necessary for $p_\gamma^* = 1$. We are inclined to believe that it is.

There is a constructive proof that $p_\gamma^* = 1$ in [93] under the assumption that $\sum_{j=1}^\infty \gamma_j^{1/3} < \infty$. The optimal exponent is achieved by a weighted Smolyak algorithm using hyperbolic points, see Section 6.

3.4.1 Limiting Discrepancy

We end this subsection by a remark on the *limiting* discrepancy which is formally defined for $d = \infty$, see [73]. For points $t_i^{(\infty)} = [t_{i,1}, t_{i,2}, \ldots,] \in [0,1]^\infty$ let $t_i^{(d)} = [t_{i,1}, \ldots, t_{i,d}] \in [0,1]^d$ denote their d-dimensional projections. Then $\mathrm{disc}_{2,\gamma}(\{t_i^{(d)}, \{n^{-1}\})$ is a non-decreasing function of d. The limiting discrepancy is then defined as

$$\mathrm{disc}_{2,\gamma}(\{t_i^{(\infty)}\}) = \lim_{d \to \infty} \mathrm{disc}_{2,\gamma}(\{t_i^{(d)}\}, \{n^{-1}\}).$$

We have

$$\mathrm{disc}_{2,\gamma}(\{t_i^{(\infty)}\}) < \infty \quad \text{iff} \quad \sum_{j=1}^\infty \gamma_j < \infty.$$

We stress that the last condition holds independently of the sample points $t_i^{(\infty)}$. We see once more that the condition $\sum_{j=1}^\infty \gamma_j < \infty$ is needed to have a finite limiting discrepancy. Error bounds for quasi-Monte Carlo algorithms in an infinite dimensional setting are also studied in [89].

3.5 L_2-Discrepancy and Errors for Integration

We now show how the different L_2-discrepancies are related to uniform integration for Sobolev spaces and the Wiener sheet measure.

3.5.1 Spaces with Boundary Values

Consider the classical Sobolev space $W_2^1([0,1])$ of absolute continuous univariate functions $f : [0,1] \to \mathbb{R}$ for which $f' \in L_2([0,1])$ and $f(0) = 0$.

The space $W_2^1([0,1])$ is equipped with the inner product

$$\langle f, g \rangle = \int_0^1 f'(x) g'(x) \, dx.$$

The Sobolev space $W_2^1([0,1])$ is a reproducing kernel Hilbert space with the reproducing kernel

$$K_1(x,t) = \min(x,t).$$

Take now F_d as the tensor product of $W_2^1([0,1])$,

$$F_d = W_2^{1,\dots,1}([0,1]^d) := W_2^1([0,1]) \otimes \cdots \otimes W_2^1([0,1]), \quad (d \text{ times}).$$

Then F_d is the Sobolev space of multivariate functions $f : [0,1]^d \to \mathbb{R}$ that are differentiable once with respect to each variable, and for which $f(x) = 0$ if at least one component of x is zero. It is a reproducing kernel Hilbert space with the reproducing kernel

$$K_d(x,t) = \prod_{j=1}^d \min(x_j, t_j), \qquad (3.22)$$

and the inner product is given by

$$\langle f, g \rangle = \int_{[0,1]^d} \frac{\partial^d f}{\partial x_1 \cdots \partial x_d}(x) \, \frac{\partial^d g}{\partial x_1 \cdots \partial x_d}(x) \, dx.$$

Since F_d is a Hilbert space and I_d is a continuous and linear functional we have by Riesz's theorem,

$$I_d(f) = \langle f, h_d \rangle, \quad \text{with} \quad h_d(x) = \int_{[0,1]^d} K_d(x,t) \, dt.$$

Clearly, the initial error satisfies $e(Q_{0,d}) = \|h_d\|$ and we have

$$e(Q_{0,d}) = \left(\int_{[0,1]^{2d}} K_d(x,t) \, dx \, dt \right)^{1/2} = 3^{-d/2}.$$

Hence, the initial error is the same as the initial L_2-discrepancy.

Let $Q_{n,d}(f) = \sum_{i=1}^n a_i f(t_i)$ be an algorithm for approximating $I_d(f)$. Consider now the worst case error of $Q_{n,d}$, which is given by (2.12) with K_d defined by (3.22). By simple integration we get

$$e^2(Q_{n,d}) = \frac{1}{3^d} - 2 \sum_{i=1}^n a_i \prod_{k=1}^d (t_{i,k} - t_{i,k}^2/2) + \sum_{i,j=1}^n a_i a_j \prod_{k=1}^d \min(t_{i,k}, t_{j,k}).$$

Since for any numbers a, b we have

$$\min(a, b) = 1 - \max(1 - a, 1 - b) \quad \text{and} \quad a - a^2/2 = \left(1 - (1 - a)^2\right)/2,$$

comparing the last formula with the L_2-discrepancy formula (3.1), we immediately obtain

$$e(Q_{n,d}, F_d) = \text{disc}_2(\{1 - t_i\}, \{a_i\}),$$

where $1 - t_i = [1 - t_{i,1}, \ldots, 1 - t_{i,d}]$. In this way, we re-proved the classical Koksma–Hlawka inequality, see e.g., [19] and [46], which says that the worst case error of $Q_{n,d}$ is equal to the L_2-discrepancy of points $1 - t_i$ and coefficients a_i. Clearly, we also have

$$e\left(n, W_2^{1,\ldots,1}([0, 1]^d), \text{QMC}\right) = \overline{\text{disc}}_2(n, d), \qquad (3.23)$$

$$e\left(n, W_2^{1,\ldots,1}([0, 1]^d), \text{LIN}\right) = \text{disc}_2(n, d).$$

From this and from the absolute bounds on the L_2-discrepancy we conclude that integration for the Sobolev space $F_d = W_2^{1,\ldots,1}([0, 1]^d)$ is *strongly tractable* with the exponent at most 1.4779 for the absolute error. On the other hand, from the normalised bounds on the L_2-discrepancy we conclude that integration for $F_d = W_2^{1,\ldots,1}([0, 1]^d)$ is *intractable* for the normalised error, i.e., for reduction of the initial error. This shows the drastic difference between tractability with respect to absolute and normalised errors.

3.5.2 Average Errors

The L_2-discrepancy is also related to the *average* case error of algorithms $Q_{n,d}$ as shown in [98]. Namely, let $F_d = C([0, 1]^d)$ be the space of continuous functions f defined over $[0, 1]^d$. The space is equipped with the Wiener sheet measure μ which is the zero mean Gaussian measure with the covariance function

$$K_\mu(x, t) := \int_{C([0,1]^d)} f(x) f(t) \, \mu(df) = \prod_{j=1}^{d} \min(x_j, t_j).$$

That is, the covariance function K_μ is the same as the reproducing kernel of the space $W_2^{1,\ldots,1}([0, 1]^d)$. Then $f(x) = 0$ with probability one if at least one component of x is zero. We have

$$\left(\int_{C([0,1]^d)} (I_d(f) - Q_{n,d}(f))^2 \, \mu(df)\right)^{1/2} = \text{disc}_2(\{1 - t_i\}, \{a_i\}).$$

Hence, the bounds on L_2-discrepancy can be also used to study integration and its tractability in the average case setting.

The relation between L_2-discrepancy and the average case setting is an example of a general relation between the worst case setting for the unit ball of a Hilbert space $H(K)$ with a reproducing kernel K and the average case setting of a Banach space equipped with a Gaussian measure with covariance function K. This subject is beyond the scope of this paper and the interested reader is referred to [68, 84].

3.5.3 Spaces without Boundary Values and Weighted Norms

So far, we have assumed that functions satisfy boundary conditions $f(x) = 0$ for x with at least one component equal to zero. We now discuss a more general weighted case without such boundary conditions. As before, we start with univariate functions, but we now drop the assumption that $f(0) = 0$. By $W_{2,\gamma}^1([0,1])$ we denote the space of absolute continuous univariate functions $f : [0,1] \to \mathbb{R}$ for which $f' \in L_2([0,1])$. The inner product is given by

$$\langle f, g \rangle = f(0) g(0) + \gamma^{-1} \int_0^1 f'(x) g'(x) dx.$$

Here γ is a positive parameter. The case $\gamma = 1$ corresponds to the classical Sobolev space. The parameter γ is not important for the univariate case. For the multivariate case we will take the parameters γ_j corresponding to successive components and their role will be crucial, as was for the weighted L_2-discrepancy.

The space $W_{2,\gamma}^1([0,1])$ is a reproducing kernel Hilbert space with the kernel

$$K_{1,\gamma}(x,t) = 1 + \gamma \min(x,t).$$

For the multivariate case, we take

$$\gamma_1 \geq \cdots \geq \gamma_d > 0,$$

and consider $F_{d,\gamma}$ as the tensor product of $W_{2,\gamma_i}^1([0,1])$,

$$F_{d,\gamma} = W_{2,\gamma}^{1,\dots,1}([0,1]^d) := W_{2,\gamma_1}([0,1]) \otimes \cdots \otimes W_{2,\gamma_d}([0,1]).$$

The space $W_{2,\gamma}^{1,\dots,1}([0,1]^d)$ is a reproducing kernel Hilbert space with the kernel

$$K_{d,\gamma}(x,t) = \prod_{k=1}^d \left(1 + \gamma_k \min(x_k, t_k)\right).$$

The inner product in the space $W_{2,\gamma}^{1,\dots,1}([0,1]^d)$ is given by

$$\langle f, g \rangle = \sum_{u \in U_d} \gamma_u^{-1} \int_{[0,1]^{|u|}} \frac{\partial^{|u|} f}{\partial x_u}(x_u, 0) \frac{\partial^{|u|} g}{\partial x_u}(x_u, 0) \, dx_u.$$

For $\gamma_j = 1$ we obtain the classical (tensor product) Sobolev spaces whereas for decreasing γ_j's the weighted spaces consist of functions with diminishing dependence on the jth variable.

It now comes as no surprise that for $Q_{n,d}(f) = \sum_{i=1}^n a_i f(t_i)$ we have

$$e(Q_{n,d}) = \mathrm{disc}_{2,\gamma}(\{1 - t_i\}, \{a_i\}).$$

This immediately leads to

$$e(n, F_{d,\gamma}, \mathrm{QMC}) = \overline{\mathrm{disc}}_{2,\gamma}(n, d), \quad e(n, F_{d,\gamma}, \mathrm{LIN}) = \mathrm{disc}_{2,\gamma}(n, d).$$
$$(3.24)$$

From the bounds on the weighted L_2-discrepancy presented in the previous subsection, we conclude that

- integration for $W_{2,\gamma}^{1,\dots,1}([0,1]^d)$ is strongly tractable (for the normalised error) iff

$$\sum_{j=1}^{\infty} \gamma_j < \infty.$$

If $\sum_{j=1}^{\infty} \gamma_j < \infty$ then the strong exponent $p^* \in [1,2]$ and $\sum_{j=1}^{\infty} \gamma_j^{1/2} < \infty$ implies that $p^* = 1$.

- integration for $W_{2,\gamma}^{1,\dots,1}([0,1]^d)$ is tractable (for the normalised error) iff

$$a := \limsup_{d \to \infty} \frac{\sum_{j=1}^d \gamma_j}{\log d} < \infty.$$

If $a < \infty$ then the d-exponent of tractability is at most $a/6$ and the ε-exponent is at most 2.

We stress that as long as $\sum_{j=1}^{\infty} \gamma_j < \infty$ then absolute and normalised errors are essentially the same, and we have strong tractability in both cases. Also if $\sum_{j=1}^d \gamma_j$ grows proportionally to $\log d$, we obtain tractability for both cases. The only difference may be in the d-exponent of tractability.

3.6 L_p-Star Discrepancy

The L_p-star discrepancy is intimately related to the worst case error of multivariate integration for the Sobolev class of functions that are once differentiable in each variable with finite L_q-norm, $1/p + 1/q = 1$, see [19] and [46].

Recall that the L_p-star discrepancy of points $t_1, \ldots, t_n \in [0,1]^d$ is defined by

$$\operatorname{disc}_p^*(t_1, \ldots, t_n) = \left(\int_{[0,1]^d} \left| x_1 \cdots x_d - \frac{1}{n} \sum_{i=1}^n 1_{[0,x)}(t_i) \right|^p dx \right)^{1/p}, \quad (3.25)$$

for $1 \leq p < \infty$, and

$$\operatorname{disc}_\infty^*(t_1, \ldots, t_n) = \sup_{x \in [0,1]^d} \left| x_1 \cdots x_d - \frac{1}{n} \sum_{i=1}^n 1_{[0,x)}(t_i) \right| \quad (3.26)$$

for $p = \infty$. It is customary to denote the L_∞-star discrepancy as the $*$-discrepancy. For $n = 0$, the L_p-star discrepancy $\operatorname{disc}_p^*(0, d)$ is the initial error of multivariate integration without sampling the function. We have $\operatorname{disc}_p^*(0, d) = (p+1)^{-d/p}$, which goes to zero with d exponentially fast for all finite p. This may indicate that the multivariate integration problem is not properly scaled for finite p. For $p = \infty$, this problem disappears since we have $\operatorname{disc}_\infty^*(0, d) = 1$.

The usual bounds on the L_p-star discrepancy are for a fixed dimension d and large n. It is well known that the asymptotic behaviour of $\operatorname{disc}_p^*(n, d)$ with respect to n is of order at most $n^{-1}(\log n)^{d-1}$, see once more [19] and [46]. As before, points for which the L_p-star discrepancy has a bound proportional to $n^{-1}(\log n)^d$ are called low discrepancy points. There is a deep and still evolving theory how to construct such low discrepancy points. This theory is mostly due to Niederreiter and his collaborators.

In this subsection we discuss the L_p-star discrepancy for uniformly distributed points. We consider only even p and define the average L_p-star discrepancy as

$$\operatorname{av}_p(n, d) = \left(\int_{[0,1]^{nd}} \operatorname{disc}_p^*(t_1, \ldots, t_n)^p \, dt \right)^{1/p}, \qquad t = (t_1, \ldots, t_n).$$

As shown in [29], the average L_p-star discrepancy depends on the Stirling numbers $s(k, i)$ of the first kind, and $S(k, i)$ of the second kind, see [67]. We have

$$\mathrm{av}_p(n,d)^p = \sum_{r=p/2}^{p-1} C(r,p,d)\cdot n^{-r}, \qquad (3.27)$$

where

$$C(r,p,d) = (-1)^r \sum_{i=0}^{p-r} \binom{p}{r+i}(-1)^i \sum_{k=i}^{i+r}(p+1-r+k-i)^{-d}s(k,i)S(i+r,k).$$

$$(3.28)$$

Furthermore,

$$|C(r,p,d)| \le \frac{(r+1)(4p)^p}{(p+1-r)^d},$$

and

$$\mathrm{av}_p(n,d) \le 4\sqrt{2}\, p\,(1+p/2)^{-d/p}\,n^{-1/2}\left(\sum_{i=0}^{p/2-1} n^{-i}\left(\frac{1+p/2}{1+p/2-i}\right)^d\right)^{1/p}$$

3.7 *-Discrepancy

For $p = 2$, there are a number of negative results for the (normalised) discrepancy, see Section 3.3. The question of dependence on d for $p = \infty$ was raised by [36] who asked whether there exists an $a > 1$ such that $\mathrm{disc}_\infty^*(\lceil a^d\rceil, d)$ tends to 1 as d goes to infinity, and also asked whether, in particular, $\mathrm{disc}_\infty^*(2^d, d)$ goes to 1 as d goes to infinity. Based on the results for $p = 2$ one may be inclined to believe that the answer to at least one of these questions is affirmative.

It was surprising for us that this is *not* the case and that a positive result holds. In [29], it is proven that $\mathrm{disc}_\infty^*(n, d)$ depends only *polynomially* on d and n^{-1}, establishing the upper bound

$$\mathrm{disc}_\infty^*(n,d) \le C\,d^{1/2}\,n^{-1/2} \qquad \forall n,d = 1,2,\dots, \qquad (3.29)$$

with an unknown multiplicative factor C.

Hence, the *-discrepancy depends at most on $d^{1/2}$. We shall see later that this dependence on d cannot be improved. We do not know whether the dependence on n in (3.29) is sharp. As already mentioned, the asymptotic behaviour of the *-discrepancy on n (for a fixed d) is much better, but it does not necessarily mean that the uniform bound which is valid for all d and n cannot be of order $n^{-1/2}$. This problem is open and seems very difficult.

The proof of (3.29) follows directly from deep results from the theory of empirical processes. In particular, we use a result of [79] combined with a result of [27], and the Vapnik–Červonenkis dimension of the family of rational cubes $[0, x)$, see e.g., [18]. The proof is non-constructive, and we do not know for which points the bound (3.29) holds.

The slightly worse upper bound

$$\text{disc}^*_\infty(n, d) \leq 2\sqrt{2}\, n^{-1/2} \left(d \log \left(\left\lceil \frac{d n^{1/2}}{2(\log 2)^{1/2}} \right\rceil + 1 \right) + \log 2 \right)^{1/2}$$

(3.30)

follows from Hoeffding's inequality and is quite elementary. Also this proof is non-constructive, but we know that many points almost satisfy the bounds (3.29) and (3.30), see [29] for details. It would be very interesting to *construct* such points.

One can also use the results on the average behaviour of L_p-star discrepancy for an even integer p to obtain upper bounds for the *-discrepancy, see [29]. For concrete values of d and ε, these upper bounds seem to be better than those of (3.30).

For the lower bounds, we extend the formula (3.26) by allowing arbitrary a_i's instead of n^{-1}, defining

$$d^*_\infty(t_1, \ldots t_n; a_1, \ldots, a_n) = \sup_{x \in [0,1]^d} \left| x_1 \cdots x_d - \sum_{i=1}^n a_i 1_{[0,x)}(t_i) \right|$$

and

$$n_\infty(d, \varepsilon) = \min \left\{ n \; : \; \inf_{t_i, a_i} d^*_\infty(t_1, \ldots t_n; a_1, \ldots, a_n) \leq \varepsilon \right\}.$$

Of course,

$$n_\infty(d, \varepsilon) \leq n^*_\infty(d, \varepsilon).$$

In [29], we prove that there exists a positive number c such that for all d and all $\varepsilon \in (0, 1/64]$,

$$n_\infty(d, \varepsilon) \geq c\, d \log \varepsilon^{-1}. \tag{3.31}$$

From (3.29) and (3.31) we conclude that the *-discrepancy depends linearly on the dimension. We also prove that for any $\lambda \in (0, 1)$ there exists a positive c_λ such that

$$n_\infty(d, \varepsilon) \geq c_\lambda\, d^\lambda\, \varepsilon^{-(1-\lambda)}. \tag{3.32}$$

3.8 Extreme or Unanchored Discrepancy

We already discussed several different discrepancies but always considered boxes of the form $[0, x)$ as our test-sets, see, for example, the classical L_2-discrepancy (3.2). A more symmetric notion of discrepancy can be defined using all boxes $[x, y)$ for $x, y \in [0, 1]^d$ with $x \le y$ (in each coordinate). Indeed, instead of (3.1) we consider weighted sums

$$\text{disc}(x, y) = (y_1 - x_1) \cdots (y_d - x_d) - \sum_{i=1}^{n} a_i 1_{[x, y)}(t_i). \qquad (3.33)$$

The *extreme or unanchored L_2-discrepancy* of points $t_1, \ldots, t_n$ and coefficients $a_1, \ldots, a_n$ is just the L_2-norm of the error function (3.33),

$$\text{disc}_2^{\text{ex}}(\{t_i\}, \{a_i\}) = \left(\int_{[0,1]^{2d}, \ x \le y} \text{disc}^2(x, y) \, dx \, dy \right)^{1/2}. \qquad (3.34)$$

By direct integration, we have the explicit formula for the extreme L_2-discrepancy,

$$\text{disc}_2^{\text{ex}}(\{t_i\}, \{a_i\})^2 = \frac{1}{12^d} - \frac{1}{2^{d-1}} \sum_{i=1}^{n} a_i \prod_{k=1}^{d} t_{i,k}(1 - t_{i,k}) \qquad (3.35)$$

$$+ \sum_{i=1}^{n} \sum_{j=1}^{n} a_i a_j \prod_{k=1}^{d} (\min(t_{i,k}, t_{j,k}) - t_{i,k} t_{j,k}).$$

Morokoff and Caflisch introduce the discrepancy (3.34) in [41] and notice that the "L_2-discrepancy over all rectangles had not been previously defined nor used". These authors prefer the new discrepancy because it is "symmetric" and does not prefer a particular vertex, like the classical star discrepancy. So far it was not known that this kind of discrepancy is also directly related to numerical integration, i.e., if it is an error bound for a suitable class of functions. We will see that the extreme discrepancy is the error for periodic functions with a boundary condition.

We begin, however, with periodic functions without a boundary condition and summarise the analysis of [30] and [31] for this case. For $d = 1$ and $p = q = 2$, we consider the Sobolev space of periodic functions $F_{1,2} = \{f \in W_2^1([0, 1]) \mid f(0) = f(1)\}$ with the norm

$$\|f\|^2 = f(1)^2 + \|f'\|_{L_2}^2.$$

This is a rank-1 modification of the space $W_2^1([0, 1])$ and we obtain the kernel

$$K_1(x,y) = 1 + (\min(x,y) - xy).$$

The analysis can be extended, by Hölder's inequality, to arbitrary q. The norm in $F_{1,q}$ is given by $\|f\|^q = |f(1)|^q + \|f'\|_{L_q}^q$. For arbitrary $d > 1$, we define $F_{d,q}$ by tensor products of factors $F_{1,q}$ with tensor product norms.

We now discuss the respective error (or discrepancy) of a QMC method $Q_{n,d}$ for the space $F_{d,2}$. The error of $Q_{n,d}$ is given by

$$e(Q_{n,d}) = \left(\sum_{u \subset S} \int_{[0,1)^{2|u|}, \, x_u \leq y_u} \mathrm{disc}^2\left((x_u, 0), (y_u, 1)\right) dx_u dy_u \right)^{1/2}.$$

For arbitrary $Q_{n,d} \in \mathrm{LIN}$, we obtain

$$e(Q_{n,d})^2 = \frac{13^d}{12^d} - \sum_{i=1}^n 2a_i \prod_{k=1}^d \left(1 + \frac{t_{i,k}(1 - t_{i,k})}{2} \right) +$$

$$\sum_{i=1}^n \sum_{j=1}^n a_i a_j \prod_{k=1}^d (1 + \min(t_{i,k}, t_{j,k}) - t_{i,k} t_{j,k}).$$

Again we can modify this to spaces with a boundary condition. We start with $d = 1$ and $p = q = 2$, and we take the space $F_{1,2} = \{f \in W_2^1([0,1]) \mid f(0) = f(1) = 0\}$. The kernel for this subspace is given by

$$K_1(x,y) = \min(x,y) - xy.$$

For $d > 1$, we use tensor product kernels and norms. For $p = 2$, the error $e(Q_{n,d})$ of any $Q_{n,d} \in \mathrm{LIN}$ is

$$e(Q_{n,d}) = \left(\int_{[0,1)^{2d}, \, x \leq y} \mathrm{disc}^2(x,y) \, dx \, dy \right)^{1/2} = \mathrm{disc}_2^{\mathrm{ex}}(\{t_i\}, \{a_i\}). \tag{3.36}$$

Hence, we see that the L_2-extreme discrepancy is also an error bound, this time for a class of periodic functions with a boundary condition. To prove the error bound (3.36) for the kernel

$$K_d(x,y) = \prod_{j=1}^d \left(\min(x_j, y_j) - x_j y_j \right),$$

we can simply use the general result (2.12), together with formula (3.35). We end this subsection with a note on the classical *extreme discrepancy* for the space $F_{d,1}$, which corresponds to $p = \infty$. For $d = 1$, the norm of $F_{1,1}$ is given by $\|f\| = 2\|f'\|_1$. We have

$$\text{disc}^{\text{ex}}_\infty(\{t_i\}, \{a_i\}) = \sup_{x \leq y} |\text{disc}(x, y)|. \tag{3.37}$$

The extreme discrepancy (3.37) is tractable, see [29]. It is enough to use equal weights $1/n$ for which we have an upper bound of the form

$$\inf_{t_1, \ldots, t_n} \text{disc}^{\text{ex}}_\infty(\{t_i\}, \{1/n\}) \leq C \cdot d^{1/2} \cdot n^{-1/2}, \tag{3.38}$$

where the positive C does not depend on n or d. This bound is the same as for the star discrepancy, we only may have a larger constant C.

3.9 Centred Discrepancy

We now consider the space $F_1 = W^1_{1/2}([0,1])$ which is the Sobolev space of absolutely continuous functions whose first derivatives are in $L_2([0,1])$ and whose function values are zero at the point $1/2$. That is,

$$F_1 = \{f : [0,1] \to \mathbb{R} : f(1/2) = 0, \ f \text{ abs. cont. and } f' \in L_2([0,1])\}$$

with the inner product $\langle f, g \rangle_{F_1} = \int_0^1 f'(t)g'(t)\,dt$. It can be checked that this Hilbert space has the reproducing kernel

$$K_1(x,t) = 1_M(x,t) \cdot \min(|x - 1/2|, |t - 1/2|), \tag{3.39}$$

where $M = [0, 1/2] \times [0, 1/2] \cup [1/2, 1] \times [1/2, 1]$. We may also write

$$K_1(x,t) = \tfrac{1}{2}(|x - 1/2| + |t - 1/2| - |x - t|). \tag{3.40}$$

The kernel K_1 is *decomposable* in the sense of [62], i.e., $K(x,t) = 0$ for all $x \leq 1/2 \leq t$.

For $d > 1$, we obtain $F_d = W^{1,1,\ldots,1}_{1/2}([0,1]^d) = W^1_{1/2}([0,1]) \otimes \cdots \otimes W^1_{1/2}([0,1])$, d times, as the Sobolev space of smooth functions f defined over $D_d = [0,1]^d$ such that $f(x) = 0$ if at least one component of x is $1/2$. The inner product of F_d is given by

$$\langle f, g \rangle_{F_d} = \int_{[0,1]^d} \frac{\partial^d}{\partial x_1 \cdots \partial x_d} f(x) \frac{\partial^d}{\partial x_1 \cdots \partial x_d} g(x)\,dx.$$

It can be checked, see [62], that the error of an algorithm $Q_{n,d}(f) = \sum_{i=1}^n a_i f(z_i)$ is now given by

$$e^2(Q_{n,d}) = \tag{3.41}$$

$$\int_{[0,1]^d} \left| \prod_{j=1}^d \min(x_j, 1 - x_j) - \sum_{i=1}^n a_i \cdot 1_{J(b(x),x)}(z_i) \right|^2 dx,$$

where $J(b(x), x)$ is the cube generated by x and the vertex $b(x)$ of $[0,1]^d$ that is closest, in the sup-norm, to x. Essentially the same formulas are presented by [30], who considers spaces similar to F_d without assuming the condition $f(\frac{1}{2}) = 0$ and considers algorithms with $a_i = 1/n$. The error of the algorithm $Q_{n,d}$ with $a_i = 1/n$ is called by Hickernell the *centred discrepancy* and denoted by $d_2^c(Q_{n,d})$. We will be using the same name for arbitrary coefficients a_i. For the space F_d with the condition $f(\frac{1}{2}) = 0$ we denote the centred discrepancy by $\tilde{d}_2^c(n, d)$. Later we remove this condition and consider the centred discrepancy $d_2^c(Q_{n,d})$ (without the tilde), as originally studied by [30].

The minimal centred discrepancy is defined as

$$\tilde{d}_2^c(n, d)^2 = \tag{3.42}$$

$$\inf_{a_i, z_i, i=1,2,\dots,n} \int_{[0,1]^d} \left| \prod_{j=1}^d \min(x_j, 1 - x_j) - \sum_{i=1}^n a_i \cdot 1_{J(b(x), x)}(z_i) \right|^2 dx.$$

In [62] we prove the lower bound

$$\tilde{d}_2^c(n, d) \geq \left(1 - n\, 2^{-d}\right)_+^{1/2} \tilde{d}_2^c(0, d) \quad \text{with} \quad \tilde{d}_2^c(0, d) = 12^{-d/2}. \tag{3.43}$$

The centred discrepancy (and uniform integration) may be also considered for the space $L_q([0,1]^d)$ with $q \in [1, \infty]$. More precisely, for $d = 1$ we take the space $F_{1,q}$ as the Sobolev space of absolutely continuous functions whose first derivatives are in $L_q([0,1])$ and that vanish at $\frac{1}{2}$. The norm in $F_{1,q}$ is given by $\|f\|_{F_{1,q}} = \left(\int_0^1 |f'(t)|^q dt\right)^{1/q}$ for $q < \infty$, and $\|f\|_{F_{1,\infty}} = \operatorname{ess\,sup}_{t \in [0,1]} |f'(t)|$ for $q = \infty$. For $d > 1$, the space $F_{d,q}$ is taken as a tensor product of factors $F_{1,q}$. Then functions from $F_{d,q}$ vanish at x whenever at least one component of x is $\frac{1}{2}$ and their norm is

$$\|f\|_{F_{d,q}} = \|D^{\vec{1}} f\|_{L_q([0,1]^d)} = \left(\int_{[0,1]^d} |D^{\vec{1}} f(x)|^q dx\right)^{1/q},$$

where $\vec{1} = [1, 1, \dots, 1]$ and $D^{\vec{1}} = \partial^d / \partial x_1 \cdots \partial x_d$. We have

$$I_d(f) - Q_{n,d}(f) = \int_{[0,1]^d} D^{\vec{1}} f(t)\, D^{\vec{1}} \left(h_d - \sum_{i=1}^n a_i K_d(\cdot, z_i) \right)(t)\, dt,$$

with

$$h_d(x) = 2^{-d} \prod_{j=1}^d \left(|x_j - \tfrac{1}{2}| - |x_j - \tfrac{1}{2}|^2 \right),$$

$$K_d(x,t) \;=\; 2^{-d} \prod_{j=1}^{d} \left(|x_j - \tfrac{1}{2}| + |t_j - \tfrac{1}{2}| - |x_j - t_j| \right).$$

From this we conclude that

$$e(Q_{n,d}) \;:=\; \sup_{f \in F_{d,q}: \|f\|_{F_{d,q}} \le 1} \left| I_d(f) - Q_{n,d}(f) \right| \;=\; \tilde{d}_p^c(Q_{n,d}),$$

where $1/p + 1/q = 1$ and $\tilde{d}_p^c(Q_{n,d})$ is the centred p-discrepancy given by

$$\tilde{d}_p^c(Q_{n,d}) = \left(\int_{[0,1]^d} \left| \prod_{j=1}^{d} \min(x_j, 1 - x_j) - \sum_{i=1}^{n} a_i \cdot 1_{J(b(x),x)}(z_i) \right|^p dx \right)^{1/p}.$$

If $q = 1$ then $p = \infty$ and, as usual, the integral is replaced by the essential supremum in the formula above.

Let $e(n, F_{d,q}) = \tilde{d}_p(n,d)$ denote the minimal error, or equivalently the minimal centred p-discrepancy, that can be achieved by using n function values. The initial error, or the initial centred discrepancy, is now given by

$$e(0, F_{d,q}, \mathrm{LIN}) = \tilde{d}_p^c(0,d) = \frac{2^{-d}}{(p+1)^{d/p}}$$

for $q > 1$, and $e(0, F_{d,1}) = 2^{-d}$. Hence, for all values of q, the initial centred discrepancy is at most 2^{-d}.

The following result is from [62]. For $n < 2^d$ and $p < \infty$, we have

$$\tilde{d}_p^c(n,d) \ge \left(1 - n\, 2^{-d} \right)^{1/p} \tilde{d}_p^c(0,d).$$

Hence, uniform integration is intractable in $F_{d,q}$ since

$$n(\varepsilon, F_{d,q}, \mathrm{LIN}) \ge (1 - \varepsilon^p) 2^d \quad \text{and} \quad \lim_{d \to \infty} \frac{\tilde{d}_p^c(d^k, d)}{\tilde{d}_p^c(0,d)} = 1 \quad \forall k = 1, 2, \ldots.$$

Observe that for $q = 1$ we have $p = \infty$, and since $\tilde{d}_p^c(n,d)$ is a nondecreasing function of p we have

$$\tilde{d}_\infty^c(n,d) = \tilde{d}_\infty^c(0,d) = 2^{-d} \quad \forall n < 2^d.$$

Hence, unlike the $*$-discrepancy and the extreme discrepancy, we find that uniform integration and the centred discrepancy are intractable for $q = 1$ and $p = \infty$.

It is known that $\tilde{d}_\infty^c(n,d)$ goes to zero at least like $n^{-1}(\log n)^{d-1}$. In view of the previous property we must wait, however, exponentially long in d to see this rate of convergence.

3.9.1 Spaces without Boundary Values

We now remove the condition $f(\frac{1}{2}) = 0$. As before, we study the more general case of the weighted L_q norm. That is, $D_1 = [0,1]$ and $F_{1,q,\gamma}$ is the Sobolev space $W_q^1([0,1])$ with the norm

$$\|f\|_{F_{1,q,\gamma}} = \left(|f(\tfrac{1}{2})|^q + \gamma^{-q/2} \int_0^1 |f'(x)|^q \, dx \right)^{1/q},$$

where $\gamma > 0$. Observe that for $q = \infty$ we have

$$\|f\|_{F_{1,\infty,\gamma}} = \max \left(|f(\tfrac{1}{2})|, \gamma^{1/2} \operatorname*{ess\,sup}_{t \in [0,1]} |f'(t)| \right).$$

For $q = 2$, we have the Hilbert space with the kernel

$$K_{1,\gamma}(x,t) = 1 + \gamma \, 1_M(x,t) \min(|x - \tfrac{1}{2}|, |t - \tfrac{1}{2}|).$$

For $d > 1$, we take $F_{d,q,\gamma} = W_q^{(1,1,\ldots,1)}([0,1]^d)$ as the tensor product of $W_q^1([0,1])$. The norm in $F_{d,q,\gamma}$ is given by

$$\|f\|_{F_{d,q,\gamma}} = \left(\sum_{u \subset \{1,2,\ldots,d\}} \gamma_u^{-q/2} \int_{[0,1]^{|u|}} \left| \frac{\partial^{|u|}}{\partial x_u} f(x_u, 1/2) \right|^q dx_u \right)^{1/q}.$$

$$(3.44)$$

The formula for the error of the algorithm $Q_{n,d}(f) = \sum_{i=1}^n a_i f(z_i)$ takes the form, see [30],

$$I_d(f) - Q_{n,d}(f) =$$

$$\sum_{u \subset \{1,2,\ldots,d\}} \int_{[0,1]^{|u|}} \frac{\partial^{|u|}}{\partial x_u} f(x_u, \tfrac{1}{2}) \frac{\partial^{|u|}}{\partial x_u} \left(h_d - \sum_{i=1}^n a_i K_d(\cdot, z_i) \right) (x_u, \tfrac{1}{2}) \, dx_u,$$

where h_d and the kernel K_d are given as before. Applying the Hölder inequality for integrals and sums to $I_d(f) - Q_{n,d}(f)$ we conclude that

$$e(Q_{n,d}) := \sup_{f \in F_{d,q,\gamma} : \|f\|_{F_{d,q,\gamma}} \le 1} \left| I_d(f) - Q_{n,d}(f) \right| = d_{p,\gamma}^c(Q_{n,d}),$$

where $1/p + 1/q = 1$ and the weighted centred p-discrepancy $d_{p,\gamma}^c(Q_{n,d})$ is given by

$$d_{p,\gamma}^c(Q_{n,d}) = \left(\sum_{u \subset \{1,2,\ldots,d\}} \gamma_u^{p/2} \int_{[0,1]^{|u|}} |\mathrm{disc}(n,d)^c(x_u, 1/2)|^p \, dx_u \right)^{1/p},$$

$$(3.45)$$

with

$$\text{disc}(n,d)^c(x_u, 1/2) = \prod_{\ell \in u} \min(x_\ell, 1 - x_\ell) - \sum_{i=1}^{n} a_i \cdot 1_{J(a(x_u), x_u)}(t_i)_u.$$

Let $e(n, F_{d,q,\gamma}, \text{LIN}) = d_{p,\gamma}(n,d)$ denote the minimal error, or equivalently the minimal weighted centred p-discrepancy, that can be achieved by using n function values. The initial error, or the initial centred weighted p-discrepancy, is now given by

$$e(0, F_{d,q}, \text{LIN}) = d_{p,\gamma}^c(0,d) =$$

$$\left(\sum_{u \subset \{1,2,\ldots,d\}} \gamma_u^{p/2} \left(\frac{2^{-p}}{p+1} \right)^{|u|} \right)^{1/p} = \prod_{j=1}^{d} \left(1 + \frac{2^{-p}}{p+1} \gamma_j^{p/2} \right)^{1/p}.$$

For $q = 1$, we have $p = \infty$ and

$$e(0, F_{d,1}, \text{LIN}) = d_{\infty,\gamma}^c(0,d) = \max_{k=0,1,\ldots,d} \left(2^{-k} (\gamma_1 \gamma_2 \cdots \gamma_k)^{1/2} \right).$$

It is proved in [62] that

$$d_{p,\gamma}^c(n,d) \geq \left(\sum_{k=0}^{d} C_{d,p,k} \left(\frac{2^{-p}}{p+1} \right)^k \cdot (1 - n\, 2^{-k})_+ \right)^{1/p},$$

where

$$C_{d,p,k} = \sum_{u \subset \{1,2,\ldots,d\} : |u| = k} \gamma_u^{p/2}.$$

For $n \leq 2^m < 2^d$, this can be rewritten for $p < \infty$ as

$$d_{p,\gamma}^c(n,d) \geq 2^{-1/p} \left(1 - \frac{\sum_{k=0}^{m} C_{d,p,k} \left(\frac{2^{-p}}{p+1} \right)^k}{\sum_{k=0}^{d} C_{d,p,k} \left(\frac{2^{-p}}{p+1} \right)^k} \right)^{1/p} d_{p,\gamma}^c(0,d),$$

and for $p = \infty$ as

$$d_{\infty,\gamma}^c \geq 2^{-(m+1)} \left(\gamma_1 \gamma_2 \cdots \gamma_{m+1} \right)^{1/2}.$$

For $q > 1$, i.e., $p < \infty$, we can check the lack of strong tractability and tractability of I_d in $F_{d,q,\gamma}$, see [62], and obtain the following. If $\sum_{j=1}^{\infty} \gamma_j^{p/2} = \infty$ then

$$\lim_{d \to \infty} \frac{d_{p,\gamma}^c(n,d)}{d_{p,\gamma}^c(0,d)} = 1 \quad \forall n,$$

and uniform integration in $F_{d,q,\gamma}$ is not strongly tractable. If

$$\lim{}^*_{d\to\infty} \frac{\sum_{j=1}^d \gamma_j^{p/2}}{\ln d} = \infty$$

then

$$\lim{}^*_{d\to\infty} \frac{d^c_{p,\gamma}(d^k,d)}{d^c_{p,\gamma}(0,d)} = 1 \quad \forall\, k = 1,2,\ldots,$$

and uniform integration in $F_{d,q,\gamma}$ is intractable, where $\lim^*$ is lim or lim sup.

4 Further Intractability Results

4.1 Hilbert Spaces with Positive Kernels

Sometimes it is relatively easy to prove lower bounds for the class POS of positive quadrature formulas whereas for the class LIN of all quadrature formulas, it is either harder to prove similar lower bounds or it is unknown where such bounds hold, see also Sections 4.2 and 5. Positive quadrature formulas were studied in [51, 52, 73, 102]. Here we present a general lower bound of [102] concerning positive quadrature formulas for Hilbert spaces with (pointwise) positive kernels.

Hence we assume that F_d is a reproducing kernel Hilbert space with the reproducing kernel $K_d : D_d \times D_d \to \mathbb{R}$ such that

$$K_d(x,t) \geq 0 \quad \forall\, x,t \in D_d. \tag{4.1}$$

We consider the linear functional I_d given by

$$I_d(f) = \int_{D_d} f(t)\, \varrho_d(t)\, dt = \langle f, h_d \rangle,$$

where $h_d \in F_d$ and

$$h_d(x) = \int_{D_d} K_d(x,t) \varrho_d(t)\, dt \quad \forall\, x \in D_d,$$

see Section 2.3. The lower bound will be expressed in terms of K_d and h_d by the quantity

$$\kappa_d := \sup_{x \in D_d} \frac{|h_d(x)|}{(K_d(x,x) \int_{D_d} h_d(y) \varrho_d(y)\, dy)^{1/2}}. \tag{4.2}$$

One can conclude that $\kappa_d \leq 1$; it may happen that $\kappa_d = 1$. If $\kappa_d = 1$, however, then multivariate integration is trivial since $I_d(f)$ can be approximated with an arbitrary small error using only one function value, see [59]. The lower bound

$$e^2(Q_{n,d}) \geq e^2(Q_{0,d})\,(1 - n\,\kappa_d^2) \tag{4.3}$$

is proved in [102] for any $Q_{n,d} \in \text{POS}$ and therefore

$$n(\varepsilon, F_d, \text{POS}) \geq (1 - \varepsilon^2)\,\kappa_d^{-2}. \tag{4.4}$$

For a discussion of this result, see [102]. Here we mention only one example.

4.1.1 Example: Tensor Product Spaces

Assume that $F_d = F_1 \otimes F_1 \otimes \ldots \otimes F_1$ is the tensor product of d copies of the same space F_1. Then $D_d = D_1^d$ and $K_d(x,t) = \prod_{j=1}^{d} K_1(x_j, t_j)$. We assume that $K_1 \geq 0$ and that the weight ϱ_d is also of product form, $\varrho_d(t) = \prod_{j=1}^{d} \varrho_1(t_j)$. Then it follows that $\kappa_d = \kappa_1^d$. If univariate integration is not trivial, i.e., $\kappa_1 < 1$, then

$$n(\varepsilon, F_d, \text{POS}) \geq (1 - \varepsilon^2)\,\kappa_1^{-2d}, \tag{4.5}$$

i.e., the minimal number of samples of positive quadrature formulas goes exponentially fast to infinity with d.

We apply this result to the kernel

$$K_1(x,t) = 1 + \min(x,t)$$

on $[0,1]^2$, which corresponds to the Sobolev space $W_2^1([0,1])$. In this case we have

$$h_1(x) = \int_0^1 K_1(x,t)\,dt = 1 + x - x^2/2$$

and

$$\kappa_1 := \sup_{x \in [0,1]} \frac{h_1(x)}{(K_1(x,x) \int_0^1 h_1(y)\,dy)^{1/2}}.$$

A trivial computation yields

$$\kappa_1 = \frac{1}{3}\frac{1 + 2\sqrt{7}}{\sqrt{2 + \sqrt{7}}} \approx 0.97298 \quad \text{and} \quad \kappa_1^{-2} \approx 1.0563.$$

This proves formula (3.20) in Section 3.4.

The bound (4.5) can also be applied to "very small" classes of analytic functions. For example we can consider $I_d(f) = \int_{[0,1]^d} f(x)\,dx$ and the norm

$$\|f\|^2 = \sum_{\alpha \in \mathbb{N}_0^d} \|D^\alpha f\|_{L_2}^2,$$

see [52]. For this last example we do not know whether arbitrary linear algorithms are more efficient than positive algorithms.

4.2 Classes of Smoother Functions

In Section 3, we presented lower bounds for different discrepancies and for related integration problems. The integrands of these classes were not very smooth, since the highest existing derivative was $f^{(1,1,\dots,1)}$. Positive quadrature formulas for smooth functions were previously discussed in Section 4.1. In this section, we present lower bounds for smoother functions for the class LIN of all quadrature formulas.

In [62] we studied multivariate (uniform and Gaussian) integration defined for integrand spaces F_d, such as weighted Sobolev spaces of functions of d variables with smooth mixed derivatives. The weight γ_j moderates the behaviour of functions with respect to the jth variable. For $\gamma_j = 1$, we obtain the classical Sobolev spaces whereas for decreasing γ_j's the weighted Sobolev spaces consist of functions with diminishing dependence on the jth variable.

The results of [62] are obtained for arbitrary linear tensor product functionals I_d defined over weighted tensor product reproducing kernel Hilbert spaces F_d of functions of d variables.

The kernel of F_d is denoted by K_d, and is fully determined by the kernel K_1 of the space F_1 of univariate functions. Our point of departure is the notion of a *decomposable* kernel. We say that the kernel K_1 is decomposable if $K_1(x,t) = 0$ for all $x \le a \le t$ for some a. For example, kernels of the Sobolev spaces of smooth functions vanishing at a are decomposable. For such kernels, the problem of approximating the linear functional I_d naturally decomposes into 2^d subproblems. These subproblems are not trivial iff the two subproblems for $d = 1$ are not trivial. This is usually the case when the point a belongs to the interior of the function domain. If so, we find that $n(\varepsilon, F_d, \mathrm{LIN})$ depends exponentially on d, and so the problem is intractable.

We then extend our analysis to certain non-decomposable kernels. We assume first that the kernel $K_1 = R_1 + \gamma R_2$ where R_i are kernels of Hilbert spaces $H(R_i)$ such that $H(R_1) \cap H(R_2) = \{0\}$, and R_2 is decomposable. For the d dimensional case, we take the weighted tensor product Hilbert space whose kernel is

$$K_d(x,t) = \prod_{j=1}^{d} \left(R_1(x_j, t_j) + \gamma_j R_2(x_j, t_j) \right),$$

where x_j and t_j denote the successive components of x and t. We stress that we may use different weights γ_j for different components. Assuming, as before, that I_d has non-trivial subproblems with respect to $H(R_2)$, we prove that $\sum_{j=1}^{\infty} \gamma_j = \infty$ implies that I_d is not strongly tractable, and that $\limsup_d \sum_{j=1}^{d} \gamma_j / \ln d = \infty$ implies that I_d is intractable.

Sobolev spaces without the condition that functions vanish at a have reproducing kernels of this form, provided that a does not belong to the boundary of the domain. This holds for the centred discrepancy, which is also analysed in the Banach case where the norm is taken in the L_p sense with $p \in [1, \infty]$, see Section 3.9.

Finally, we consider the case when $K_1 = R_1 + \gamma R_2$, where R_2 is not necessarily decomposable but is the sum of two kernels with one decomposable term. This last assumption holds for the Sobolev spaces with or without boundary conditions when the kernel R_2 is modified by an operator of finite rank. In particular, this decomposition holds for the L_2-star discrepancy. In this case, we obtain the same sufficient conditions for the lack of strong tractability and intractability of I_d.

The approach in [62] requires us to identify a part of the reproducing kernel that is decomposable. As demonstrated by many examples in this paper, this can be done for many classical spaces. Decomposability means, roughly speaking, that F_1 contains two orthogonal subspaces $F_{(0)}$ and $F_{(1)}$ such that

$$f(x) = 0 \quad \text{for} \quad x \notin D_{(0)} \ \forall f \in F_{(0)},$$

$$f(x) = 0 \quad \text{for} \quad x \notin D_{(1)} \ \forall f \in F_{(1)},$$

where the set $D_{(0)} \cap D_{(1)}$ has at most one element and both the $D_{(i)}$ have positive Lebesgue measure. If the spaces $F_{(i)}$ are non-trivial (i.e., they contain elements different from zero) then our results are non-trivial and lead to intractability.

This theory cannot be applied, however, to spaces of analytic functions, since these functions cannot be decomposed as above. For example, the kernel

$$K_1(x,t) = \sum_{j=1}^{\infty} \frac{x^j t^j}{(j!)^2},$$

corresponds to a Hilbert space of entire functions. In general, for spaces of analytic functions, tractability issues of linear functionals are still open.

4.2.1 Sobolev Spaces

In this subsection, we illustrate the results of [62] for smooth Sobolev spaces with and without boundary conditions. Let r be an arbitrary (maybe large) natural number. We consider the Sobolev space $F_1 = W_0^r([0,1])$ with the boundary condition $f(0) = \ldots = f^{(r-1)}(0) = 0$ and the inner product

$$\langle f, g \rangle_{F_1} = \int_0^1 f^{(r)}(t) g^{(r)}(t)\, dt.$$

The reproducing kernel is

$$K_1(x,t) = \int_0^1 \frac{(x-u)_+^{r-1}}{(r-1)!} \frac{(t-u)_+^{r-1}}{(r-1)!}\, du.$$

An explicit form of K_1 is

$$K_1(x,t) = \frac{x^r}{r!} \sum_{j=0}^{r-1} \frac{r}{r+j} \frac{x^j}{j!} \frac{(t-x)^{r-1-j}}{(r-1-j)!},$$

for $x \le t$. This means that for a fixed x, the function $g(t) = K_1(x,t)$ is a polynomial in $t \in [x,1]$ of degree at most $r-1$.

For an arbitrary $a^* \in (0,1)$, consider the subspace F_{a^*} of F_1 consisting of functions f for which $f^{(i)}(a^*) = 0$ for $i = 0,1,\ldots,r-1$. Observe that $f^{(i)}(a^*) = \left\langle f, K_1^{(i,0)}(a^*, \cdot) \right\rangle_{F_1}$, where $K_1^{(i,0)}$ denotes i times differentiation with respect to the first variable. Hence, $f \in F_{a^*}$ iff f is orthogonal to

$$A_{r-1} = \mathrm{span}(K^{(0,0)}(a^*, \cdot), K^{(1,0)}(a^*, \cdot), \ldots, K^{(r-1,0)}(a^*, \cdot)).$$

Let $\{g_0, \ldots, g_{r-1}\}$ be an orthonormal basis of A_{r-1}. Since each function $K_1^{(i,0)}(a^*, \cdot)$ is a polynomial of degree at most $r-1$, the same holds for g_i in the interval $[a^*, 1]$.

Let K_{a^*} be the reproducing kernel of F_{a^*}. It is known that

$$K_{a^*}(x,t) = K_1(x,t) - \sum_{j=0}^{r-1} g_j(x)g_j(t).$$

One can check that K_{a^*} is decomposable with a^*. That is, for $x \leq a^* \leq t$ we have $K_{a^*}(x,t) = 0$. Hence, we have $K_1 = R_{1,a^*} + R_{2,a^*}$, where $R_{1,a^*}(x,t) = \sum_{j=1}^{r-1} g_j(x)g_j(t)$ and $R_{2,a^*} = K_{a^*}$ is decomposable.

Consider now uniform integration, $I_1(f) = \int_0^1 f(t)\,dt$. Then

$$h_1(x) = \int_0^x \frac{(1-u)^r}{r!} \frac{(x-u)^{r-1}}{(r-1)!}\,du$$

is a polynomial of degree $2r$. Since $h_{1,2}$ differs from h_1 by a polynomial of degree $r-1$, the function $h_{1,2}$ is a polynomial of degree $2r$. Its restrictions $h_{1,2,(0)}$ and $h_{1,2,(1)}$ are also polynomials of degree $2r$ over $[0,a^*]$ and $[a^*,1]$ and therefore their norms must be non-zero. It follows from the general results of [62] that uniform integration is intractable.

We now remove the boundary conditions and consider the Sobolev space $F_{1,\gamma} = W^r([0,1]^d)$ with the inner product

$$\langle f,g \rangle_{F_{1,\gamma}} = \sum_{j=0}^{r-1} f^{(j)}(0)\, g^{(j)}(0) + \gamma^{-1} \int_0^1 f^{(r)}(t)\, g^{(r)}(t)\,dt.$$

The kernel is now

$$K_{1,\gamma}(x,t) = R_1(x,t) + \gamma \int_0^1 \frac{(x-u)_+^{r-1}}{(r-1)!} \frac{(t-u)_+^{r-1}}{(r-1)!}\,du$$

with

$$R_1(x,t) = \sum_{j=0}^{r-1} \frac{x^j}{j!}\frac{t^j}{j!}.$$

The kernel $K_{1,\gamma}$ can be written as

$$K_{1,\gamma}(x,t) = R_1(x,t) + \gamma \sum_{j=1}^{r-1} g_j(x)g_j(t) + \gamma K_{a^*}(x,t).$$

For uniform integration we have

$$h_1(x) = \sum_{j=0}^{r-1} \frac{x^j}{j!(j+1)!} + \gamma \int_0^x \frac{(1-u)^r}{r!} \frac{(x-u)^{r-1}}{(r-1)!}\,du.$$

So h_1 is still a polynomial of degree $2r$. We summarise the results for this case.

Uniform integration for the Sobolev space $F_d = W_0^{r,r,\ldots,r}([0,1]^d)$ is intractable for any $r \geq 1$. In particular,

$$\lim_{d \to \infty} \frac{e(d^q, F_d)}{e(0, F_d)} = 1 \quad \forall q = 1, 2, \ldots, .$$

If $\sum_{j=1}^{\infty} \gamma_j = \infty$ then uniform integration in the Sobolev space $F_{d,\gamma} = W^{r,r,\ldots,r}([0,1]^d)$ is not strongly tractable and

$$\lim_{d \to \infty} \frac{e(n, F_{d,\gamma})}{e(0, F_{d,\gamma})} = 1 \quad \forall n.$$

If $\lim_{d \to \infty}^{*} \sum_{j=1}^{d} \gamma_j / \ln d = \infty$ then uniform integration in the Sobolev space $F_{d,\gamma} = W^{r,r,\ldots,r}([0,1]^d)$ is intractable and

$$\lim_{d \to \infty}^{*} \frac{e(d^q, F_{d,\gamma})}{e(0, F_{d,\gamma})} = 1 \quad \forall q = 1, 2, \ldots,$$

where $\lim^* \in \{\lim, \lim \sup\}$.

If $\gamma_j \geq \gamma_1 > 0$ for all j then uniform integration in the Sobolev space $F_{d,\gamma} = W^{r,r,\ldots,r}([0,1]^d)$ is intractable and

$$\lim_{d \to \infty} \frac{e(\lfloor b^d \rfloor, F_{d,\gamma})}{e(0, F_{d,\gamma})} = 1$$

for $b \in (1, b^*)$ with $b^* > 1$. The proof of this result gives a bound for b^*. This allows to prove the lower bound in (3.13) and in (3.21).

5 On the Power of Negative Weights

In Section 4.1 we studied linear functionals defined on Hilbert spaces and obtained the general lower bounds (4.3) and (4.4) for quadrature formulas modulo two assumptions:

- We assumed that the kernel of the Hilbert space is (pointwise) positive.
- We assumed that the quadrature formula only uses positive weights.

We may ask whether the second assumption is really necessary. One might be inclined to believe that for positive functionals, such as the unweighted Lebesgue integral, positive quadrature formulas should be optimal. This guess is supported by the results reported in Section 3.4: the (normalised) weighted L_2-discrepancy is strongly tractable (or tractable) for the class POS iff it is strongly tractable (or tractable) for the class LIN.

On the other hand, there is the very powerful algorithm of Smolyak for general multivariate tensor product problems which uses negative weights, see Section 6. In some cases the Smolyak algorithm is better than all *explicitly known* quadrature formulas with positive weights. Of course this does not prove that general linear algorithms are better than positive ones. Proving the optimality of positive algorithms for natural function spaces is a challenging problem.

One way to define such natural function spaces is as follows. We assume that F_d is a tensor product Hilbert space consisting of functions f : $[0,1]^d \to \mathbb{R}$. We consider quadrature formulas for the usual (unweighted) integral and assume that the kernel of F_d is positive. Moreover, we assume that F_d is a subspace of the Sobolev space $W^{1,1,\cdots,1}([0,1]^d)$. We will see that these assumptions are not enough to prove the optimality of positive algorithms. In fact, it may happen that general linear algorithms are exponentially better than positive algorithms. Our example is rather artificial, however, and we still do not know whether optimal algorithms for classes such as the $F_{d,\gamma}$ from Section 3.5.3 use negative weights. We begin with a result from [59]. We consider a Hilbert space F_d of functions defined on $[0,1]^d$ that is a tensor product, $F_d = F_1 \otimes \ldots \otimes F_1$. The space F_1 is two-dimensional and generated by the orthonormal functions e_1 and e_2. We may take $e_1(x) = |x - 1/2|$ and $e_2(x) = (x - 1/2) \cdot g(x)$, where $|g(x)| \leq 1$ is symmetric about $1/2$, $g(1/2 - x) = g(1/2 + x)$, and takes infinitely many values. We can take g of the form $g(x) = 1/2 + \alpha (x - 1/2)^2$. Note that

$$K_1(x,t) = e_1(x)e_1(t) + e_2(x)e_2(t)$$

is pointwise non-negative.

Then the (partially defined) function e_1/e_2 has a range that is infinite but not dense in $\mathbb{R}$, since $|e_1(x)/e_2(x)| \geq 1$. Let I_1 be a linear functional satisfying $e(1, F_1, \mathrm{LIN}) > 0$. Note that $I_1(f) = \int_0^1 f(t)\,dt$ is such a functional. For any such I_1, let $I_d = I_1 \otimes \ldots \otimes I_1$ be the d-fold tensor product functional. Then

$$e(d, F_d, \mathrm{LIN}) > 0 \quad \text{and} \quad e(d+1, F_d, \mathrm{LIN}) = 0.$$

Hence,

$$n(\varepsilon, F_d, \mathrm{LIN}) \leq d+1 \quad \forall \varepsilon \in [0,1).$$

This means that all tensor product functionals on F_d are tractable. If we only allow positive quadrature formulas then we obtain a completely different result. Since all the assumptions of Section 4.1.1 hold, we obtain

$$n(\varepsilon, F_d, \mathrm{POS}) \geq (1 - \varepsilon^2)\,\kappa_1^{-2d},$$

where $\kappa_1 < 1$, i.e., the minimal number of samples of positive quadrature formulas goes exponentially fast to infinity with d.

This example proves that, at least for some spaces, allowing negative weights breaks intractability of integration using positive quadrature formulas. It would be of practical importance to characterise for which spaces F_d this phenomenon occurs.

6 Smolyak-Type Algorithms

The known constructions of polynomial time algorithms for multivariate tensor product problems are based on a construction of Smolyak (1963) and on certain modifications of this algorithm, see also [4, 11, 16, 17, 20, 21, 24, 25, 26, 53, 54, 55, 56, 57, 58, 64, 65, 66, 76, 77, 78, 80, 90, 93, 95]. Even this long list is not complete since similar algorithms (under names such as *blending algorithm* or *sparse grid algorithm* or *Boolean algorithm* or *hyperbolic cross points algorithm*) were studied by many authors in different contexts.

Explicit cost bounds and error estimates (not including any unknown constants) for Smolyak formulas are proved in [90]. These estimates are quite general. For a given problem, one only has to know suitable algorithms and error bounds for the univariate case $d = 1$. Then the Smolyak algorithm and the bounds from [90] automatically give the algorithm and the error bounds for the tensor product problem in any dimension $d > 1$. The Smolyak algorithm has been modified in different papers, see the papers which we already mentioned.

For simplicity, we describe the Smolyak algorithm only for multivariate integration

$$I_d(f) = \int_{[0,1]^d} f(x)\, dx.$$

We assume that for $d = 1$ quadrature formulas†

$$U^i(f) = \sum_{j=1}^{m_i} a_j^i\, f(t_j^i)$$

are given. Here we have $m_{i+1} > m_i$. For $d > 1$ a *(tensor) product*

---

† Hence the a_j^i are real numbers. In case of interpolation or operator equations the a_i^j are functions.

formula is given by

$$(U^{i_1} \otimes \cdots \otimes U^{i_d})(f) = \sum_{j_1=1}^{m_{i_1}} \cdots \sum_{j_d=1}^{m_{i_d}} a_{j_1}^{i_1} \cdots a_{j_d}^{i_d} \, f(t_{j_1}^{i_1}, \ldots, t_{j_d}^{i_d}).$$

To evaluate this formula one needs

$$n = \prod_{\ell=1}^{d} m_{i_\ell}$$

function values. The algorithm of Smolyak is a suitable linear combination of such tensor products with the aim of only using a small number of function values. It is given by

$$A(q, d) = \sum_{q-d+1 \leq |i| \leq q} (-1)^{q-|i|} \cdot \binom{d-1}{q-|i|} \cdot (U^{i_1} \otimes \cdots \otimes U^{i_d}),$$

where $q \geq d$. Here $i = [i_1, i_2, \ldots, i_d]$ with integers $i_j \geq 1$, and $|i| = i_1 + i_2 + \cdots + i_d$.

The information used by the algorithm $A(q, d)$ consists of the function values $f(t_{j_1}^{i_1}, \ldots, t_{j_d}^{i_d})$ with $j_k \leq m_{i_k}$. Assume that $m_i \leq M^i$ for some $M > 1$. Then the indices of the sample points $t_{j_1}^{i_1}, \ldots, t_{j_d}^{i_d}$ satisfy

$$j_1 j_2 \cdots j_d \leq m_{i_1} m_{i_2} \cdots m_{i_d} \leq M^{|i|} \leq M^q.$$

Hence, the indices $j = [j_1, j_2, \ldots, j_d]$ satisfy the so called *hyperbolic cross* inequality, and therefore the information used by $A(q, d)$ is called *hyperbolic cross* information.

We now discuss the (worst case) error of the Smolyak algorithm $A(q, d)$ for the unit ball of a tensor product Hilbert space $F_d = F_1 \otimes F_1 \otimes \cdots \otimes F_1$ (d times). The error of $A(q, d)$ depends on the error of the one-dimensional algorithms U^i for the unit ball of the Hilbert space F_1. If we assume that *nested* information is used, i.e., the sample points of U^i are also used by U^{i+1}, and the weights a_j^i are chosen optimally, then the algorithm $A(q, d)$ remains optimal for all d. This holds independently of the choice of nested sample points t_j^i. Obviously, the smaller the one-dimensional error of U^i is, the smaller the error of $A(q, d)$ will be. The error of $A(q, d)$ can be expressed in terms of the errors of U^i, and the explicit formulas can be found in [90].

We now discuss the cost $n(\varepsilon, F_d)$ of the Smolyak algorithm $A(q, d)$ which is defined as the minimal number of function values for which the worst case error $\|I_d - A(q, d)\|$ is at most ε. Assume that the error of

U^i is polynomial in m_i. That is, there exist positive C and p such that

$$\|I_1 - U^i\| \leq C\, m_i^{-1/p} \qquad \forall\, i.$$

Then, see [90],

$$n(\varepsilon, F_d) \leq \beta_1 \left(\beta_2 + \beta_3 \frac{\log \varepsilon^{-1}}{d-1}\right)^{\beta_4(d-1)} \left(\frac{1}{\varepsilon}\right)^p \tag{6.1}$$

for some positive β_i which are fully determined by the error of the one-dimensional algorithms U^i.

We stress that the leading factor of (6.1) is ε^{-p} and the exponent of ε^{-1} does *not* depend on d and is the same as for $d = 1$. The dependence on d is through $\log \varepsilon^{-1}$. Note that for a fixed d, we have

$$C_d := \limsup_{\varepsilon \to 0} \frac{n(\varepsilon, F_d)}{(\log \varepsilon^{-1})^{\beta_4(d-1)}\,\varepsilon^{-p}} \leq \beta_1 \left(\frac{\beta_3}{d-1}\right)^{\beta_4(d-1)}$$

Hence, the asymptotic constant C_d goes to zero super-exponentially fast with d.

For a fixed ε and varying d, or for both ε and d varying, the bound (6.1) may depend exponentially on d. It turns out that if $\|I_1\| < 1$, then $n(\varepsilon, F_d)$ has a bound

$$n(\varepsilon, F_d) \leq C\,\varepsilon^{-p^*} \qquad \forall\, \varepsilon > 0$$

for some positive C and p^*. (In general, $p^* > p$.) This bound is independent of d and depends polynomially on ε^{-1}, see [90]†.

Hence, we have *strong tractability* (for the absolute error) and the Smolyak algorithm is a *constructive* algorithm that computes an ε-approximation using a polynomial number of function values. We stress that the Smolyak algorithm is fully constructive. We only need to know the one-dimensional algorithms U^i with good error properties, which usually is easy to obtain for many spaces F_1. The total computing time of the Smolyak algorithm is proportional to the number of sample points; this holds for any dimension d. The additional cost (for example, the cost of computing the binomials) is negligible.‡

---

† The assumption $\|I_1\| < 1$ is also a necessary condition for strong tractability provided that I_1, instead of being an integration operator, is an arbitrary continuous linear operator of rank at least two.

‡ There exist other constructive algorithms whose errors serve as upper bounds on the complexity. For example, this holds for algorithms based on constructive versions of Tchakalov's theorem. This theorem says the following. If $V(d, k)$ denotes the space of polynomials of d variables and degree at most k then there exists a quadrature formula with positive weights and a number of sample points not larger than the dimension of $V(d, k)$ which is exact for all polynomials from

We illustrate the Smolyak algorithm by an example. Many more examples can be found in the papers cited previously. We take

$$m_1 = 1 \quad \text{and} \quad m_i = 2^{i-1} + 1 \quad \text{for } i > 1 \qquad (6.2)$$

and choose Clenshaw–Curtis formulas

$$U^i(f) = \sum_{j=1}^{m_i} a_j^i \, f(t_j^i)$$

with (6.2) and the sample points

$$t_j^i = -\cos \frac{\pi(j-1)}{m_i - 1}$$

($t_1^1 = 0$). We then obtain an algorithm which was presented in [54] and studied further in several other papers. Here one uses, for $d = 1$, *non-equidistant* sample points that are *nested*. Both properties improve the error bounds, as well as the stability of the algorithm. The $A(q, d)$ are almost optimal for many different function classes, see [54] and [55]. For all the spaces from Section 3 we obtain

$$e(A(q, d)) \preceq n^{-1} \cdot (\log n)^{2(d-1)},$$

where n is the number of sample points of $A(q, d)$. The weights of $A(q, d)$ are relatively small since $\|A(q, d)\|_\infty \preceq (\log n)^{d-1}$. The formula $A(k + d, d)$ is exact for all polynomials of degree $2k + 1$ and uses, for large d, about $n \approx 2^k \, d^k / k!$ sample points, see [56].

The Smolyak algorithm (together with the midpoint rule for $d = 1$) can be used to prove *constructive* upper bounds for the star-discrepancy. From the results of [90] we obtain

$$n(\varepsilon, F_d, \mathrm{LIN}) \leq C_\varepsilon \cdot 62^d. \qquad (6.3)$$

This upper bound is very bad if compared with the best non-constructive upper bound $n(\varepsilon, F_d, \mathrm{LIN}) \leq C_\varepsilon \cdot d$. To prove (6.3) we use the formulas of [90], p. 32. For the nested midpoint rule U_i in one dimension we use $n_i = (3^i - 1)/2$ sample points to obtain the error 3^{-i}. Using the notation of [90], we have $\alpha_1 = \alpha_2 \cdot \log(3e/\log 3)$, with $\alpha_2 = e^2/((e-1)\log 3)$. Hence we obtain $\beta_2 = \alpha_1 < 7.85$ and (6.3) follows from formula (50) of [90] since $C = 1$ and $\beta_4 = 2$.

A major modification of the original Smolyak algorithm for the case

$V(d, k)$. The computation of this quadrature formula is, however, so expensive that it cannot be implemented for, say, $d > 10$.

of weighted tensor product problems together with explicit error bounds was given in [93]. Using the *WTP-algorithm* from this paper one can constructively prove strong tractability of weighted problems provided the weights go to zero sufficiently fast. Such a result was already mentioned in Section 3.4.

7 Path Integration

Many applications require approximate values of path integrals. Often path integrals are approximated by a high dimensional integrals and then a *Monte Carlo* (or randomised) algorithm is applied. Typical Monte Carlo algorithms need roughly ε^{-2} integrand evaluations and the error bound ε is guaranteed only in a stochastic sense. Do we really need to use randomised algorithms for path integrals? In this section we present results concerning *deterministic algorithms* together with *worst case error (and complexity) bounds*.

7.1 *Tractability of Path Integration*

We consider path integration with respect to a Gaussian measure μ for different classes F of integrands. That is, we want to compute an ε-approximation to

$$I(f) = \int_X f(x)\, d\,\mu(x), \qquad f \in F,$$

where X is a Banach (usually infinite dimensional) space equipped with a Gaussian measure μ.

Tractability of path integration means that the complexity (the minimal number of integrand evaluations) depends polynomially on ε^{-1}, where ε is the worst case error. The results depend on the Gaussian measure, mainly through the eigenvalues λ_i of the covariance operator, and on the class F of integrands. We present two results from [91].

For the first result we assume that $r \in \mathbb{N}$, and F^r is the class of all $f : X \to \mathbb{R}$ such that $f^{(r)}$ is continuous with $\|f^{(k)}(x)\| \le 1$ for all $x \in X$ and $k = 0, 1, \ldots, r$. Then the tractability of path integration depends on the eigenvalues λ_i, which we assume to be ordered,

$$\lambda_1 \ge \lambda_2 \ge \ldots \ge 0.$$

·There are two cases.

First case: If $r \ge 1$ and only k eigenvalues are positive, i.e., $\lambda_k > 0$ and

$\lambda_{k+1} = 0$, then the path integration problem is tractable with exponent k/r, i.e.,

$$n(\varepsilon, F^r) \asymp \varepsilon^{-k/r}.$$

Second case: If $r = 0$ or if infinitely many eigenvalues are positive, then the path integration problem is intractable.

Observe that the first case is not very typical. In this case the measure is concentrated on a k-dimensional space, which contradicts the essence of the path integration problem since we really have a finite dimensional integration problem. Hence the second case is typical and we have a negative result on tractability of path integration. This result indicates that the class F^r is simply too large.

To obtain a positive result on tractability of path integration, one may study classes of entire integrands. In [91] the authors assume additionally that one can compute the derivatives of integrands at zero. For a class of entire integrands the optimal algorithm is given in [91], together with error bounds that prove the tractability of the problem. It turns out that the complexity of path integration is at most of the order ε^{-p}, where p depends on the Gaussian measure and is always less than or equal to 2. For the classical Wiener measure, $\lambda_k \asymp k^{-2}$, the worst case upper bound is of the order $\varepsilon^{-2/3}$.

7.2 A Smolyak Type algorithm for Wiener Integrals

In this section we present a Smolyak type algorithm for Wiener integrals, i.e., integrals with respect to the Wiener measure W. Such integrals are usually computed with the Monte Carlo algorithm or with number theoretic (or quasi-Monte Carlo) algorithms, based on low discrepancy sequences. These algorithms usually are reliable but slow. In particular, these algorithms can hardly exploit the smoothness properties of the integrand. Here we discuss the computation of Wiener integrals by a Smolyak type algorithm following [58] and [78]. The new algorithm is superior if the integrand is sufficiently smooth.

We begin with the *Lévy Ciesielski decomposition* of the Wiener measure by means of the *Schauder or hat functions*.

The Schauder functions $S_{j,k}$ are defined for $k = 1, \ldots, m(j)$, where $m(j) = 2^{j-2}$ for $j \geq 2$. Consider independent standard normal random variables $Y_{j,k}$ and variances $\sigma^2(1) = 1$, $\sigma^2(j) = 2^{-j}$ for $j \geq 2$. Then for

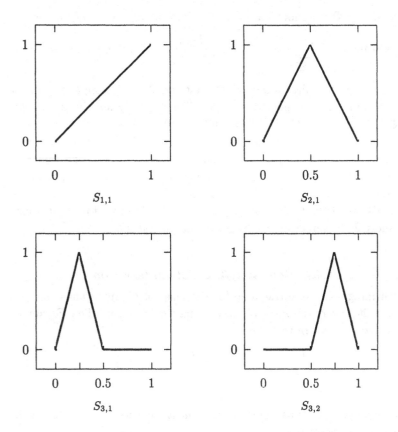

$S_{1,1}$ $S_{2,1}$

$S_{3,1}$ $S_{3,2}$

$d = 2^{s-1}$ let

$$W^{(d)} = \sum_{j=1}^{s} \sigma(j) \sum_{k=1}^{m(j)} Y_{j,k} \cdot S_{j,k}$$

be the *s-th Lévy Ciesielski approximation* of the Wiener measure. For $s \to \infty$ we get weak convergence to the Wiener measure. We approximate Wiener integrals

$$E(f) = \int_{C[0,1]} f(x)\, dW(x)$$

by the finite dimensional expectations $E(f(W^{(d)}))$. Here the influence of the variables $Y_{j,k}$ decreases with increasing j.

For approximation of $E(f(W^{(d)}))$ we use

$$A(s)(f) = A(s-1+d, d)(f \circ W^{(d)}), \quad d = 2^{s-1} \tag{7.1}$$

where $A(q,d)$ is a modification of the Smolyak algorithm of the form

$$A(q,d) = \sum_{q-d+1 \le |i| \le q} (-1)^{q-|i|} \cdot \binom{d-1}{q-|i|} \cdot (U^{i_1-\beta_1} \otimes \cdots \otimes U^{i_d-\beta_d})$$

with Gauss Hermite formulas U^k that use $2k-1$ sample points for $k \in \mathbb{N}$ and 1 sample point for $k \le 0$. The β_j are defined by $\beta_j = n$ for $j \in \{2^n+1, \ldots, 2^{n+1}\}$ if $n \in \mathbb{N}_0$ and $\beta_1 = 0$.

Worst case error bounds for the algorithm (7.1) and a class F of analytic functions are proved in [78]. The class F contains all Lipschitz continuous functionals with the property that all even partial derivatives in the direction of Schauder functions are bounded by 1. The error bounds are of the form $e(A(s)) \le 2^{-s/2}$ and it is proved that $A(s)$ uses at most 2^{2s-1} integrand evaluations (sample paths).

7.3 Feynman–Kac Path Integration

Path integration is intractable for the class of C^r-integrands, see 7.1. Often, however, the integrals have a particular form. The *Feynman–Kac integrals* are of the form

$$S(v,V) = \int_{C[0,1]} v(x(1)) \exp \left(\int_0^1 V(x(s)) \, ds \right) dw(x), \qquad (7.2)$$

where v and V belong to a Banach space F. Recall that Feynman–Kac integrals occur as the solution of heat equations by the use of the Feynman–Kac formula.

We report some results from [66], where a slightly more general problem is studied. For example, the function exp in (7.2) can be replaced by any smooth function. Observe that the solution operator S depends linearly on the function v, but nonlinearly on the function V. In [66] it is assumed that the functions v and V belong, in particular, to a subclass F of $C^r(\mathbb{R})$ functions. The authors present an algorithm A_ε that computes, for inputs v and V from F, the number $S(v,V)$ up to the error bound ε. The algorithm A_ε is again based on the Smolyak algorithm and uses the values of v and V at deterministically chosen sample points derived from from a certain approximation problem for the class F. An explicit bound on the cost of the algorithm A_ε is given. By the cost, we mean the total number of function values of v and V plus all arithmetic operations needed to compute $A_\varepsilon(v,V)$.

For the given subclass F of C^r-functions the cost is roughly of the order $\varepsilon^{-1/r}$ and this is almost optimal. This should be compared with the classical Monte Carlo algorithm whose cost is at least of the order ε^{-2}. Hence there is a great improvement, even for a modest smoothness $r = 2$ or $r = 3$.

To apply the new algorithm A_ε, one must precompute certain constants that are needed by the algorithm. Precomputation is typical for many algorithms. Consider, for example, the classical Gaussian quadrature formula

$$Q_n(f) = \sum_{i=1}^{n} a_i\, f(x_i) \quad \text{for} \quad \int_{-1}^{1} f(x)\, dx. \tag{7.3}$$

To compute the sample points x_i and the weights a_i one has to solve certain equations. This is usually done just once (for example with Maple or Mathematica and high precision) and the results of this precomputation are stored in a file that defines the algorithm. The cost of precomputation is usually ignored since it is done only once for all functions f for which we use (7.3). The cost of the algorithm (7.3) is then of order n.

The problem with the new algorithm A_ε is that it requires calculation of many multivariate integrals. Hence it is by itself a difficult problem. So far the precomputation has been done for $r = 1$, 2, and 4 and the precomputed values have an error of the order 10^{-4}. This is enough for the very fast computation of $S(v, V)$ if the needed accuracy is not too large. Nevertheless, for higher accuracy or for higher r, the problem of precomputation currently limits the use of this algorithm.

Acknowledgments

We thank Klaus Ritter, Joseph Traub and Arthur Werschulz for valuable comments. We also thank Arieh Iserles for editing our paper for the FOCM proceedings. The first author was supported by a Heisenberg scholarship of the German Research Council (DFG) and by a scholarship of the KU Leuven. The second author was partly supported by the National Science Foundation.

References

[1]   Aronszajn, N. (1950). Theory of reproducing kernels, *Trans. Amer. Soc.* **68**, 337–404.
[2]   Bakhvalov, N.S. (1959). On approximate computation of integrals, *Vestnik MGU, Ser. Math. Mech. Astron. Phys. Chem.* **4**, 3–18 (Russian).

[3] Bakhvalov, N.S. (1971). On the optimality of linear methods for operator approximation in convex classes of functions. USSR Comput. Maths. Math. Phys. **11**, 244–249.

[4] Barthelmann, V., Novak, E. and Ritter, K. (1998). High dimensional polynomial interpolation on sparse grids. *Adv. in Comput. Math.*, to appear.

[5] Beck, J. and Chen, W. (1987). *Irregularities of Distribution* (Cambridge University Press, Cambridge).

[6] Beck, J. and Sós, V.T. (1995). Discrepancy theory, in *Handbook of Combinatorics* (R. Graham, M. Grötschel and L. Lovász, eds) (Elsevier, Amsterdam).

[7] Bellman, R. (1957). *Dynamic Programming* (University Press, Princeton NJ).

[8] Blum, L., Shub, M. and Smale, S. (1989). On a theory of computation and complexity over the real numbers: NP-completeness, recursive functions and universal machines, *Bull. AMS* **21**, 1–46.

[9] Blum, L., Cucker, F., Shub, M. and Smale, S. (1998). *Complexity and Real Computation* (Springer-Verlag, New York).

[10] Boender, C.G.E. and Romeijn, H.E. (1995). Stochastic methods. in *Handbook of Global Optimization* (R. Horst, P. M. Pardalos, eds) (Kluwer, Dordrecht), 829–869.

[11] Bungartz, H.-J. and Griebel, M. (1999). A note on the complexity of solving Poisson's equation for spaces of bounded mixed derivatives, *J. Complexity* **15**, 167–199.

[12] Chen, W.W.L. (1985). On irregularities of distribution and approximate evaluation of certain functions, *Quarterly J. Math. Oxford* **36**, 173–182.

[13] Chen, W.W.L. (1987). On irregularities of distribution and approximate evaluation of certain functions, II, in *Analytic Number Theory and Diophantine Problems, Proceedings of a Conference at Oklahoma State University in 1984* (Birkhäuser, Basel), 75–86.

[14] van der Corput, J.G. (1935). Verteilungsfunktionen. I, *Proc. Kon. Ned. Akad. v. Wetensch.* **38**, 813–821.

[15] van der Corput, J.G. (1935). Verteilungsfunktionen. II, *Proc. Kon. Ned. Akad. v. Wetensch.* **38**, 1058–1066.

[16] Cools, R., Novak, E. and Ritter, K. (1999). Smolyak's construction of cubature formulas of arbitrary trigonometric degree, *Computing* **62**, 147–162.

[17] Delvos, F.-J. and Schempp, W. (1989). *Boolean Methods in Interpolation and Approximation*, Pitman Research Notes in Mathematics Series **230** (Longman, Essex).

[18] Dudley, R.M. (1984). A course on empirical processes (École d'Été de Probabilités de Saint-Flour XII-1982), LNiM 1097 (Springer-Verlag, New York), 2–141,

[19] Drmota, M. and Tichy, R.F. (1997). *Sequences, Discrepancies and Applications*, LNiM **1651** (Springer-Verlag, Berlin).

[20] Frank, K. and Heinrich, S. (1996). Computing discrepancies of Smolyak quadrature rules, *J. Complexity* **12**, 287–314.

[21] Frank, K., Heinrich, S. and Pereverzev, S. (1996). Information complexity of multivariate Fredholm integral equations in Sobolev classes. *J. Complexity* **12**, 17–34.

[22] Frolov, K.K. (1980). Upper bounds on the discrepancy in L_p, $2 \le p < \infty$, *Dokl. Akad. Nauk USSR* **252**, 805–807.

[23] Gal, S. and Micchelli, C.A. (1980). Optimal sequential and non-sequential procedures for evaluating a functional, *Appl. Anal.* **10**, 105–120.

[24] Genz, A.C. (1986). Fully symmetric interpolatory rules for multiple integrals, *SIAM J. Numer. Anal.* **23**, 1273–1283.

[25] Gerstner, Th. and Griebel, M. (1998). Numerical integration using sparse grids, *Num. Algorithms* **18**, 209–232.

[26] Griebel, M., Schneider, M. and Zenger, Ch. (1992). A combination technique for the solution of sparse grid problems, in *Iterative Methods in Linear Algebra* (R. Beauwens R. and P. de Groen, eds) (Elsevier, Amsterdam), 263–281.

[27] Haussler, D. (1995). Sphere packing numbers for subsets of the Boolean n-cube with bounded Vapnik-Červonenkis dimension, *J. Combinatorial Theory* A **69**, 217–232.

[28] Heinrich, S. (1995). Efficient algorithms for computing the L_2-discrepancy, *Math. Comput.* **65**, 1621–1633.

[29] Heinrich, S., Novak, E., Wasilkowski, G.W. and Woźniakowski, H. (1999). The inverse of the star-discrepancy depends linearly on the dimension, *Acta Arithmetica*, to appear.

[30] Hickernell, F.J. (1998). A generalized discrepancy and quadrature error bound, *Math. Comput.* **67**, 299–322.

[31] Hickernell, F.J. (1998). What affects the accuracy of quasi-Monte Carlo quadrature? Preprint.

[32] Hickernell, F.J. and Woźniakowski, H. (1999). Integration and approximation in arbitrary dimension, *Adv. Comput. Math.* **12**, 25–58.

[33] Joe, S. (1997). Formulas for the computation of the weighted L_2-discrepancy. Manuscript.

[34] Kon, M.A. and Novak, E. (1990). The adaption problem for approximating linear operators, *Bull. AMS* **23**, 159–165.

[35] Larcher, G. (1998a). Digital point sets: analysis and application, in *Random and quasi-random point sets* (P. Hellekalek, G. Larcher, eds), LNiS **138** (Springer-Verlag, Berlin) 167–222.

[36] Larcher, G. (1998b). Talk at Dagstuhl-Seminar.

[37] Matoušek, J. (1998). The exponent of discrepancy is at least 1.0669, *J. Complexity* **14**, 448–453.

[38] Matoušek, J. (1999). *Geometric Discrepancy* (Springer-Verlag, Berlin).

[39] Meer, K. and Michaux, C. (1997). A survey on real structural complexity theory, *Bull. Belg. Math. Soc.* **4**, 113–148.

[40] Morokoff, W.J. (1998). Generating quasi-random paths for stochastic processes, *SIAM Review* **40**, 765–788.

[41] Morokoff, W.J. and Caflisch, R.E. (1994). Quasi-random sequences and their discrepancies, *SIAM J. Sci. Comput.* **15**, 1251–1279.

[42] Nemirovsky, A. (1996). Polynomial time methods in convex programming, in *The Mathematics of Numerical Analysis, 1995 AMS–SIAM Summer Seminar in Applied Mathematics* (J. Renegar, M. Shub and S. Male, eds), AMS Lect. Appl. Math. **32**, 543–589.

[43] Nemirovsky, A.S. and Yudin, D.B. (1983). *Problem Complexity and Method Efficiency in Optimization* (Wiley-Interscience, New York).

[44] Nesterov, Yu. and Nemirovskii, A. (1994). *Interior-Point Polynomial*

Algorithms in Convex Programming, SIAM Studies in Applied Mathematics (SIAM, Philadelphia).

[45] Neumayer, A. (1997). Molecular modeling of proteins and mathematical prediction of protein structure, *SIAM Review* **39**, 407–460.

[46] Niederreiter, H. (1992). *Random Number Generation and Quasi-Monte Carlo Methods* (SIAM, Philadelphia).

[47] Niederreiter, H. and Xing, C. (1998). Nets, (t, s)-sequences, and algebraic geometry, in *Random and quasi-random point sets* (P. Hellekalek, G. Larcher, eds), LNiS **138** (Springer-Verlag, Berlin), 267–302.

[48] Novak, E. (1988). *Deterministic and Stochastic Error Bounds in Numerical Analysis*, LNiM **1349** (Springer-Verlag, Berlin).

[49] Novak, E. (1995). The real number model in numerical analysis, *J. Complexity* **11**, 57–73.

[50] Novak, E. (1996). On the power of adaption, *J. Complexity* **12**, 199–237.

[51] Novak, E. (1999). Intractability results for positive quadrature formulas and extremal problems for trigonometric polynomials, *J. Complexity* **15**, 299–316.

[52] Novak, E. (1999). Is there a curse of dimension for integration? in *Eleventh Int. Conf. on Domain Decomposition Methods* (C.-H. Lai, P. E. Bjørstad, M. Cross and O. B. Widlund, eds), 89–96.

[53] Novak, E. and Ritter, K. (1996). Global optimization using hyperbolic cross points, in *State of the Art in Global Optimization: Computational Methods and Applications* (C. A. Floudas and P. M. Pardalos, eds) (Kluwer, Dordrecht), 19–33.

[54] Novak, E. and Ritter, K. (1996). High dimensional integration of smooth functions over cubes, *Numer. Math.* **75**, 79–97.

[55] Novak, E. and Ritter, K. (1997). The curse of dimension and a universal method for numerical integration, in *Multivariate Approximation and Splines* (G. Nürnberger, J. W. Schmidt and G. Walz, eds), ISNM **125** (Birkhäuser, Basel), 177–188.

[56] Novak, E. and Ritter, K. (1999). Simple cubature formulas with high polynomial exactness, *Construct. Approx.* **15**, 499–522.

[57] Novak, E., Ritter, K., Schmitt, R. and Steinbauer, A. (1999). On an interpolatory method for high dimensional integration, *J. Comput. Appl. Math.* **112**, 215–228.

[58] Novak, E., Ritter, K. and Steinbauer, A. (1998). A multiscale method for the evaluation of Wiener integrals, in *Approximation Theory IX, Volume 2* (C.K. Chui and L.L. Schumaker, eds), 251–258.

[59] Novak, E., Sloan, I.H. and Woźniakowski, H. (1997). Tractability of tensor product linear operators, *J. Complexity* **13**, 387–418.

[60] Novak, E. and Woźniakowski, H. (1996). Topological complexity of zero finding, *J. Complexity* **12**, 380–400.

[61] Novak, E. and Woźniakowski, H. (1999). On the cost of uniform and nonuniform algorithms, *Theoretical Comp. Sc.* **219**, 301–318.

[62] Novak, E. and Woźniakowski, H. (1999). Intractability results for integration and discrepancy, *J. Complexity*, to appear.

[63] Novak, E. and Woźniakowski, H. (2000). Complexity of linear problems with a fixed output basis, *J. Complexity* **16**, 333–362.

[64] Petras, K. (2000). On the Smolyak cubature error for analytic functions, *Adv. in Comput. Math.* **12**, 71–93.

[65] Plaskota, L. (2000). The exponent of discrepancy of sparse grids is at least 2.1933, *Adv. in Comput. Math.* **12**, 3–24.

[66] Plaskota, L., Wasilkowski, G.W. and Woźniakowski, H. (1999). A new algorithm and worst case complexity for Feynman-Kac path integration. Preprint.

[67] Riordan, J. (1958). *An Introduction to Combinatorial Analysis*, (Wiley and Sons, New York).

[68] Ritter, K. (1999). *Average Case Analysis of Numerical Problems*, LNiM (Springer-Verlag, Berlin), to appear.

[69] Roth, K.F. (1954). On irregularities of distributions, *Mathematika* **1**, 73–79.

[70] Roth, K.F. (1980). On irregularities of distributions IV, *Acta Arithm.* **37**, 67–75.

[71] Sloan, I.H. and Joe, S. (1994). *Lattice Methods for Multiple Integration* (Clarendon Press, Oxford).

[72] Sloan, I.H. and Woźniakowski, H. (1997). An intractability result for multiple integration, *Math. Comp.* **66**, 1119–1124.

[73] Sloan, I.H. and Woźniakowski, H. (1998). When are quasi-Monte Carlo algorithms efficient for high dimensional integrals? *J. Complexity* **14**, 1-33.

[74] Schäffler, S. (1995). Unconstrained global optimization using stochastic integral equations, *Optimization* **35**, 43–60.

[75] Smale, S. (1998). Mathematical problems for the next century, *Math. Intell.* **20**, 7–15.

[76] Smolyak, S.A. (1963). Quadrature and interpolation formulas for tensor products of certain classes of functions, *Sov. Math. Doklady* **4**, 240–243.

[77] Sprengel, F. (1997). Interpolation and wavelets on sparse Gauss–Chebyshev grids, in *Multivariate Approximation* (W. Haussmann et al., eds), Math. Res. **101** (Akademie Verlag, Berlin), 269–286.

[78] Steinbauer, A. (1999). Quadrature formulas for the Wiener measure, *J. Complexity* **15**, 476–498.

[79] Talagrand, M. (1994). Sharper bounds for Gaussian and empirical processes, *Ann. Probability* **22**, 28–76.

[80] Temlyakov, V.N. (1994). *Approximation of Periodic Functions* (Nova Science, New York).

[81] Tezuka, S. (1995). *Uniform Random Numbers: Theory and Practice* (Kluwer Academic Publishers, Boston).

[82] Tezuka, S. (1998), Financial applications of Monte Carlo and Quasi-Monte Carlo methods, in *Random and Quasi-random Point Sets* (P. Hellekalek, G. Larcher, eds), LNiM **138** (Springer-Verlag, Berlin).

[83] Thiémard, E. (1999). Computing bounds for the star discrepancy. Preprint.

[84] Traub, J.F., Wasilkowski, G.W. and Woźniakowski, H. (1988). *Information-Based Complexity* (Academic Press, New York).

[85] Traub, J.F. and Werschulz, A.G. (1998). *Complexity and Information* (Cambridge University Press, Cambridge).

[86] Traub, J.F. and Woźniakowski, H. (1980). *A General Theory of Optimal Algorithms* (Academic Press, New York).

[87] Traub, J.F. and Woźniakowski, H. (1982). Complexity of linear programming, *Oper. Res. Lett.* **1**, 59–62.

[88] Wahba, G. (1990). *Spline Models for Observational Data*, SIAM-NSF Regional Conference Series in Appl. Math. **59** (SIAM, Philadelphia).

[89] Wang, X. and Hickernell, F.H. (1999). Error bounds and tractability of quasi-Monte Carlo algorithms in infinite dimension. Preprint.

[90] Wasilkowski, G.W. and Woźniakowski, H. (1995). Explicit cost bounds of algorithms for multivariate tensor product problems, *J. Complexity* **11**, 1–56.

[91] Wasilkowski, G.W. and Woźniakowski, H. (1996). On tractability of path integration, *J. Math. Phys.* **37**, 2071–2088.

[92] Wasilkowski, G.W. and Woźniakowski, H. (1997). The exponent of discrepancy is at most 1.4778..., *Math. Comput.* **66**, 1125–1132.

[93] Wasilkowski, G.W. and Woźniakowski, H. (1999). Weighted tensor-product algorithms for linear multivariate problems, *J. Complexity* **15**, 402–447.

[94] Weihrauch, K. (1998). A refined model of computation for continuous problems, *J. Complexity* **14**, 102–121.

[95] Werschulz, A.G. (1996). The complexity of the Poisson problem for spaces of bounded mixed derivatives, in *The Mathematics of Numerical Analysis* (J. Renegar, M. Shub, S. Smale, eds.), Lect. in Appl. Math. **32** (AMS, Providence, RI) 895–914.

[96] Weyl, H. (1916). Über die Gleichverteilung von Zahlen mod Eins, *Math. Ann.* **77**, 313–352.

[97] Winkler, G. (1995). *Image Analysis, Random Fields and Dynamic Monte Carlo Methods* (Springer-Verlag, Berlin).

[98] Woźniakowski, H. (1991). Average case complexity of multivariate integration, *Bull. AMS (new series)* **24**, 185–191.

[99] Woźniakowski, H. (1994). Tractability and strong tractability of linear multivariate problems, *J. Complexity* **10** (1994), 96–128.

[100] Woźniakowski, H. (1994). Tractability and strong tractability of multivariate tensor product problems, *J. of Computing and Information* **4**, 1–19.

[101] Woźniakowski, H. (1998). Why does information-based complexity use the real number model? *Theoretical Comp. Sc.* **219**, 451–466.

[102] Woźniakowski, H. (1999). Efficiency of quasi-Monte Carlo algorithms for high dimensional integrals, in *Monte Carlo and Quasi-Monte Carlo Methods 1998* (H. Niederreiter and J. Spanier, eds.) (Springer-Verlag, Berlin), 114–136.

Moving Frames — in Geometry, Algebra, Computer Vision, and Numerical Analysis

Peter J. Olver

School of Mathematics
University of Minnesota
Minneapolis, MN 55455, USA
Email: olver@ima.umn.edu
Url: www.math.umn.edu/~olver

Abstract

This paper surveys the new, algorithmic theory of moving frames developed by the author and M. Fels. Applications in geometry, computer vision, classical invariant theory, and numerical analysis are indicated.

1 Introduction

The method of moving frames ("repères mobiles") was forged by Élie Cartan, [13, 14], into a powerful and algorithmic tool for studying the geometric properties of submanifolds and their invariants under the action of a transformation group. However, Cartan's methods remained incompletely understood and the applications were exclusively concentrated in classical differential geometry; see [22, 23, 26]. Three years ago, [20, 21], Mark Fels and I formulated a new approach to the moving frame theory that can be systematically applied to general transformation groups. The key idea is to formulate a moving frame as an equivariant map to the transformation group. All classical moving frames can be reinterpreted in this manner, but the new approach applies in far wider generality. Cartan's normalization procedure for the explicit construction of a moving frame relies on the choice of a cross-section to the group orbits. Building on these two simple ideas, one may algorithmically construct moving frames and complete systems of invariants for completely general group actions.

The existence of a moving frame requires freeness of the underlying group action. Classically, non-free actions are made free by prolonging to a jet space of sufficiently high order, leading to differential invariants and the solution to equivalence and symmetry problems via the differential invariant signature. More recently, the moving frame method was also applied to Cartesian product "prolongations" of group actions, leading to classification of joint invariants and joint differential invariants, [42]. The combination of jet and Cartesian product actions known as multi-space was proposed in [43] as a framework for the geometric analysis of numerical approximations, and, via the application of the moving frame method, to the systematic construction of invariant numerical algorithms.

New and significant applications of these results have been developed in a wide variety of directions. In [40, 1, 29], the theory was applied to produce new algorithms for solving the basic symmetry and equivalence problems of polynomials that form the foundation of classical invariant theory. In [32], the differential invariants of projective surfaces were classified and applied to generate integrable Poisson flows arising in soliton theory. In [20], the moving frame algorithm was extended to include infinite-dimensional pseudo-group actions. Faugeras, [19], initiated the applications of moving frames in computer vision. In [11], the characterization of submanifolds via their differential invariant signatures was applied to the problem of object recognition and symmetry detection, [4, 5, 7, 45]. The moving frame method provides a direct route to the classification of joint invariants and joint differential invariants, [21, 42], establishing a geometric counterpart of what Weyl, [51], in the algebraic framework, calls the first main theorem for the transformation group. In computer vision, joint differential invariants have been proposed as noise-resistant alternatives to the standard differential invariant signatures, [6, 10, 16, 36, 49, 50]. The approximation of higher order differential invariants by joint differential invariants and, generally, ordinary joint invariants leads to fully invariant finite difference numerical schemes, first proposed in [12, 11, 3, 43]. Finally, a complete solution to the calculus of variations problem of directly constructing differential invariant Euler-Lagrange equations from their differential invariant Lagrangians has been recently effected, [30].

2 Moving frames

We begin by outlining the basic moving frame construction in [21]. Let G be an r-dimensional Lie group acting smoothly on an m-dimensional manifold M. Let $G_S = \{g \in G : g \cdot S = S\}$ denote the *isotropy subgroup* of a subset $S \subset M$, and $G_S^* = \cap_{z \in S} G_z$ its *global isotropy subgroup*, which consists of those group elements which fix all points in S. We always assume, without any significant loss of generality, that G acts *effectively on subsets*, and so $G_U^* = \{e\}$ for any open $U \subset M$, i.e., there are no group elements other than the identity which act completely trivially on an open subset of M.

The crucial idea is to decouple the moving frame theory from reliance on any form of frame bundle. In other words,

$$\text{Moving frames} \neq \text{Frames!}$$

A careful study of Cartan's analysis of the case of projective curves, [13], reveals that Cartan was well aware of this fact. However, this important and instructive example did not receive the attention it deserves. Building on this example and the constructions in [22, 26], we arrive at the proper definition of a moving frame for a general transformation group.

Definition 1 *A* moving frame *is defined as a smooth, G-equivariant map $\rho : M \to G$.*

The group G acts on itself by left or right multiplication. If $\rho(z)$ is any right-equivariant moving frame then $\tilde{\rho}(z) = \rho(z)^{-1}$ is left-equivariant and conversely. All classical moving frames are left-equivariant, but, in many cases, the right versions are easier to compute. In many geometrical situations, one can identify our left moving frames with the usual frame-based versions, but these identifications break down for more general transformation groups.

Theorem 1 *A moving frame exists in a neighborhood of a point $z \in M$ if and only if G acts freely and regularly near z.*

Recall that G acts *freely* if the isotropy subgroup of each point is trivial, $G_z = \{e\}$ for all $z \in M$. This implies that the orbits all have the same dimension as G itself. *Regularity* requires that, in addition, each point $x \in M$ has a system of arbitrarily small neighborhoods whose intersection with each orbit is connected, cf. [38]. Regularity prevents global topological pathologies such as the irrational flow on the torus.

The practical construction of a moving frame is based on Cartan's method of *normalization*, [28, 13], which requires the choice of a (local) *cross-section* to the group orbits.

Theorem 2 *Let G act freely and regularly on M, and let $K \subset M$ be a cross-section. Given $z \in M$, let $g = \rho(z)$ be the unique group element*

*that maps z to the cross-section: $g \cdot z = \rho(z) \cdot z \in K$. Then $\rho: M \to G$
is a right moving frame for the group action.*

Given local coordinates $z = (z_1, \ldots, z_m)$ on M, let $w(g, z) = g \cdot z$ be
the explicit formulae for the group transformations. The right† moving
frame $g = \rho(z)$ associated with a *coordinate cross-section* $K = \{\, z_1 = c_1, \ldots, z_r = c_r \,\}$ is obtained by solving the *normalization equations*

$$w_1(g, z) = c_1, \qquad \ldots \qquad w_r(g, z) = c_r, \qquad (2.1)$$

for the group parameters $g = (g_1, \ldots, g_r)$ in terms of the coordinates $z = (z_1, \ldots, z_m)$. Substituting the moving frame formulae into the remaining
transformation rules leads to a complete system of invariants for the
group action.

Theorem 3 *If $g = \rho(z)$ is the moving frame solution to the normalization equations (2.1), then the functions*

$$I_{r+1}(z) = w_{r+1}(\rho(z), z), \qquad \ldots \qquad I_m(z) = w_m(\rho(z), z), \qquad (2.2)$$

form a complete system of functionally independent invariants.

Definition 2 *The* invariantization *of a scalar function $F: M \to \mathbb{R}$ with
respect to a right moving frame ρ is the the invariant function $I = \iota(F)$
defined by $I(z) = F(\rho(z) \cdot z)$.*

Invariantization amounts to restricting F to the cross-section, $I \mid K = F \mid K$, and then requiring that I be constant along the orbits. Invari-
antizing the coordinate functions $I_i(z) = \iota(z_i)$ yields the fundamen-
tal invariants (2.2) along with the normalization constants $c_i = I_i(z)$,
$i = 1, \ldots, r$. More generally, the invariantization of $F(z_1, \ldots, z_m)$ is

$$I(z) = F(I_1(z), \ldots, I_m(z)) = F(c_1, \ldots, c_r, I_{r+1}(z), \ldots, I_m(z)).$$

In particular, if $I(z)$ is an invariant, then $\iota(I) = I$, so invariantization
defines a canonical projection, depending on the moving frame, that
maps functions to invariants.

Of course, most interesting group actions are *not* free, and therefore
do not admit moving frames in the sense of definition 1. There are two
common methods for converting a non-free (but effective) action into a
free action. In the traditional moving frame theory, [13, 23, 26], this is
accomplished by prolonging the action to a jet space J^n of suitably high

---

† The left version can be obtained directly by replacing g by g^{-1} throughout the
construction.

order; the consequential invariants are the classical differential invariants for the group, [21, 38]. Alternatively, one may consider the product action of G on a sufficiently large Cartesian product $M^{\times(n+1)}$; here, the invariants are joint invariants, [42], of particular interest in classical algebra, [40, 51]. In neither case is there a general theorem guaranteeing the freeness and regularity of the prolonged or product actions — indeed, there are counterexamples in the product case — but such pathologies never occur in practical examples. In our approach to invariant numerical approximations, we will amalgamate the two methods by prolonging to an appropriate multi-space, as defined below.

3 Prolongation and differential invariants

Traditional moving frames are obtained by prolonging the group action to the nth order (extended) jet bundle $J^n = J^n(M, p)$ consisting of equivalence classes of p-dimensional submanifolds $S \subset M$ modulo n^{th} order contact at a single point; see [38] (Chapter 3) for details. Since G preserves the contact equivalence relation, it induces an action on the jet space J^n, known as its nth order *prolongation* and denoted by $G^{(n)}$.

An nth *order moving frame* $\rho^{(n)} \colon J^n \to G$ is an equivariant map defined on an open subset of the jet space. In practical examples, for n sufficiently large, the prolonged action $G^{(n)}$ becomes regular and free on a dense open subset $V^n \subset J^n$, the set of *regular jets*. In [41] it is rigorously proved that, for $n \gg 0$ sufficiently large, if G acts effectively on subsets, then $G^{(n)}$ acts locally freely on an open subset $V^n \subset J^n$.

Theorem 4 *An nth order moving frame exists in a neighborhood of a point $z^{(n)} \in J^n$ if and only if $z^{(n)} \in V^n$ is a regular jet.*

Our normalization construction will produce a moving frame and a complete system of differential invariants in the neighborhood of any regular jet. Local coordinates $z = (x, u)$ on M — considering the first p components $x = (x^1, \ldots, x^p)$ as independent variables, and the latter $q = m - p$ components $u = (u^1, \ldots, u^q)$ as dependent variables — induce local coordinates $z^{(n)} = (x, u^{(n)})$ on J^n with components u_J^α representing the partial derivatives of the dependent variables with respect to the independent variables, [38, 39]. We compute the prolonged transformation formulae

$$w^{(n)}(g, z^{(n)}) = g^{(n)} \cdot z^{(n)}, \qquad \text{or} \qquad (y, v^{(n)}) = g^{(n)} \cdot (x, u^{(n)}),$$

P.J. Olver

by implicit differentiation of the v's with respect to the y's. For simplicity, we restrict to a coordinate cross-section by choosing $r = \dim G$ components of $w^{(n)}$ to normalize to constants:

$$w_1(g, z^{(n)}) = c_1, \quad \ldots \quad w_r(g, z^{(n)}) = c_r. \tag{3.1}$$

Solving the normalization equations (3.1) for the group transformations leads to the explicit formulae $g = \rho^{(n)}(z^{(n)})$ for the right moving frame. As in theorem 3, substituting the moving frame formulae into the unnormalized components of $w^{(n)}$ leads to the *fundamental n^{th} order differential invariants*

$$I^{(n)}(z^{(n)}) = w^{(n)}(\rho^{(n)}(z^{(n)}), z^{(n)}) = \rho^{(n)}(z^{(n)}) \cdot z^{(n)}. \tag{3.2}$$

Once the moving frame is established, the *invariantization* process will map general differential functions $F(x, u^{(n)})$ to differential invariants $I = \iota(F)$. As before, invariantization defines a projection, depending on the moving frame, from the space of differential functions to the space of differential invariants. The fundamental differential invariants $I^{(n)}$ are obtained by invariantization of the coordinate functions

$$\begin{aligned}
H^i(x, u^{(n)}) &= \iota(x^i) = y^i(\rho^{(n)}(x, u^{(n)}), x, u), \\
I_K^\alpha(x, u^{(k)}) &= \iota(u_J^\alpha) = v_K^\alpha(\rho^{(n)}(x, u^{(n)}), x, u^{(k)}).
\end{aligned} \tag{3.3}$$

In particular, those corresponding to the normalization components (3.1) of $w^{(n)}$ will be constant, and are known as the *phantom differential invariants*. The nonphantom differential invariants form a complete system of functionally independent n^{th} order differential invariants.

Theorem 5 *Let $\rho^{(n)}: J^n \to G$ be a moving frame of order $\leq n$. Every n^{th} order differential invariant can be locally written as a function $J = \Phi(I^{(n)})$ of the fundamental n^{th} order differential invariants (3.3). The function Φ is unique provided it does not depend on the phantom invariants.*

Example 1 Let us illustrate the theory with a very simple, well-known example: curves in the Euclidean plane. The orientation-preserving Euclidean group SE(2) acts on $M = \mathbb{R}^2$, mapping a point $z = (x, u)$ to

$$y = x\cos\theta - u\sin\theta + a, \quad v = x\sin\theta + u\cos\theta + b. \tag{3.4}$$

For a general parametrized† curve $z(t) = (x(t), u(t))$, the prolonged

† While the local coordinates $(x, u, u_x, u_{xx}, \ldots)$ on the jet space assume that the

group transformations

$$v_y = \frac{dv}{dy} = \frac{\dot{x}\sin\theta + \dot{u}\cos\theta}{\dot{x}\cos\theta - \dot{u}\sin\theta}, \qquad v_{yy} = \frac{d^2v}{dy^2} = \frac{\dot{x}\ddot{u} - \ddot{x}\dot{u}}{(\dot{x}\cos\theta - \dot{u}\sin\theta)^3},$$

$$(3.5)$$

and so on, are found by successively applying the implicit differentiation operator

$$\frac{d}{dy} = \frac{1}{\dot{x}\cos\theta - \dot{u}\sin\theta} \frac{d}{dt} \qquad (3.6)$$

to v. The classical Euclidean moving frame for planar curves, [23], follows from the cross-section normalizations

$$y = 0, \qquad v = 0, \qquad v_y = 0. \qquad (3.7)$$

Solving for the group parameters $g = (\theta, a, b)$ leads to the right-equivariant moving frame

$$\theta = -\tan^{-1}\frac{\dot{u}}{\dot{x}},$$

$$a = -\frac{x\dot{x} + u\dot{u}}{\sqrt{\dot{x}^2 + \dot{u}^2}} = \frac{z \cdot \dot{z}}{\|\dot{z}\|}, \qquad (3.8)$$

$$b = \frac{x\dot{u} - u\dot{x}}{\sqrt{\dot{x}^2 + \dot{u}^2}} = \frac{z \wedge \dot{z}}{\|\dot{z}\|}.$$

The inverse group transformation $g^{-1} = (\tilde{\theta}, \tilde{a}, \tilde{b})$ is the classical left moving frame, [13, 23]: one identifies the translation component $(\tilde{a}, \tilde{b}) = (x, u) = z$ as the point on the curve, while the columns of the rotation matrix $\tilde{R}(\tilde{\theta}) = (\mathbf{t}, \mathbf{n})$ are the unit tangent and unit normal vectors. Substituting the moving frame normalizations (3.8) into the prolonged transformation formulae (3.5), results in the fundamental differential invariants

$$v_{yy} \longmapsto \kappa = \frac{\dot{x}\ddot{u} - \ddot{x}\dot{u}}{(\dot{x}^2 + \dot{u}^2)^{3/2}} = \frac{\dot{z} \wedge \ddot{z}}{\|\dot{z}\|^3},$$

$$(3.9)$$

$$v_{yyy} \longmapsto \frac{d\kappa}{ds}, \qquad v_{yyyy} \longmapsto \frac{d^2\kappa}{ds^2} + 3\kappa^3,$$

where $d/ds = \|\dot{z}\|^{-1}\, d/dt$ is the arc length derivative — which is itself found by substituting the moving frame formulae (3.8) into the implicit

curve is given as the graph of a function $u = f(x)$, the moving frame computations also apply, as indicated in this example, to general parametrized curves. Two parametrized curves are equivalent if and only if one can be mapped to the other under a suitable reparametrization.

differentiation operator (3.6). A complete system of differential invariants for the planar Euclidean group is provided by the curvature and its successive derivatives with respect to arc length: $\kappa, \kappa_s, \kappa_{ss}, \ldots$.

The one caveat is that the first prolongation of SE(2) is only locally free on J^1 since a 180° rotation preserves the tangent line to a curve and so has trivial first prolongation. The even derivatives of κ with respect to s change sign under a 180° rotation, and so only their absolute values are fully invariant. The ambiguity can be removed by including the second order constraint $v_{yy} > 0$ in the derivation of the moving frame. Extending the analysis to the full Euclidean group E(2) adds in a second sign ambiguity which can only be resolved at third order. See [42] for complete details.

Example 2 Let $n \neq 0, 1$. In classical invariant theory, the planar actions

$$y = \frac{\alpha x + \beta}{\gamma x + \delta}, \qquad \bar{u} = (\gamma x + \delta)^{-n} u, \tag{3.10}$$

of $G = GL(2)$ play a key role in the equivalence and symmetry properties of binary forms, when $u = q(x)$ is a polynomial of degree $\leq n$, [24, 40, 1, 29]. We identify the graph of the function $u = q(x)$ as a plane curve. The prolonged action on such graphs is found by implicit differentiation:

$$v_y = \frac{\sigma u_x - n\gamma u}{\Delta \sigma^{n-1}}, \qquad v_{yy} = \frac{\sigma^2 u_{xx} - 2(n-1)\gamma \sigma u_x + n(n-1)\gamma^2 u}{\Delta^2 \sigma^{n-2}},$$

$$v_{yyy} = \frac{\sigma^3 u_{xxx} - 3(n-2)\gamma \sigma^2 u_{xx} + 3(n-1)(n-2)\gamma^2 \sigma u_x - n(n-1)(n-2)\gamma^3 u}{\Delta^3 \sigma^{n-3}},$$

and so on, where $\sigma = \gamma p + \delta$, $\Delta = \alpha\delta - \beta\gamma \neq 0$. On the regular subdomain

$$\mathcal{V}^2 = \{uH \neq 0\} \subset J^2, \qquad \text{where} \qquad H = uu_{xx} - \frac{n-1}{n} u_x^2$$

is the classical Hessian covariant of u, we can choose the cross-section defined by the normalizations

$$y = 0, \qquad v = 1, \qquad v_y = 0, \qquad v_{yy} = 1.$$

Solving for the group parameters gives the right moving frame formulae†

$$\alpha = u^{(1-n)/n}\sqrt{H}, \qquad \beta = -x\,u^{(1-n)/n}\sqrt{H},$$
$$\gamma = \tfrac{1}{n} u^{(1-n)/n} u_x, \qquad \delta = u^{1/n} - \tfrac{1}{n} x\, u^{(1-n)/n} u_x. \tag{3.11}$$

† See [1] for a detailed discussion of how to resolve the square root ambiguities.

Substituting the normalizations (3.11) into the higher order transformation rules gives us the differential invariants, the first two of which are

$$v_{yyy} \;\longmapsto\; J = \frac{T}{H^{3/2}}, \qquad v_{yyyy} \;\longmapsto\; K = \frac{V}{H^2}. \qquad (3.12)$$

Here

$$T = u^2 u_{xxx} - 3\,\frac{n-2}{n}\,uu_x u_{xx} + 2\,\frac{(n-1)(n-2)}{n^2}\,u_x^3,$$

$$V = u^3 u_{xxxx} - 4\,\frac{n-3}{n}\,u^2 u_x u_{xx} + 6\,\frac{(n-2)(n-3)}{n^2}\,uu_x{}^2 u_{xx}$$
$$- 3\,\frac{(n-1)(n-2)(n-3)}{n^3}\,u_x^4,$$

can be identified with classical covariants, which may be constructed using the basic transvectant process of classical invariant theory, cf. [24, 40]. Using $J^2 = T^2/H^3$ as the fundamental differential invariant will remove the ambiguity caused by the square root. As in the Euclidean case, higher order differential invariants are found by successive application of the normalized implicit differentiation operator $d/ds = uH^{-1/2}d/dx$ to the fundamental invariant J.

4 Equivalence and signatures

The moving frame method was developed by Cartan expressly for the solution to problems of equivalence and symmetry of submanifolds under group actions. Two submanifolds $S, \bar{S} \subset M$ are said to be *equivalent* if $\bar{S} = g \cdot S$ for some $g \in G$. A *symmetry* of a submanifold is a group transformation that maps S to itself, and so is an element $g \in G_S$. As emphasized by Cartan, [13], the solution to the equivalence and symmetry problems for submanifolds is based on the functional interrelationships among the fundamental differential invariants restricted to the submanifold.

Suppose we have constructed an nth order moving frame $\rho^{(n)} \colon \mathrm{J}^n \to G$ defined on an open subset of jet space. A submanifold S is called *regular* if its n-jet $\mathrm{j}_n S$ lies in the domain of definition of the moving frame. For any $k \geq n$, we use $J^k = I^{(k)} \mid S = I^{(k)} \circ \mathrm{j}_k S$ to denote the kth order *restricted differential invariants*. The kth order *signature* $\mathcal{S}^k = \mathcal{S}^k(S)$ is the set parametrized by the restricted differential invariants; S is called *fully regular* if J^k has constant rank $0 \leq t_k \leq p = \dim S$ for all $k \geq n$. In this case, $\mathcal{S}^k$ forms a submanifold of dimension t_k — perhaps with self-intersections. In the fully regular case, we have

$$t_n < t_{n+1} < t_{n+2} < \cdots < t_s = t_{s+1} = \cdots = t \leq p,$$

where t is the *differential invariant rank* and s the *differential invariant order* of S.

The fundamental equivalence criterion is a conequence of the Frobenius Theorem governing the existence of solutions to certain involutive systems of first order partial differential equations, [39].

Theorem 6 *Two fully regular p-dimensional submanifolds S, $\bar{S} \subset M$ are locally equivalent, $\bar{S} = g \cdot S$, if and only if they have the same differential invariant order s and their signature manifolds of order $s + 1$ are identical: $\mathcal{S}^{s+1}(\bar{S}) = \mathcal{S}^{s+1}(S)$.*

Since symmetries are the same as self-equivalences, the signature also determines the symmetry group of the submanifold.

Theorem 7 *If $S \subset M$ is a fully regular p-dimensional submanifold of differential invariant rank t, then its symmetry group G_S is an $(r - t)$-dimensional subgroup of G that acts locally freely on S.*

A submanifold that has maximal differential invariant rank $t = p$, and hence only a discrete symmetry group, is called *nonsingular*. The number of symmetries of a submanifold is determined by its *index*, which is defined as the number of points in S map to a single generic point of its signature:

$$\operatorname{ind} S = \min \{ \# (J^{s+1})^{-1}\{\zeta\} : \zeta \in \mathcal{S}^{s+1} \}.$$

Theorem 8 *If S is a nonsingular submanifold, then its symmetry group is a discrete subgroup of cardinality $\# G_S = \operatorname{ind} S$.*

At the other extreme, a rank 0 or *maximally symmetric* submanifold has all constant differential invariants, and so its signature degenerates to a single point.

Theorem 9 *A regular p-dimensional submanifold S has differential invariant rank 0 if and only if its symmetry group is a p-dimensional subgroup $H = G_S \subset G$ and an H–orbit: $S = H \cdot z_0$.*

Remark: "Totally singular" submanifolds may have even larger, non-free symmetry groups, but these are not covered by the preceding results. See [41] for details and precise characterization of such submanifolds.

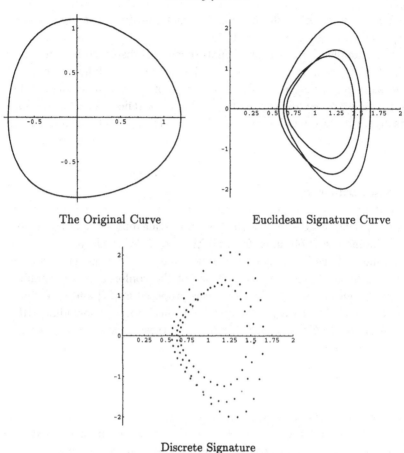

The Original Curve   Euclidean Signature Curve

Discrete Signature

Fig. 4.1. The Curve $x = \cos t + \frac{1}{5}\cos^2 t$, $y = \sin t + \frac{1}{10}\sin^2 t$

Example 3 The *Euclidean signature* for a curve in the Euclidean plane is the planar curve $\mathcal{S}(C) = \{(\kappa, \kappa_s)\}$ parametrized by the curvature invariant κ and its first derivative with respect to arc length. Two planar curves are equivalent under oriented rigid motions if and only if they have the same signature curves. The maximally symmetric curves have constant Euclidean curvature, and so their signature curve degenerates to a single point. These are the circles and straight lines, and, in accordance with theorem 9, each is the orbit of its one-parameter symmetry subgroup of SE(2). The number of Euclidean symmetries of a curve is equal to its index — the number of times the Euclidean signature is retraced as we go around the curve.

An example of a Euclidean signature curve is displayed in figure 4.1. The first figure shows the curve, and the second its Euclidean signature; the axes are κ and κ_s in the signature plot. Note in particular the approximate three-fold symmetry of the curve is reflected in the fact that its signature has winding number three. If the symmetries were exact, the signature curve would be exactly retraced three times as we go once around the original curve. The final figure gives a discrete approximation to the signature which is based on the invariant numerical algorithms to be discussed below.

In figure 4.2 we display some signature curves computed from an actual medical image — a 70×70, 8-bit gray-scale image of a cross section of a canine heart, obtained from an MRI scan. We then display an enlargement of the left ventricle. The boundary of the ventricle has been automatically segmented through use of the conformally Riemannian moving contour or snake flow that was proposed in [27] and sucessfully applied to a wide variety of 2D and 3D medical imagery, including MRI, ultrasound and CT data, [52]. Underneath these images, we display the ventricle boundary curve along with two smoothed versions which were obtained by application of the standard Euclidean-invariant curve shortening procedure. Below each curve is the associated spline-interpolated discrete signature curves for the smoothed boundary, as computed using the invariant numerical approximations to κ and κ_s discussed below. As the evolving curves approach circularity the signature curves exhibit less variation in curvature and appear to be winding more and more tightly around a single point, which is the signature of a circle of area equal to the area inside the evolving curve. Despite the rather extensive smoothing involved, except for an overall shrinkage as the contour approaches circularity, the basic qualitative features of the different signature curves, and particularly their winding behavior manifesting approximate Euclidean symmetries, appear to be remarkably robust.

Thus, the signature curve method has the potential to be of practical use in the general problem of object recognition and symmetry classification. It offers several advantages over more traditional approaches. First, it is purely local, and therefore immediately applicable to occluded objects. Second, it provides a mechanism for recognizing symmetries and approximate symmetries of the object. The design of a suitably robust "signature metric" for practical comparison of signatures is the subject of ongoing research.

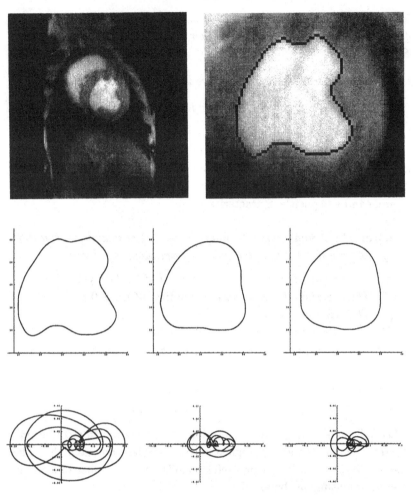

Fig. 4.2. Signature of a Canine Heart Image

Example 4 Let us next consider the equivalence and symmetry problems for binary forms. According to the general moving frame construction in example 2, the signature curve $\mathcal{S} = \mathcal{S}(q)$ of a polynomial (or more general function) $u = q(x)$ is parametrized by the covariants J^2 and K, as given in (3.12). The following solution to the equivalence problem for complex-valued binary forms, [1, 37, 40], is an immediate consequence of the general equivalence theorem 6.

Theorem 10 *Two nondegenerate complex-valued binary forms $q(x)$ and $\bar{q}(x)$ are equivalent if and only if their signature curves are identical:* $\mathcal{S}(q) = \mathcal{S}(\bar{q})$.

All equivalence maps $\bar{x} = \varphi(x)$ solve the two rational equations

$$J(x)^2 = \bar{J}(\bar{x})^2, \qquad K(x) = \overline{K}(\bar{x}). \tag{4.1}$$

In particular, the theory guarantees φ is necessarily a linear fractional transformation! The symmetries of a nonsingular form can be explicitly determined by solving the rational equations (4.1) with $\bar{J} = J$, $\overline{K} = K$. The maximally symmetric forms, which have a one-parameter symmetry group, can be explicitly characterized.

Theorem 11 *A nondegenerate binary form $q(x)$ is maximally symmetric if and only if it satisfies the following equivalent conditions:*

(a) *q is complex-equivalent to a monomial x^k, with $k \neq 0, n$.*
(b) *The covariant T^2 is a constant multiple of $H^3 \not\equiv 0$.*
(c) *The signature is just a single point.*
(d) *q admits a one-parameter symmetry group.*
(e) *The graph of q coincides with the orbit of a one-parameter subgroup of* GL(2).

A binary form $q(x)$ is nonsingular if and only if it is not complex-equivalent to a monomial if and only if it has a finite symmetry group.

See [1] for a MAPLE implementation of this method for computing discrete symmetries and classification of univariate polynomials. The method leads to the following useful bounds on the number of symmetries of a nonsingular binary form.

Theorem 12 *If $q(x)$ is a complex binary form of degree n which is not complex-equivalent to a monomial, then its symmetry group has cardinality*

$$k \leq \begin{cases} 6n^2 - 12n & \text{if } V = cH^2 \text{ for some constant } c, \text{ or} \\ 4n^2 - 8n & \text{in all other cases.} \end{cases}$$

In her thesis, Kogan, [29] extends these results to forms in several variables. In particular, a complete signature for ternary forms leads to a practical algorithm for computing discrete symmetries of such forms, including the particularly important case of ternary cubics defining elliptic curves.

5 Joint invariants and joint differential invariants

One practical difficulty with the differential invariant signature is its dependence upon high order derivatives, which makes it very sensitive to data noise. For this reason, a new signature paradigm, based on joint invariants, was proposed in [42]. We consider now the joint action

$$g \cdot (z_0, \ldots, z_n) = (g \cdot z_0, \ldots, g \cdot z_n), \qquad g \in G, \quad z_0, \ldots, z_n \in M. \quad (5.1)$$

of the group G on the $(n+1)$-fold Cartesian product manifold $M^{\times(n+1)} = M \times \cdots \times M$. An invariant $I(z_0, \ldots, z_n)$ of (5.1) is an $(n+1)$-*point joint invariant* of the original transformation group. In most cases of interest, although not in general, if G acts effectively on M, then, for $n \gg 0$ sufficiently large, the product action is free and regular on an open subset of $M^{\times(n+1)}$. Consequently, the moving frame method outlined in section 2 can be applied to such joint actions, and thereby establish complete classifications of joint invariants and, via prolongation to Cartesian products of jet spaces, joint differential invariants. We will discuss two particular examples — planar curves in Euclidean geometry and projective geometry, referring to [42] for details.

Example 5 Consider the proper Euclidean group SE(2) acting on oriented curves in the plane $M = \mathbb{R}^2$. We begin with the Cartesian product action on $M^{\times 2} \simeq \mathbb{R}^4$. The cross-section $x_0 = u_0 = x_1 = 0, u_1 > 0$, leads to the normalization equations

$$
\begin{aligned}
y_0 &= x_0 \cos\theta - u_0 \sin\theta + a = 0, \\
v_0 &= x_0 \sin\theta + u_0 \cos\theta + b = 0, \qquad (5.2) \\
y_1 &= x_1 \cos\theta - u_1 \sin\theta + a = 0.
\end{aligned}
$$

Solving, we obtain a right moving frame

$$
\begin{aligned}
\theta &= \tan^{-1}\left(\frac{x_1 - x_0}{u_1 - u_0}\right), \\
a &= -x_0 \cos\theta + u_0 \sin\theta, \qquad (5.3) \\
b &= -x_0 \sin\theta - u_0 \cos\theta,
\end{aligned}
$$

along with the fundamental interpoint distance invariant

$$v_1 = x_1 \sin\theta + u_1 \cos\theta + b \quad \longmapsto \quad I = \| z_1 - z_0 \|. \quad (5.4)$$

282   P.J. Olver

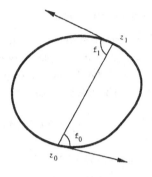
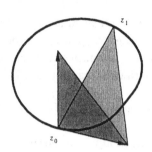

Fig. 5.1. First and Second Order Joint Euclidean Differential Invariants

Substituting (5.3) into the prolongation formulae (3.5) leads to the the normalized first and second order joint differential invariants

$$\frac{dv_k}{dy} \longmapsto J_k = -\frac{(z_1 - z_0) \cdot \dot{z}_k}{(z_1 - z_0) \wedge \dot{z}_k},$$

$$\frac{d^2 v_k}{dy^2} \longmapsto K_k = -\frac{\| z_1 - z_0 \|^3 \,(\dot{z}_k \wedge \ddot{z}_k)}{\left[(z_1 - z_0) \wedge \dot{z}_0\right]^3},$$

$$(5.5)$$

for $k = 0, 1$. Note that

$$J_0 = -\cot \phi_0, \qquad J_1 = +\cot \phi_1, \qquad (5.6)$$

where $\phi_k = \sphericalangle(z_1 - z_0, \dot{z}_k)$ denotes the angle between the chord connecting z_0, z_1 and the tangent vector at z_k, as illustrated in figure 5.1. The modified second order joint differential invariant

$$\widehat{K}_0 = -\| z_1 - z_0 \|^{-3} K_0 = \frac{\dot{z}_0 \wedge \ddot{z}_0}{\left[(z_1 - z_0) \wedge \dot{z}_0\right]^3} \qquad (5.7)$$

equals the ratio of the area of triangle whose sides are the first and second derivative vectors $\dot{z}_0$, $\ddot{z}_0$ at the point z_0 over the *cube* of the area of triangle whose sides are the chord from z_0 to z_1 and the tangent vector at z_0; see figure 5.1.

On the other hand, we can construct the joint differential invariants by invariant differentiation of the basic distance invariant (5.4). The normalized invariant differential operators are

$$D_{y_k} \longmapsto \mathcal{D}_k = -\frac{\| z_1 - z_0 \|}{(z_1 - z_0) \wedge \dot{z}_k} \frac{d}{dt_k}. \qquad (5.8)$$

Proposition 13 *Every two-point Euclidean joint differential invariant is a function of the interpoint distance $I = \| z_1 - z_0 \|$ and its invariant derivatives with respect to (5.8).*

A generic product curve $\mathbf{C} = C_0 \times C_1 \subset M^{\times 2}$ has joint differential invariant rank $2 = \dim \mathbf{C}$, and its joint signature $\mathcal{S}^2(\mathbf{C})$ will be a two-dimensional submanifold parametrized by the joint differential invariants I, J_0, J_1.

Theorem 14 *A curve C or, more generally, a pair of curves $C_0, C_1 \subset \mathbb{R}^2$, is uniquely determined up to a Euclidean transformation by its reduced joint signature $\mathcal{S}^2(\mathbf{C})$. The curve(s) have a one-dimensional symmetry group if and only if their signature is a one-dimensional curve if and only if they are orbits of a common one-parameter subgroup, i.e., concentric circles or parallel straight lines.*

For $n > 2$ points, we can use the two-point moving frame (5.3) to construct the additional joint invariants

$$y_k \longmapsto H_k = \| z_k - z_0 \| \cos \psi_k, \qquad v_k \longmapsto I_k = \| z_k - z_0 \| \sin \psi_k,$$

where $\psi_k = \sphericalangle(z_k - z_0, z_1 - z_0)$. Therefore, a complete system of joint invariants for SE(2) consists of the angles ψ_k, $k \geq 2$, and distances $\| z_k - z_0 \|$, $k \geq 1$. The other interpoint distances can all be recovered from these angles; vice versa, given the distances, and the sign of one angle, one can recover all other angles. In this manner, we establish a "First Main Theorem" for joint Euclidean differential invariants.

Theorem 15 *If $n \geq 2$, then every n-point joint E(2) differential invariant is a function of the interpoint distances $\| z_i - z_j \|$ and their invariant derivatives with respect to (5.8). For the proper Euclidean group SE(2), one must also include the sign of one of the angles, say $\psi_2 = \sphericalangle(z_2 - z_0, z_1 - z_0)$.*

Generic three-pointed Euclidean curves still require first order signature invariants. To create a Euclidean signature based entirely on joint invariants, we take four points z_0, z_1, z_2, z_3 on our curve $C \subset \mathbb{R}^2$. As illustrated in figure 5.2, there are six different interpoint distance invariants

$$a = \| z_1 - z_0 \|, \qquad b = \| z_2 - z_0 \|, \qquad c = \| z_3 - z_0 \|,$$
$$d = \| z_2 - z_1 \|, \qquad e = \| z_3 - z_1 \|, \qquad f = \| z_3 - z_2 \|, \tag{5.9}$$

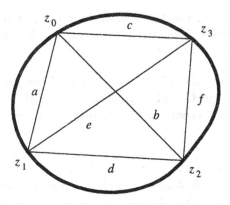

Fig. 5.2. Four-Point Euclidean Curve Invariants

which parametrize the joint signature $\widehat{S} = \widehat{S}(C)$ that uniquely characterizes the curve C up to Euclidean motion. This signature has the advantage of requiring no differentiation, and so is not sensitive to noisy image data. There are two local syzygies

$$\Phi_1(a, b, c, d, e, f) = 0, \qquad \Phi_2(a, b, c, d, e, f) = 0, \qquad (5.10)$$

among the the six interpoint distances. One of these is the universal *Cayley–Menger syzygy*

$$\det \begin{vmatrix} 2a^2 & a^2 + b^2 - d^2 & a^2 + c^2 - e^2 \\ a^2 + b^2 - d^2 & 2b^2 & b^2 + c^2 - f^2 \\ a^2 + c^2 - e^2 & b^2 + c^2 - f^2 & 2c^2 \end{vmatrix} = 0, \qquad (5.11)$$

which is valid for all possible configurations of the four points, and is a consequence of their coplanarity, cf. [2, 34]. The second syzygy in (5.10) is curve-dependent and serves to effectively characterize the joint invariant signature. Euclidean symmetries of the curve, both continuous and discrete, are characterized by this joint signature. For example, the number of discrete symmetries equals the signature index — the number of points in the original curve that map to a single, generic point in S.

A wide variety of additional cases, including curves and surfaces in two and three-dimensional space under the Euclidean, equi-affine, affine and projective groups, are investigated in detail in [42].

6 Multi-space for curves

In modern numerical analysis, the development of numerical schemes that incorporate additional structure enjoyed by the problem being approximated have become quite popular in recent years. The first instances of such schemes are the symplectic integrators arising in Hamiltonian mechanics, and related energy conserving methods, [15, 31, 48]. The design of symmetry-based numerical approximation schemes for differential equations has been studied by various authors, including Shokin, [47], Dorodnitsyn, [17, 18], Axford and Jaegers, [25], and Budd and Collins, [8]. These methods are closely related to the active area of geometric integration of ordinary differential equations on Lie groups, [9, 33]. In practical applications of invariant theory to computer vision, group-invariant numerical schemes to approximate differential invariants have been applied to the problem of symmetry-based object recognition, [3, 11, 12].

In this section, we outline the basic construction of multi-space that forms the foundation for the study of the geometric properties of discrete approximations to derivatives and numerical solutions to differential equations. We will only discuss the case of curves, which correspond to functions of a single independent variable, and hence satisfy ordinary differential equations. The more difficult case of higher dimensional submanifolds, corresponding to functions of several variables that satisfy partial differential equations, relies on a new approach to multi-dimensional interpolation theory, and hence will be the subject of a forthcoming paper, [44].

Numerical finite difference approximations to the derivatives of a function $u = f(x)$ rely on its values $u_0 = f(x_0), \ldots, u_n = f(x_n)$ at several distinct points $z_i = (x_i, u_i) = (x_i, f(x_i))$ on its graph. Thus, discrete approximations to jet coordinates on J^n are functions $F(z_0, \ldots, z_n)$ defined on the $(n+1)$-fold Cartesian product space $M^{\times(n+1)} = M \times \cdots \times M$. In order to seamlessly connect the jet coordinates with their discrete approximations, then, we need to relate the jet space for curves, $\mathrm{J}^n = \mathrm{J}^n(M, 1)$, to the Cartesian product space $M^{\times(n+1)}$. Now, as the points $z_0, \ldots, z_n$ coalesce, the approximation $F(z_0, \ldots, z_n)$ will not be well-defined unless we specify the "direction" of convergence. Thus, strictly speaking, F is not defined on all of $M^{\times(n+1)}$, but, rather, on the "off-diagonal" part, by which we mean the subset

$$M^{\diamond(n+1)} \equiv \{(z_0, \ldots, z_n) : z_i \neq z_j \text{ for all } i \neq j\} \subset M^{\times(n+1)}$$

consisting of all *distinct* $(n+1)$-tuples of points. As two or more points come together, the limiting value of $F(z_0, \ldots, z_n)$ will be governed by the derivatives (or jet) of the appropriate order governing the direction of convergence. This observation serves to motivate our construction of the nth *order multi-space* $M^{(n)}$, which shall contain both the jet space J^n and the off-diagonal Cartesian product space $M^{\diamond(n+1)}$ in a consistent manner.

Definition 3 *An* $(n+1)$-*pointed curve* $\mathbf{C} = (z_0, \ldots, z_n; C)$ *consists of a smooth curve* C *and* $n+1$ *not necessarily distinct points* $z_0, \ldots, z_n \in C$ *thereon. Given* $\mathbf{C}$, *we let* $\#i = \#\{j : z_j = z_i\}$. *Two* $(n+1)$-*pointed curves* $\mathbf{C} = (z_0, \ldots, z_n; C)$, $\widetilde{\mathbf{C}} = (\tilde{z}_0, \ldots, \tilde{z}_n; \widetilde{C})$, *have* n^{th} *order multi-contact if and only if*

$$z_i = \tilde{z}_i, \quad \text{and} \quad \mathrm{j}_{\#i-1}C|_{z_i} = \mathrm{j}_{\#i-1}\widetilde{C}|_{z_i}, \quad \text{for each } i = 0, \ldots, n.$$

Definition 4 *The* n^{th} *order multi-space, denoted* $M^{(n)}$ *is the set of equivalence classes of* $(n+1)$-*pointed curves in* M *under the equivalence relation of* n^{th} *order multi-contact. The equivalence class of an* $(n+1)$-*pointed curve* $\mathbf{C}$ *is called its* n^{th} *order multi-jet, and will be denoted as* $\mathrm{j}_n\mathbf{C} \in M^{(n)}$.

In particular, if the points on $\mathbf{C} = (z_0, \ldots, z_n; C)$ are all distinct, then $\mathrm{j}_n\mathbf{C} = \mathrm{j}_n\widetilde{\mathbf{C}}$ if and only if $z_i = \tilde{z}_i$ for all i, which means that $\mathbf{C}$ and $\widetilde{\mathbf{C}}$ have all $n+1$ points in common. Therefore, we can identify the subset of multi-jets of multi-pointed curves having distinct points with the off-diagonal Cartesian product space $M^{\diamond(n+1)} \subset J^n$. On the other hand, if all $n+1$ points coincide, $z_0 = \cdots = z_n$, then $\mathrm{j}_n\mathbf{C} = \mathrm{j}_n\widetilde{\mathbf{C}}$ if and only if $\mathbf{C}$ and $\widetilde{\mathbf{C}}$ have n^{th} order contact at their common point $z_0 = \tilde{z}_0$. Therefore, the multi-space equivalence relation reduces to the ordinary jet space equivalence relation on the set of coincident multi-pointed curves, and in this way $J^n \subset M^{(n)}$. These two extremes do not exhaust the possibilities, since one can have some but not all points coincide. Intermediate cases correspond to "off-diagonal" Cartesian products of jet spaces

$$J^{k_1} \diamond \cdots \diamond J^{k_i} \equiv \{(z_0^{(k_1)}, \ldots, z_i^{(k_i)}) \in J^{k_1} \times \cdots \times J^{k_i} : \pi(z_\nu^{(k_\nu)}) \text{ are distinct}\},$$
$$(6.1)$$

where $\sum k_\nu = n$ and $\pi: J^k \to M$ is the usual jet space projection. These *multi-jet spaces* appear in the work of Dhooghe, [16], on the theory of "semi-differential invariants" in computer vision.

Theorem 16 *If M is a smooth m-dimensional manifold, then its nth order multi-space $M^{(n)}$ is a smooth manifold of dimension $(n + 1)m$, which contains the off-diagonal part $M^{\diamond(n+1)}$ of the Cartesian product space as an open, dense submanifold, and the nth order jet space J^n as a smooth submanifold.*

The proof of theorem 16 requires the introduction of coordinate charts on $M^{(n)}$. Just as the local coordinates on J^n are provided by the coefficients of Taylor polynomials, the local coordinates on $M^{(n)}$ are provided by the coefficients of interpolating polynomials, which are the classical divided differences of numerical interpolation theory, [35, 46].

Definition 5 *Given an $(n + 1)$-pointed graph $\mathbf{C} = (z_0, \ldots, z_n; C)$, its divided differences are defined by $[\, z_j \,]_C = f(x_j)$, and*

$$[\, z_0 z_1 \ldots z_{k-1} z_k \,]_C = \lim_{z \to z_k} \frac{[\, z_0 z_1 z_2 \ldots z_{k-2} z \,]_C - [\, z_0 z_1 z_2 \ldots z_{k-2} z_{k-1} \,]_C}{x - x_{k-1}}.$$
(6.2)

When taking the limit, the point $z = (x, f(x))$ must lie on the curve C, and take limiting values $x \to x_k$ and $f(x) \to f(x_k)$.

In the non-confluent case $z_k \neq z_{k-1}$ we can replace z by z_k directly in the difference quotient (6.2) and so ignore the limit. On the other hand, when all $k+1$ points coincide, the kth order confluent divided difference converges to

$$[\, z_0 \ldots z_0 \,]_C = \frac{f^{(k)}(x_0)}{k!}.$$
(6.3)

Remark: Classically, one employs the simpler notation $[\, u_0 u_1 \ldots u_k \,]$ for the divided difference $[\, z_0 z_1 \ldots z_k \,]_C$. However, the classical notation is ambiguous since it assumes that the mesh $x_0, \ldots, x_n$ is fixed throughout. Because we are regarding the independent and dependent variables on the same footing — and, indeed, are allowing changes of variables that scramble the two — it is important to adopt an unambiguous divided difference notation here.

Theorem 17 *Two $(n + 1)$-pointed graphs $\mathbf{C}, \tilde{\mathbf{C}}$ have n^{th} order multi-contact if and only if they have the same divided differences:*

$$[\, z_0 z_1 \ldots z_k \,]_C = [\, z_0 z_1 \ldots z_k \,]_{\tilde{C}}, \qquad k = 0, \ldots, n.$$

The required local coordinates on multi-space $M^{(n)}$ consist of the independent variables along with all the divided differences

$$x_0, \ldots, x_n, \quad \begin{aligned} u^{(0)} &= u_0 = [\, z_0 \,]_C, & u^{(1)} &= [\, z_0 z_1 \,]_C, \\ u^{(2)} &= 2 \, [\, z_0 z_1 z_2 \,]_C & \cdots & \quad u^{(n)} = n! \, [\, z_0 z_1 \ldots z_n \,]_C, \end{aligned}$$
$$(6.4)$$

prescribed by $(n+1)$-pointed graphs $\mathbf{C} = (z_0, \ldots, z_n; C)$. The $n!$ factor is included so that $u^{(n)}$ agrees with the usual derivative coordinate when restricted to J^n, cf. (6.3). See [43] for a proof that these local coordinate charts endow $M^{(n)}$ with the structure of a smooth manifold.

7 Invariant numerical methods

To implement a numerical solution to a system of differential equations

$$\Delta_\nu(x, u, u^{(1)}, \ldots, u^{(n)}) = 0 \qquad (7.1)$$

by finite difference methods, one relies on suitable discrete approximations to each of its defining differential functions Δ_ν, and this requires extending the differential functions from the jet space to the associated multi-space, in accordance with the following definition.

Definition 6 *An $(n+1)$-point numerical approximation of order k to a differential function $\Delta: J^n \to \mathbb{R}$ is a function $F: M^{(n)} \to \mathbb{R}$ that, when restricted to the jet space, agrees with Δ to order k.*

The simplest illustration is provided by the divided difference coordinates (6.4). Each divided difference $u^{(n)}$ forms an $(n+1)$-point numerical approximation to the nth order derivative coordinate on J^n. According to the usual Taylor expansion, the order of the approximation is $k = 1$. More generally, any differential function $\Delta(x, u, u^{(1)}, \ldots, u^{(n)})$ can immediately be assigned an $(n+1)$-point numerical approximation $F = \Delta(x_0, u^{(0)}, u^{(1)}, \ldots, u^{(n)})$ by replacing each derivative by its divided difference coordinate approximation. However, these are by no means the only numerical approximations possible.

Now let us consider an r-dimensional Lie group G which acts smoothly on M. Since G evidently maps multi-pointed curves to multi-pointed curves while preserving the multi-contact equivalence relation, it induces an action on the multi-space $M^{(n)}$ that will be called the nth *multi-prolongation* of G and denoted by $G^{(n)}$. On the jet subset $J^n \subset M^{(n)}$ the multi-prolonged action reduced to the usual jet space prolongation. On the other hand, on the off-diagonal part $M^{\diamond(n+1)} \subset M^{(n)}$ the action coincides with the $(n+1)$-fold Cartesian product action of G on $M^{\times(n+1)}$.

We define a *multi-invariant* to be a function $K: M^{(n)} \to \mathbb{R}$ on multi-space which is invariant under the multi-prolonged action of $G^{(n)}$. The restriction of a multi-invariant K to jet space will be a differential invariant, $I = K \mid J^n$, while restriction to $M^{\diamond(n+1)}$ will define a joint invariant $J = K \mid M^{\diamond(n+1)}$. Smoothness of K will imply that the joint invariant J is an *invariant numerical approximation to the differential invariant* I. Moreover, every invariant finite difference numerical approximation arises in this manner. Thus, the theory of multi-invariants *is* the theory of invariant numerical approximations!

Furthermore, the restriction of a multi-invariant to an intermediate multi-jet subspace, as in (6.1), will define a joint differential invariant, [42] — also known as a semi-differential invariant in the computer vision literature, [16, 36]. The approximation of differential invariants by joint differential invariants is, therefore, based on the extension of the differential invariant from the jet space to a suitable multi-jet subspace (6.1). The invariant numerical approximations to joint differential invariants are, in turn, obtained by extending them from the multi-jet subspace to the entire multi-space. Thus, multi-invariants also include invariant semi-differential approximations to differential invariants as well as joint invariant numerical approximations to differential invariants and semi-differential invariants — all in one seamless geometric framework.

Effectiveness of the group action on M implies, typically, freeness and regularity of the multi-prolonged action on an open subset of $M^{(n)}$. This allows us to directly apply the basic moving frame construction. The resulting *multi-frame* $\rho^{(n)}: M^{(n)} \to G$ will lead us immediately to the required multi-invariants and hence a general, systematic construction for invariant numerical approximations to differential invariants. Any multi-frame will evidently restrict to a classical moving frame $\rho^{(n)}: J^n \to G$ on the jet space along with a suitably compatible product frame $\rho^{\diamond(n+1)}: M^{\diamond(n+1)} \to G$.

In local coordinates, we let $w_k = (y_k, v_k) = g \cdot z_k$ denote the transformation formulae for the individual points on a multi-pointed curve. The multi-prolonged action on the divided difference coordinates gives

$$y_0, \ldots, y_n, \quad \begin{matrix} v^{(0)} = v_0 = [\, w_0 \,], & v^{(1)} = [\, w_0 w_1 \,], \\ v^{(2)} = [\, w_0 w_1 w_2 \,], & \ldots & v^{(n)} = n! \, [\, w_0, \ldots, w_n \,], \end{matrix} \quad (7.2)$$

where the formulae are most easily computed via the difference quotients

$$[\, w_j \,] = v_j, \qquad \text{and}$$

$$[w_0 w_1 \ldots w_{k-1} w_k] = \frac{[w_0 w_1 w_2 \ldots w_{k-2} w_k] - [w_0 w_1 w_2 \ldots w_{k-2} w_{k-1}]}{y_k - y_{k-1}},$$

and then taking appropriate limits to cover the case of coalescing points. Inspired by the constructions in [21], we will refer to (7.2) as the *lifted divided difference invariants*.

To construct a multi-frame, we need to normalize by choosing a cross-section to the group orbits in $M^{(n)}$, which amounts to setting $r = \dim G$ of the lifted divided difference invariants (7.2) equal to suitably chosen constants. An important observation is that in order to obtain the limiting differential invariants, we must require our local cross-section to pass through the jet space $J^n \subset M^{(n)}$, and define, by intersection, a cross-section for the prolonged action on J^n. This compatibility constraint implies, in particular, that we are only allowed to normalize the first lifted independent variable $y_0 = c_0$.

With the aid of the multi-frame, the most direct construction of the requisite multi-invariants and associated invariant numerical differentiation formulae is through the invariantization of the original finite difference quotients (6.2). Substituting the multi-frame formulae for the group parameters into the lifted coordinates (7.2) provides a complete system of multi-invariants on $M^{(n)}$; this follows immediately from theorem 3. We denote the fundamental multi-invariants by

$$y_i \longmapsto H_i = \iota(x_i), \qquad v^{(n)} \longmapsto K^{(n)} = \iota(u^{(n)}), \qquad (7.3)$$

where ι denotes the invariantization map associated with the multi-frame. The fundamental differential invariants for the prolonged action of G on J^n can all be obtained by restriction, so that $I^{(n)} = K^{(n)} \,|\, J^n$. On the jet space, the points are coincident, and so the multi-invariants H_i will all restrict to the *same* constant invariant $c_0 = H = H_i \,|\, J^n$, the normalization value of y_0. On the other hand, the fundamental joint invariants on $M^{\diamond(n+1)}$ are obtained by restricting the multi-invariants $H_i = \iota(x_i)$ and $K_i = \iota(u_i)$. The multi-invariants can computed by using a multi-invariant divided difference recursion

$$[I_j] = K_j = \iota(u_j), \qquad (7.4)$$

$$[I_0 \ldots I_k] = \iota([z_0 z_1 \ldots z_k]) = \frac{[I_0 \ldots I_{k-2} I_k] - [I_0 \ldots I_{k-2} I_{k-1}]}{H_k - H_{k-1}},$$

and then relying on continuity to extend the formulae to coincident points. The multi-invariants

$$K^{(n)} = n! \,[I_0 \ldots I_n] = \iota(u^{(n)}) \qquad (7.5)$$

define the fundamental first order invariant numerical approximations to the differential invariants $I^{(n)}$. Higher order invariant approximations can be obtained by invariantization of the higher order divided difference approximations. The moving frame construction has a significant advantage over the infinitesimal approach used by Dorodnitsyn, [17, 18], in that it does not require the solution of partial differential equations in order to construct the multi-invariants.

Given a regular G-invariant differential equation

$$\Delta(x, u, u^{(1)}, \ldots, u^{(n)}) = 0, \tag{7.6}$$

we can invariantize the left hand side to rewrite the differential equation in terms of the fundamental differential invariants:

$$\iota(\Delta(x, u, u^{(1)}, \ldots, u^{(n)})) = \Delta(H, I^{(0)}, \ldots, I^{(n)}) = 0.$$

The invariant finite difference approximation to the differential equation is then obtained by replacing the differential invariants $I^{(k)}$ by their multi-invariant counterparts $K^{(k)}$:

$$\Delta(c_0, K^{(0)}, \ldots, K^{(n)}) = 0. \tag{7.7}$$

Example 6 Consider the elementary action

$$(x, u) \longmapsto (\lambda^{-1} x + a, \lambda u + b)$$

of the three-parameter similarity group $G = \mathbb{R}^2 \ltimes \mathbb{R}$ on $M = \mathbb{R}^2$. To obtain the multi-prolonged action, we compute the divided differences (7.2) of the basic lifted invariants

$$y_k = \lambda^{-1} x_k + a, \qquad v_k = \lambda u_k + b.$$

We find

$$v^{(1)} = [\, w_0 w_1 \,] = \frac{v_1 - v_0}{y_1 - y_0} = \lambda^2 \frac{u_1 - u_0}{x_1 - x_0} = \lambda^2 [\, z_0 z_1 \,] = \lambda^2 u^{(1)}.$$

More generally,

$$v^{(n)} = \lambda^{n+1} u^{(n)}, \qquad n \geq 1. \tag{7.8}$$

Note that we may compute the multi-space transformation formulae assuming initially that the points are distinct, and then extending to coincident cases by continuity. (In fact, this gives an alternative method for computing the standard jet space prolongations of group actions!) In particular, when all the points coincide, each $u^{(n)}$ reduces to the nth

order derivative coordinate, and (7.8) reduces to the prolonged action of G on J^n. We choose the normalization cross-section defined by

$$y_0 = 0, \qquad v_0 = 0, \qquad v^{(1)} = 1,$$

which, upon solving for the group parameters, leads to the basic moving frame

$$a = -x_0 \sqrt{u^{(1)}}, \qquad b = -\frac{u_0}{\sqrt{u^{(1)}}}, \qquad \lambda = \frac{1}{\sqrt{u^{(1)}}}, \qquad (7.9)$$

where, for simplicity, we restrict to the subset where $u^{(1)} = [\, z_0 z_1 \,] > 0$. The fundamental joint similarity invariants are obtained by substituting these formulae into

$$y_k \longmapsto H_k = (x_k - x_0)\sqrt{u^{(1)}} = (x_k - x_0)\sqrt{\frac{u_1 - u_0}{x_1 - x_0}},$$

$$v_k \longmapsto K_k = \frac{u_k - u_0}{\sqrt{u^{(1)}}} = (u_k - u_0)\sqrt{\frac{x_1 - x_0}{u_1 - u_0}},$$

both of which reduce to the trivial zero differential invariant on J^n. Higher order multi-invariants are obtained by substituting (7.9) into the lifted invariants (7.8), leading to

$$K^{(n)} = \frac{u^{(n)}}{(u^{(1)})^{(n+1)/2}} = \frac{n!\,[\, z_0 z_1 \ldots z_n \,]}{[\, z_0 z_1 z_2 \,]^{(n+1)/2}}.$$

In the limit, these reduce to the differential invariants

$$I^{(n)} = (u^{(1)})^{-(n+1)/2}\, u^{(n)},$$

and so the multi-invariants $K^{(n)}$ are the desired similarity-invariant, first order numerical approximations. To construct an invariant numerical scheme for any similarity-invariant ordinary differential equation

$$\Delta(x, u, u^{(1)}, u^{(2)}, \ldots u^{(n)}) = 0,$$

we merely invariantize the defining differential function, leading to the general similarity–invariant numerical approximation

$$\Delta(0, 0, 1, K^{(2)}, \ldots, K^{(n)}) = 0.$$

Example 7 For the action (3.4) of the proper Euclidean group SE(2) on $M = \mathbb{R}^2$, the multi-prolonged action is free on $M^{(n)}$ for $n \geq 1$. We can thereby determine a first order multi-frame and use it to completely classify Euclidean multi-invariants. The first order transformation formulae are

$$y_0 = x_0 \cos\theta - u_0 \sin\theta + a, \quad v_0 = x_0 \sin\theta + u_0 \cos\theta + b,$$

$$y_1 = x_1 \cos\theta - u_1 \sin\theta + a, \quad v^{(1)} = \frac{\sin\theta + u^{(1)}\cos\theta}{\cos\theta - u^{(1)}\sin\theta}, \tag{7.10}$$

where $u^{(1)} = [z_0 z_1]$. Normalization based on the cross-section $y_0 = v_0 = v^{(1)} = 0$ results in the right moving frame

$$a = -x_0 \cos\theta + u_0 \sin\theta = -\frac{x_0 + u^{(1)} u_0}{\sqrt{1 + (u^{(1)})^2}},$$

$$b = -x_0 \sin\theta - u_0 \cos\theta = \frac{x_0 u^{(1)} - u_0}{\sqrt{1 + (u^{(1)})^2}}, \qquad \theta = -\tan^{-1} u^{(1)}.$$

$$\tag{7.11}$$

Substituting the moving frame formulae (7.11) into the lifted divided differences results in a complete system of (oriented) Euclidean multi-invariants. These are easily computed by beginning with the fundamental joint invariants $I_k = (H_k, K_k) = \iota(x_k, u_k)$, where

$$y_k \longmapsto H_k = \frac{(x_k - x_0) + u^{(1)}(u_k - u_0)}{\sqrt{1 + (u^{(1)})^2}} = (x_k - x_0)\frac{1 + [z_0 z_1][z_0 z_k]}{\sqrt{1 + [z_0 z_1]^2}},$$

$$v_k \longmapsto K_k = \frac{(u_k - u_0) - u^{(1)}(x_k - x_0)}{\sqrt{1 + (u^{(1)})^2}} = (x_k - x_0)\frac{[z_0 z_k] - [z_0 z_1]}{\sqrt{1 + [z_0 z_1]^2}}.$$

The multi-invariants are obtained by forming divided difference quotients

$$[I_0 I_k] = \frac{K_k - K_0}{H_k - H_0} = \frac{K_k}{H_k} = \frac{(x_k - x_1)[z_0 z_1 z_k]}{1 + [z_0 z_k][z_0 z_1]},$$

where, in particular, $I^{(1)} = [I_0 I_1] = 0$. The second order multi-invariant

$$I^{(2)} = 2[I_0 I_1 I_2] = 2\frac{[I_0 I_2] - [I_0 I_1]}{H_2 - H_1}$$

$$= \frac{2[z_0 z_1 z_2]\sqrt{1 + [z_0 z_1]^2}}{\left(1 + [z_0 z_1][z_1 z_2]\right)\left(1 + [z_0 z_1][z_0 z_2]\right)}$$

$$= \frac{u^{(2)}\sqrt{1 + (u^{(1)})^2}}{\left[1 + (u^{(1)})^2 + \frac{1}{2}u^{(1)}u^{(2)}(x_2 - x_0)\right]\left[1 + (u^{(1)})^2 + \frac{1}{2}u^{(1)}u^{(2)}(x_2 - x_1)\right]}$$

provides a Euclidean–invariant numerical approximation to the Euclidean curvature:

$$\lim_{z_1, z_2 \to z_0} I^{(2)} = \kappa = \frac{u^{(2)}}{(1 + (u^{(1)})^2)^{3/2}}.$$

Similarly, the third order multi-invariant

$$I^{(3)} = 6 \left[I_0 I_1 I_2 I_3 \right] = 6 \frac{\left[I_0 I_1 I_3 \right] - \left[I_0 I_1 I_2 \right]}{H_3 - H_2}$$

will form a Euclidean–invariant approximation for the normalized differential invariant $\kappa_s = \iota(u_{xxx})$, the derivative of curvature with respect to arc length, [11, 21].

To compare these with the invariant numerical approximations proposed in [11, 12, 3], we reformulate the divided difference formulae in terms of the geometrical configurations of the four distinct points z_0, z_1, z_2, z_3 on our curve. We find

$$H_k = \frac{(z_1 - z_0) \cdot (z_k - z_0)}{\| z_1 - z_0 \|} = r_k \cos \phi_k,$$

$$[I_0 I_k] = \tan \phi_k,$$

$$K_k = \frac{(z_1 - z_0) \wedge (z_k - z_0)}{\| z_1 - z_0 \|} = r_k \sin \phi_k,$$

where

$$r_k = \| z_k - z_0 \|, \qquad \phi_k = \sphericalangle (z_k - z_0, z_1 - z_0),$$

denotes the distance and the angle between the indicated vectors. Therefore,

$$I^{(2)} = 2 \frac{\tan \phi_2}{r_2 \cos \phi_2 - r_1},$$

$$I^{(3)} = 6 \frac{(r_2 \cos \phi_2 - r_1) \tan \phi_3 - (r_3 \cos \phi_3 - r_1) \tan \phi_2}{(r_2 \cos \phi_2 - r_1)(r_3 \cos \phi_3 - r_1)(r_3 \cos \phi_3 - r_2 \cos \phi_2)}. \tag{7.12}$$

Interestingly, $I^{(2)}$ is *not* the same Euclidean approximation to the curvature that was used in [11, 12]. The latter was based on the Heron formula for the radius of a circle through three points:

$$I^\star = \frac{4\Delta}{abc} = \frac{2 \sin \phi_2}{\| z_1 - z_2 \|}. \tag{7.13}$$

Here Δ denotes the area of the triangle connecting z_0, z_1, z_2 and

$$a = r_1 = \| z_1 - z_0 \|, \qquad b = r_2 = \| z_2 - z_0 \|, \qquad c = \| z_2 - z_1 \|,$$

are its side lengths. The ratio between these two curvature approximations tends to a limit $I^\star / I^{(2)} \to 1$ as the points coalesce. The geometrical approximation (7.13) has the advantage that it is symmetric under permutations of the points; one can achieve the same thing by symmetrizing the divided difference version $I^{(2)}$. Furthermore, $I^{(3)}$ is an invariant approximation for the differential invariant κ_s, that, like the approximations constructed by Boutin, [3], converges properly for arbitrary spacings of the points on the curve.

References

[1] Berchenko, I.A. and Olver, P.J. (2000). Symmetries of polynomials, *J. Symb. Comp.*, to appear.

[2] Blumenthal, L.M. (1953). *Theory and Applications of Distance Geometry* (Oxford Univ. Press, Oxford).

[3] Boutin, M. (1999). Numerically invariant signature curves, preprint (University of Minnesota).

[4] Bruckstein, A.M., Holt, R.J., Netravali and A.N., Richardson, T.J. (1993). Invariant signatures for planar shape recognition under partial occlusion, *CVGIP: Image Understanding* **58**, 49–65.

[5] Bruckstein, A.M. and Netravali, A.N. (1995). On differential invariants of planar curves and recognizing partially occluded planar shapes, *Ann. Math. Artificial Intel.* **13**, 227–250.

[6] Bruckstein, A.M., Rivlin, E. and Weiss, I. (1997). Scale space semi-local invariants, *Image Vision Comp.* **15**, 335–344.

[7] Bruckstein, A.M. and Shaked, D. (1998). Skew-symmetry detection via invariant signatures, *Pattern Recognition* **31**, 181–192.

[8] Budd, C.J. and Collins, C.B. (1998). Symmetry based numerical methods for partial differential equations, in *Numerical Analysis 1997* (D.F. Griffiths, D.J. Higham and G.A. Watson, eds), Pitman Res. Notes Math., vol. 380 (Longman, Harlow), pp. 16–36.

[9] Budd, C.J. and Iserles, A. (1999). Geometric integration: numerical solution of differential equations on manifolds, *Phil. Trans. Roy. Soc. London A* **357**, 945–956.

[10] Carlsson, S., Mohr, R., Moons, T., Morin, L., Rothwell, C., Van Diest, M., Van Gool, L., Veillon, F. and Zisserman, A. (1996). Semi-local projective invariants for the recognition of smooth plane curves, *Int. J. Comput. Vision* **19**, 211–236.

[11] Calabi, E., Olver, P.J., Shakiban, C., Tannenbaum, A. and Haker, S. (1998). Differential and numerically invariant signature curves applied to object recognition, *Int. J. Computer Vision* **26**, 107–135.

[12] Calabi, E., Olver, P.J. and Tannenbaum, A. (1996). Affine geometry, curve flows, and invariant numerical approximations, *Adv. in Math.* **124**, 154–196.

[13] Cartan, É. (1935). La Méthode du Repère Mobile, la Théorie des Groupes Continus, et les Espaces Généralisés, Exposés de Géométrie No. 5 (Hermann, Paris).

[14] Cartan, É. (1937). La Théorie des Groupes Finis et Continus et la Géométrie Différentielle Traitées par la Méthode du Repère Mobile, Cahiers Scientifiques, Vol. 18 (Gauthier–Villars, Paris).

[15] Channell, P.J. and Scovel, C. (1990). Symplectic integration of Hamiltonian systems, *Nonlinearity* **3**, 231–259.

[16] Dhooghe, P.F. (1996). Multilocal invariants, in *Geometry and Topology of Submanifolds, VIII* (F. Dillen, B. Komrakov, U. Simon, I. Van de

Woestyne, and L. Verstraelen, eds), (World Sci. Publishing, Singapore), 121–137.

[17] Dorodnitsyn, V.A. (1991). Transformation groups in net spaces, *J. Sov. Math.* **55**, 1490–1517 .

[18] Dorodnitsyn, V.A. (1994). Finite difference models entirely inheriting continuous symmetry of original differential equations, *Int. J. Mod. Phys. C* **5**, 723–734.

[19] Faugeras, O. (1994). Cartan's moving frame method and its application to the geometry and evolution of curves in the euclidean, affine and projective planes, in *Applications of Invariance in Computer Vision* (J.L. Mundy, A. Zisserman and D. Forsyth, eds), Lecture Notes in Computer Science, Vol. 825, (Springer-Verlag, Berlin), 11–46.

[20] Fels, M. and Olver, P.J. (1998). Moving coframes. I. A practical algorithm, *Acta Appl. Math.* **51**, 161–213.

[21] Fels, M., Olver, P.J. (1999). Moving coframes. II. Regularization and theoretical foundations, *Acta Appl. Math.* **55**, 127–208.

[22] Griffiths, P.A. (1974). On Cartan's method of Lie groups and moving frames as applied to uniqueness and existence questions in differential geometry, *Duke Math. J.* **41**, 775–814.

[23] Guggenheimer, H.W. (1963). *Differential Geometry* (McGraw–Hill, New York).

[24] Hilbert, D. (1993). *Theory of Algebraic Invariants* (Cambridge Univ. Press, New York).

[25] Jaegers, P.J. (1994). Lie group invariant finite difference schemes for the neutron diffusion equation, Los Alamos National Lab Report, LA–12791–T.

[26] Jensen, G.R. (1977). Higher order contact of submanifolds of homogeneous spaces, Lecture Notes in Math., No. 610, (Springer–Verlag, New York).

[27] Kichenassamy, S., Kumar, A., Olver, P.J., Tannenbaum, A. and Yezzi, A. (1996). Conformal curvature flows: from phase transitions to active vision, *Arch. Rat. Mech. Anal.* **134**, 275–301.

[28] Killing, W. (1890). Erweiterung der Begriffes der Invarianten von Transformationgruppen, *Math. Ann.* **35**, 423–432.

[29] Kogan, I.A. (2000). Inductive approach to moving frames and applications in classical invariant theory, Ph. D. Thesis (University of Minnesota).

[30] Kogan, I.A. and Olver, P.J. (2000). The invariant variational bicomplex, preprint (University of Minnesota).

[31] Lewis, D. and Simo, J.C. (1994). Conserving algorithms for the dynamics of Hamiltonian systems on Lie groups, *J. Nonlin. Sci.* **4**, 253–299.

[32] Marí-Beffa, G. and Olver, P.J. (1999). Differential invariants for parametrized projective surfaces, *Commun. Anal. Geom.* **7**, 807–839.

[33] McLachlan, R.I., Quispel, G.R.W. and Robidoux, N. (1999). Geometric integration using discrete gradients, *Phil. Trans. Roy. Soc. London A* **357**, 1021–1045.

[34] Menger, K. (1928). Untersuchungen über allgemeine Metrik, *Math. Ann.* **100**, 75–163.

[35] Milne–Thompson, L.M. (1951). *The Calculus of Finite Differences* (Macmillan, London).

[36]   Moons, T., Pauwels, E., Van Gool, L. and Oosterlinck, A. (1995). Foundations of semi-differential invariants, *Int. J. Comput. Vision* **14**, 25–48.

[37]   Olver, P.J. (1990). Classical invariant theory and the equivalence problem for particle Lagrangians. I. Binary Forms, *Adv. in Math.* **80** 39–77.

[38]   Olver, P.J. (1993). *Applications of Lie Groups to Differential Equations* (2nd ed.), GTM 107 (Springer-Verlag, New York).

[39]   Olver, P.J. (1995). *Equivalence, Invariants, and Symmetry* (Cambridge University Press, Cambridge). .

[40]   Olver, P.J. (1999). *Classical Invariant Theory*, London Math. Soc. Student Texts, vol. 44 (Cambridge University Press, Cambridge).

[41]   Olver, P.J. (1999). Moving frames and singularities of prolonged group actions. *Selecta Math.*, to appear.

[42]   Olver, P.J. (2000), Joint invariant signatures, *Found. Comp. Math.*, to appear.

[43]   Olver, P.J. (2000). Geometric foundations of numerical algorithms and symmetry, *Appl. Alg. Engin. Comp. Commun.*, to appear.

[44]   Olver, P.J. (1999). Multi-space, in preparation.

[45]   Pauwels, E., Moons, T., Van Gool, L.J., Kempenaers, P. and Oosterlinck, A. (1995). Recognition of planar shapes under affine distortion, *Int. J. Comput. Vision* **14**, 49–65.

[46]   Powell, M.J.D. (1981). *Approximation Theory and Methods* (Cambridge University Press, Cambridge).

[47]   Shokin, Y.I. (1983). *The Method of Differential Approximation* (Springer-Verlag, New York).

[48]   van Beckum, F.P.H. and van Groesen, E. (1987). Discretizations conserving energy and other constants of the motion, in *Proc. ICIAM 87*, *Paris*, 17–35.

[49]   Van Gool, L., Brill, M.H., Barrett, E.D., Moons, T. and Pauwels, E. (1992). Semi-differential invariants for nonplanar curves in *Geometric Invariance in Computer Vision* (J.L. Mundy and A. Zisserman, eds), (MIT Press, Cambridge, Mass), 293–309.

[50]   Van Gool, L., Moons, T., Pauwels, E. and Oosterlinck, A. (1992). Semi-differential invariants, in *Geometric Invariance in Computer Vision* (J.L. Mundy and A. Zisserman, eds), (MIT Press, Cambridge, Mass.), 157–192.

[51]   Weyl, H. (1946). *Classical Groups* (Princeton Univ. Press, Princeton, N.J.).

[52]   Yezzi, A., Kichenassamy, S., Kumar, A., Olver and P.J., Tannenbaum, A. (1997). A geometric snake model for segmentation of medical imagery, *IEEE Trans. Medical Imaging* **16**, 199–209.

Harmonic Map Flows and Image Processing

Guillermo Sapiro

Electrical and Computer Engineering
University of Minnesota
Minneapolis, MN 55455
Email: guille@ece.umn.edu

1 Introduction

In a a number of applications in image processing, computer vision, and computer graphics, the data of interest is defined on non-flat manifolds and maps onto non-flat manifolds. A classical and important example is directional data, including gradient directions, optical flow directions, surface normals, principal directions, and chroma. Frequently, this data is available in a noisy fashion, and there is a need for noise removal. In addition, it is often desired to obtain a multiscale-type representation of the directional data, similar to those representations obtained for gray-level images, [2, 31, 36, 37, 55]. Addressing the processing of non-flat data is the goal of this chapter. We will illustrate the basic ideas with directional data and probability distributions. In the first case, the data maps onto an hypersphere, while on the second one it maps onto a semi-hyperplane.

Image data, as well as directions and other sources of information, are not always defined on the $\mathbb{R}^2$ plane or $\mathbb{R}^3$ space. They can be, for example, defined over a surface embedded in $\mathbb{R}^3$. It is important then to define basic image processing operation for general data defined on general (not-necessarily flat) manifolds. In other words, we want to deal with maps between two general manifolds, and be able for example to isotropically and anisotropically diffuse them with the goal of noise removal. This will make it possible for example to denoise data defined on 3D surfaces. Although we are particularly interested in this chapter in directions defined on $\mathbb{R}^2$ (non-flat data defined on a flat manifold), the framework here presented applies to the general case as well. We will also briefly illustrate the ideas for data defined on semi-hyperplanes, mainly posterior probabilities.

299

An $\mathbb{R}^n$ direction defined on an image in $\mathbb{R}^2$ is given by a vector $I(x, y, 0) : \mathbb{R}^2 \to \mathbb{R}^n$ such that the Euclidean norm of $I(x, y, 0)$ is equal to one, that is,

$$\sqrt{\sum_{i=1}^{n} I_i^2(x, y, 0)} = 1,$$

where $I_i(x, y, 0) : \mathbb{R}^2 \to \mathbb{R}$ are the components of the vector. The notation can be simplified by considering $I(x, y, 0) : \mathbb{R}^2 \to S^{n-1}$, where S^{n-1} is the unit ball in $\mathbb{R}^n$. This implicitly includes the unit norm constraint. (Any non-zero vector can be transformed into a direction by normalizing it. For zero vectors, the unit norm constraint has to be relaxed, and a norm less or equal to one needs to be required.) When smoothing the data, or computing a multiscale representation $I(x, y, t)$ of a direction $I(x, y, 0)$ (t stands for the scale), it is crucial to maintain the unit norm constraint, which is an intrinsic characteristic of directional data. That is, the smoothed direction $\hat{I}(x, y, 0) : \mathbb{R}^2 \to \mathbb{R}^n$ must also satisfy

$$\sqrt{\sum_{i=1}^{n} \hat{I}_i^2(x, y, 0)} = 1.$$

Or, $\hat{I}(x, y, 0) : \mathbb{R}^2 \to S^{n-1}$. The same constraint holds for a multiscale representation $I(x, y, t)$ of the original direction $I(x, y, 0)$. This is what makes the smoothing of directions different from the smoothing of ordinary vectorial data as in [8, 43, 44, 53, 54, 56]. The smoothing is performed in S^{n-1} instead of $\mathbb{R}^n$.

Directions can also be represented by the angle(s) the vector makes with a given coordinate system, denoted in this Chapter as *orientation(s)*. In the 2D case for example, the direction of a vector (I_1, I_2) can be given by the angle θ that this vector makes with the x axis (we consider $\theta \in [0, 2\pi)$): $\theta = \arctan(I_2/I_1)$ (with the corresponding sign considerations to have the map in the $[0, 2\pi)$ interval). There is of course a one-to-one map between a direction vector $I(x, y) : \mathbb{R}^2 \to S^1$ and the angle function $\theta(x, y)$. Using this relation, Perona, [36], well motivated the necessity for orientation and direction diffusion and transformed the problem of 2D direction diffusion into a 1D problem of angle or orientation diffusion. Perona then proposed PDE's based techniques for the isotropic smoothing of 2D orientations; see also [26, 52] and the general discussion of these methods in [36]. Smoothing orientations instead of directions solves the unit norm constraint, but adds a periodicity constraint. Perona showed that a simple heat flow (Laplacian or Gaussian filtering) applied to the $\theta(x, y)$ image, together with special numerical attention, can address this periodicity issue. This pioneering approach theoretically applies only to smooth data (indeed to vectors with coordinates in the Sobolev space $W^{1,2}$), and thereby disqualifying edges. The straightforward extension of this to S^{n-1} would be to consider $n - 1$ angles, and smooth each one of these as a scalar image. The natural coupling is then missing, obtaining a set of decoupled PDE's.

As Perona pointed out in his work, directional data is just one example of the diffusion of images representing data beyond flat manifolds. Extensions to Perona's work, using intrinsic metrics on the manifold can be found in [11, 44]. In [11] the authors explicitly deal with orientations (no directions) and present the L_1 norm as well as many additional new features, contributions on discrete formulations, and connections with the now described approach. The work [44] does not deal with orientations or directions, although the framework is valid to approach this problem as well. In [40] the authors also mention the minimization of the L_1 norm of the divergence of the normalized image gradient (curvature of the level-sets). This is done in the framework of image denoising, without addressing the regularization and analysis of directional data or presenting examples. Note that not only the data (e.g., directions) can go beyond flat manifolds, but its domain can also be non-flat (e.g., the data can be defined on a surface). The harmonic framework here described addresses the non-flatness of both manifolds. That is, with the general framework here introduced we can obtain isotropic and anisotropic diffusion and scale-spaces for any function mapping two manifolds (see also [44], a framework also permitting this). We can for example diffuse (and denoise) data on a 3D surface, or diffuse posterior probability vectors. In this chapter we present the general framework and detail the case of directions defined on the plane (and briefly data onto hyperplanes), while other cases are described elsewhere.

From the original unit norm vectorial image $I(x, y, 0) : \mathbb{R}^2 \to S^{n-1}$ we construct a family of unit norm vectorial images $I(x, y, t) : \mathbb{R}^2 \times [0, \tau) \to S^{n-1}$ that provides a multiscale representation of directions. The method intrinsically takes care of the normalization constraint, eliminating the need to consider orientations and develop special periodicity preserving numerical approximations. Discontinuities in the directions are also allowed by the algorithm. The approach follows results from the literature on harmonic maps in liquid crystals, and $I(x, y, t)$ is obtained

from a system of coupled partial differential equations that reduces a
given (harmonic) energy. Energies giving both isotropic and anisotropic
flows will be described. Due to the large amount of literature in the subject of harmonic maps applied to liquid crystals, a number of relevant
theoretical results can immediately be obtained.

3 Harmonic maps and direction diffusion

Let $I(x, y, 0) : \mathbb{R}^2 \to S^{n-1}$ be the original image of directions. That is,
this is a collection of vectors from $\mathbb{R}^2$ to $\mathbb{R}^n$ such that their unit norm
is equal to one, i.e., $\| I(x, y, 0) \| = 1$, where $\| \cdot \|$ indicates Euclidean
length. $I_i(x, y, 0) : \mathbb{R}^2 \to \mathbb{R}$ stand for each one of the n components
of $I(x, y, 0)$. We search for a family of images, a multiscale representation, of the form $I(x, y, t) : \mathbb{R}^2 \times [0, \tau) \to S^{n-1}$, and once again we
use $I_i(x, y, t) : \mathbb{R}^2 \to \mathbb{R}$ to represent each one of the components of this
family. Let us define the *component gradient* ∇I_i as

$$\nabla I_i := \frac{\partial I_i}{\partial x} \vec{x} + \frac{\partial I_i}{\partial y} \vec{y}, \tag{3.1}$$

where $\vec{x}$ and $\vec{y}$ are the unit vectors in the x and y directions respectively.
From this,

$$\| \nabla I_i \| = \left(\left(\frac{\partial I_i}{\partial x} \right)^2 + \left(\frac{\partial I_i}{\partial y} \right)^2 \right)^{1/2}, \tag{3.2}$$

gives the absolute value of the component gradient.

The *component Laplacian* is given by

$$\Delta I_i = \frac{\partial^2 I_i}{\partial x^2} + \frac{\partial^2 I_i}{\partial y^2}. \tag{3.3}$$

We are also interested in the *absolute value of the image gradient*,
given by

$$\| \nabla I \| := \left(\sum_{i=1}^{n} \left(\left(\frac{\partial I_i}{\partial x} \right)^2 + \left(\frac{\partial I_i}{\partial y} \right)^2 \right) \right)^{1/2}. \tag{3.4}$$

Having this notation, we are now ready to formulate the framework.
The problem of *harmonic maps in liquid crystals* is formulated as the
search for the solution to

$$\min_{I:\mathbb{R}^2 \to S^{n-1}} \int \int_{\Omega} \| \nabla I \|^p \, dx dy, \tag{3.5}$$

where Ω stands for the image domain and $p \geq 1$. This variational formulation can be re-written as

$$\min_{I:\mathbb{R}^2 \to \mathbb{R}^n} \int \int_\Omega \parallel \nabla I \parallel^p dx dy, \qquad (3.6)$$

such that

$$\parallel I \parallel = 1. \qquad (3.7)$$

This is a particular case of the search for maps I between Riemannian manifolds (M, g) and (N, h) which are critical points (that is, minimizers) of the *harmonic energy*

$$E_p(I) = \int_M \parallel \nabla_M I \parallel^p \text{dvol} M, \qquad (3.8)$$

where $\parallel \nabla_M I \parallel$ is the length of the differential in M. In our particular case, M is a domain in $\mathbb{R}^2$ and $N = S^{n-1}$, and $\parallel \nabla_M I \parallel$ reduces to (3.4). The critical points of (3.8) are called *p-harmonic maps* (or simply *harmonic maps* for $p = 2$). This is in analogy to the critical points of the Dirichlet energy $\int_\Omega \parallel \nabla f \parallel^2$ for real valued functions f, which are called *harmonic functions*.

The general form of the harmonic energy, normally from a 3D surface (M) to the plane (N) with $p = 2$ (the most classical case, e.g., [20, 21]), was successfully used for example in computer graphics to find smooth maps between two given (triangulated) surfaces (again, normally a surface and the complex or real plane); e.g. [19, 3, 27, 58]. In this case, the search is indeed for the critical point, that is, for the harmonic map between the surfaces. This can be done for example via finite elements [3, 27]. In our case, the problem is different. We already have a candidate map, the original (normally noisy or with many details at all scales) image of directions $I(x, y, 0)$, and we want to compute a multiscale representation or regularized/denoised version of it. That is, we are not (just) interested in the harmonic map between the domain in $\mathbb{R}^2$ and S^{n-1} (the critical point of the energy), but are interested in the process of computing this map via partial differential equations. More specifically, we are interested in the gradient-descent type flow of the harmonic energy (3.8). This is partially motivated by the fact that, as we have already seen, the basic diffusion equations for multiscale representations and denoising of gray-valued images are obtained as well as gradient descent flows acting on real-valued data; see for example [7, 37, 41, 57]. Isotropic diffusion (linear heat flow) is just the gradient descent of the L_2 norm of the image gradient, while anisotropic diffusion can be interpreted as the gradient descent flow of more robust functions acting on the image gradient.

Since we have an energy formulation, it is straightforward to add additional data-dependent constraints on the minimization process, e.g., preservation of the original average; see for example [41] for examples for gray-valued images. In this case we might indeed be interested in the critical point of the modified energy, which can be obtained as the steady-state solution of the corresponding gradient descent flow. Since the goal of this Chapter is to describe the general framework for direction diffusion, we will not add these type of constraints in the examples in §4. These constraints are normally closely tied to both the specific problem and the available information about the type of noise present in the image. In [11], data-terms are added.

For the most popular case of $p = 2$, the Euler-Lagrange equation corresponding to (3.8) is a simple formula based on Δ_M, the Laplace-Beltrami operator of M, and $A_N(I)$, the second fundamental form of N (assumed to be embedded in $\mathbb{R}^k$) evaluated at I; e.g., [20, 21, 45]:

$$\Delta_M I + A_N(I)\langle \nabla_M I, \nabla_M I \rangle = 0. \tag{3.9}$$

This leads to a gradient-descent type of flow, that is,

$$\frac{\partial I}{\partial t} = \Delta_M I + A_N(I)\langle \nabla_M I, \nabla_M I \rangle. \tag{3.10}$$

In the following sections, we will present the gradient descent flows for our particular energy (3.5), that is, for M being a domain in $\mathbb{R}^2$ and N equal to S^{n-1}. We concentrate on the cases of $p = 2$, isotropic, and $p = 1$, anisotropic (or in general $1 \leq p < 2$). The use of $p = 2$ corresponds to the classical heat flow from the linear scale-space theory [31, 55], while the case $p = 1$ corresponds to the *total variation* flow studied in [41]. For data like surface normals, principal directions, or simple gray values on 3D, M is a surface in 3D and the general flow (3.10) is used. This flow can be implemented using classical numerical techniques to compute ∇_M, Δ_M, $A_N(I)$ on triangulated surfaces, e.g., [3, 27, 30]. The harmonic maps framework can also be extended to data defined on *implicit* surfaces adapting the variational level-sets approach [4], and to data mapping onto arbitrary implicit surfaces [32].

Most of the literature on harmonic maps deals with $p = 2$ in (3.8) or (3.5), the linear case. Some more recent results are available for $1 < p < \infty$, $p \neq 2$, [14, 18], and very few results deal with the case $p = 1$ [25]. A number of theoretical results, both for the variational formulation and its corresponding gradient descent flow, which are relevant to the multiscale representation of directions, will be given in the following sections as well. The papers [20, 21] are an excellent source of information for regular harmonic maps, while [28] contains a comprehensive review of singularities of harmonic maps (check also [45], a classic on harmonic maps). A classical paper for harmonic maps in liquid crystals, that is, the particular case of (3.5) (or in general, M being a domain in $\mathbb{R}^n$ and $N = S^{n-1}$), is [9].

3.1 Isotropic diffusion of directions

It is easy to show that for $p = 2$, the gradient descent flow corresponding to (3.6) with the constraint (3.7) is given by the set of coupled PDE's:

$$\frac{\partial I_i}{\partial t} = \Delta I_i + I_i \parallel \nabla I \parallel^2, \ 1 \leq i \leq n. \tag{3.11}$$

This system of coupled PDE's defines the isotropic multiscale representation of $I(x, y, 0)$, which is used as initial data to solve (3.11). (Boundary conditions are also added in the case of finite domains.)

This result can also be obtained directly from the general Euler-Lagrange equations (3.9) and (3.10). The Laplace-Beltrami operator Δ_M and manifold gradient ∇_M become the regular Laplace and gradient respectively, since we are working on $\mathbb{R}^2$ (the same for $\mathbb{R}^n$). The second fundamental form $A_N(I)$ of the sphere (in any dimension) is I.

The first part of (3.11) comes from the variational form, while the second one comes from the constraint (see for example [46]). As expected, the first part is decoupled between components I_i, and linear, while the coupling and non-linearity come from the constraint.

If $n = 2$, that is, we have 2D directions, then it is easy to show that for (smooth data) $I(x, y) = (\cos \theta(x, y), \sin \theta(x, y))$, the energy in (3.5) becomes $E_p(\theta) := \int \int_\Omega (\theta_x^2 + \theta_y^2)^{p/2} dx dy$. For $p = 2$ we then obtain the linear heat flow on θ ($\theta_t = \Delta\theta$) as the corresponding gradient descent flow, as expected from the results in [36]. The precise equivalence between the formulation of the energy in terms of directions, $E_p(I)$, and the formulation in terms of orientations, $E_p(\theta)$, is only true for $p \geq 2$ [5, 25]. In spite of this connection, and as was pointed out in [25], using directions and orientations is not fully equivalent. Observe for example the map $v(X) = \frac{X}{\parallel X \parallel}$ defined on the unit ball B^2 of $\mathbb{R}^2$. On one hand, this map has finite E_p energy if, and only if, $1 \leq p < 2$. On the other hand, the map cannot be defined with an angle function $\theta(x, y)$, θ being in the Sobolev space $W^{1,p}(B^2)$. As pointed out in [25], the only obstruction to this representation is the index of the vector of directions. Thus, smooth directions can lead to non-smooth orientations (Perona's goal in the periodic formulations he proposes is in part to address this issue). This problem then gives an additional advantage to selecting a directions-based representation instead of an orientations one.

For the isotropic case, $p = 2$, we have the following important results from the literature on harmonic maps:

Existence: Existence results for harmonic mappings were already reported in [22] for a particular selection of the target manifold N (nonpositive curvature). Struwe [45] showed, in one of the classical papers in the area, that for initial data with finite energy (as measured by (3.8)), M a two dimensional manifold with $\partial M = \emptyset$ (manifold without boundary), and $N = S^{n-1}$, there is a unique solution to the general gradient-descent flow. Moreover, this solution is regular with the exception of a finite number of isolated points and the harmonic energy is decreasing in time. If the initial energy is small, the solution is completely regular and converges to a constant value. (These results actually hold for any compact N.) These uniqueness result was later extended to manifolds with smooth $\partial M \neq \emptyset$ and for weak solutions [24]. Recapping, there is a unique weak solution to (3.11) (weak solutions defined in natural spaces, $H^{1,2}(M, N)$ or $W^{1,2}(M, N)$), and the set of possible singularities is finite. These solutions decrease the harmonic energy. The result is not completely true for M with dimension greater than 2, and this was investigated for example in [13]. Global weak solutions exist for example for $N = S^{n-1}$, although there is no uniqueness for the general initial value problem [17]. Results on the regularity of the solution, for a restricted suitable class of weak solutions, to the harmonic flow for high dimensional manifolds M into S^{n-1} have been recently reported [15, 23]. In this case, it is assumed that the weak solutions hold a number of given energy constraints.

Singularities in 2D: If $N = S^1$, and the initial and boundary conditions are well behaved (smooth, finite energy), then the solution of the harmonic flow is regular. This is the case for example for smooth 2D image gradients and 2D optical flow.

Singularities in 3D: Unfortunately, for $n = 3$ in (3.11) (that is $N = S^2$, 3D vectors), smooth initial data can lead to singularities in finite time [12]. Chang *et al.* showed examples where the flow (3.11), with initial data $I(x, y, 0) = I_0(x, y) \in C^1(D^2, S^2)$ (D^2 is the unit disk on the plane) and boundary conditions $I(x, y, t)|_{\partial D^2} = I_0|_{\partial D^2}$, develops singularities in finite time. The idea is to use as original data I_0 a function that covers S^2 more than once in a certain region. From the point of view of the harmonic energy, the solution is "giving up on" regularity in order to reduce energy.

Singularities topology: Since singularities can occur, it is then interesting to study them [9, 28, 39]. For example, Brezis *et al.* studied the value of the harmonic energy when the singularities of the critical point are prescribed (the map is from R^3 to S^2 in this case). (Perona suggested to look at this line of work to analyze the singularities of the orientation diffusion flow.) Qing characterized the energy at the singularities. A recent review on the singularities of harmonic maps was prepared by Hardt [28]. (Singularities for more general energies are studied for example in [38].) The results there reported can be used to characterize the behavior of the multiscale representation of high dimensional directions, although these results mainly address the shape of the harmonic map, that is, the critical point of the harmonic energy and not the flow. Of course, for the case of M being of dimension two, which corresponds to (3.11), we have Struwe's results mentioned above.

3.2 Anisotropic diffusion of directions

The picture becomes even more interesting for the case $1 \leq p < 2$. Now the gradient descent flow corresponding to (3.6), in the range $1 < p < 2$ (and formally for $p = 1$), with the constraint (3.7) is given by the set of coupled PDE's:

$$\frac{\partial I_i}{\partial t} = \text{div}\left(\parallel \nabla I \parallel^{p-2} \nabla I_i\right) + I_i \parallel \nabla I \parallel^p, \ 1 \leq i \leq n. \qquad (3.12)$$

This system of coupled PDE's defines the anisotropic multiscale representation of $I(x, y, 0)$, which is used as initial datum to solve (3.11). In contrast with the isotropic case, now both terms in (3.12) are non-linear and include coupled components. Formally, we can also explicitly write the case $p = 1$, giving

$$\frac{\partial I_i}{\partial t} = \text{div}\left(\frac{\nabla I_i}{\parallel \nabla I \parallel}\right) + I_i \parallel \nabla I \parallel, \ 1 \leq i \leq n, \qquad (3.13)$$

although the formal analysis and interpretation of this case is much more delicate.

The case of $p \neq 2$ in (3.8) has been less studied in the literature.

When M is a domain in $\mathbb{R}^m$, and $N = S^{n-1}$, the function $v(X) := \frac{X}{\|X\|}$, $X \in \mathbb{R}^m$, is a critical point of the energy for $p \in \{2, 3, ..., m-1\}$, for $p \in [m-1, m)$ (this interval includes the energy case that leads to (3.12)), and for $p \in [2, m - 2\sqrt{m-1}]$ [28]. For $n = 2$ and $p = 1$, the variational problem has also been investigated in [25], where the authors addressed, among other things, the correct spaces to perform the minimization (in the scalar case, $BV(\Omega, \mathbb{R})$ is used), and the existence of minimizers. Of course, we are more interested in the results for the flow (3.12), and not just in its corresponding energy. Some results exist for $1 < p < \infty$, $p \neq 2$, showing in a number of cases the existence of local solutions which are not smooth. To the best of our knowledge, the case of $1 \leq p < 2$, and in particular $p = 1$, has not been fully studied for the evolution equation.

Following the framework for robust anisotropic diffusion we can generalize (3.5) and study problems of the form

$$\min_{I:\mathbb{R}^2 \to S^{n-1}} \int \int_\Omega \rho(\| \nabla I \|) dx dy, \qquad (3.14)$$

where ρ is now a robust function like the Tukey biweight. Results on this approach will be reported elsewhere.

4 Examples of direction diffusion

In this section we present a number of illustrative examples for the harmonic flows for $p = 2$ (isotropic) and $p = 1$ (anisotropic) presented above. These examples will mainly show that the proposed framework produces for directional data the same qualitative behavior that is well known and studied for scalar images. One of the advantages of directional diffusion is that, although advanced specialized numerical techniques to solve (3.5) and its corresponding gradient-descent flow, have been developed, e.g., [1], as a first approximation we can basically use the algorithms developed for isotropic and anisotropic diffusion without the unit norm constraint to implement (3.11) and (3.12) [16]. Although, as stated before, these PDE's preserve the unit norm (that is, the solutions are vectors in S^{n-1}), numerical errors might violate the constraint (recent developments by Osher and Vese might solve this numerical problem [34]). Therefore, between every two steps of the numerical implementation of this equations we add a renormalization step [16]. Basically, a simple time-step iteration is used for (3.11), while for (3.12) we incorporate the edge capturing technique developed in [41] (we always use the maximal time step that ensures stability).

For the examples we show below, a number of visualization techniques are used (note that the visualization of vectors is an active research area in the graphics community):

1. Arrows. Drawing arrows indicating the vector direction is very illustrative, but can be done only for sparse images and they are not very informative for dense data like gradients or optical flow. Therefore, we use arrows either for toy examples or to show the behavior of the algorithm in local areas.

2. HSV color mapping. We use the HSV color map (applied to orientation) to visualize whole images of directions while being able to also illustrate details like small noise.

3. Line integral convolution (LIC) [10]. LIC is based on locally integrating at each pixel the values of a random image. The integration is done in the line corresponding to the direction at the pixel. The LIC technique gives the general form of the flow, while the color map is useful to detect small noise in the direction (orientation) image.

All these visualization techniques are used for vectors in S^1. We also show examples for vectors in S^2. In this case, we consider these vectors as RGB vectors, and color is used to visualize the results.

Figure 4.1 shows a number of toy examples to illustrate the general ideas introduced in this Chapter. The first row shows, using LIC, an image with two regions having two different (2D) orientations on the left (original), followed by the results of isotropic diffusion for 200, 2000, and 8000 iterations (scale-space). Three different steps of the scale-space are shown in the next row using arrows. Note how the edge in the directional data is being smoothed out. The horizontal and vertical directions are being smoothed out to converge to the diagonal average. This is in contrast with the edges in the third row. This row shows the result of removing noise in the directional data. The original noisy image is shown first, followed by the results with isotropic and anisotropic smoothing. Note how the anisotropic flow gets rid of the noise (outliers) while preserving the rest of the data, while the isotropic flow also affects the data itself while removing the noise. Note that since the discrete theory developed in [36] applies only to small changes in orientation, theoretically it can not be applied to the images we have seen so far, all of them contain sharp discontinuities in the directional data. The last two rows of Figure 4.1 deal with 3D directions. In this case, we interpret the vector as RGB coordinates, and use color to visualize them. First we show two steps in the scale space for the isotropic flow (original on the left). Then, the last row shows the original image followed by the results of isotropic and anisotropic diffusion. Note how the colors, and then the corresponding directions, get blurred with the isotropic flow, while the edges in direction (color) are well preserved with the anisotropic process.

Figure 4.2 shows results for simulated optical flow. This simulated optical flow data was computed with the public domain software described in [6] (down-loaded from M. Black's home page), stopping at early annealing stages to have enough noise for experimentation. This data is used here for illustration purposes only. Alternatively, we could include the harmonic energy as a regularization term inside the optical flow computation, that is, combined with the optical flow constraint. The first figure on the first row shows a frame of the famous Yosemite movie. Next, in the same row, we show from left to right the original optical flow direction, the result of the isotropic flow, the result of the anisotropic flow, and the result of the isotropic flow for a large number of iterations. The HSV color map is used. In the next row, we use LIC to visualize again the three middle figures of the first row. In the third row, arrows are used to show a blown-up of the marked region corresponding to the isotropic flow for 20, 60, and 500 iterations. Note how the noise in the optical flow directions is removed.

Figure 4.3 presents two examples for color images, that is, 3D directions defined on $\mathbb{R}^2$. Both rows show the original image, followed by the noisy image and the enhanced one. In the first row, noise is added only to the 3D RGB directions representing the chroma (the RGB vector normalized to a unit norm vector),† while the magnitude of the vector (brightness) was kept untouched. The enhanced image was then obtained applying the proposed direction diffusion flow to the chroma, while keeping the original magnitude. This experiment shows that when the magnitude is preserved (or well reconstructed, see below), direction diffusion for chroma denoising produces an image practically indistinguishable from the original one. In the second row, the noise was added to the original image, resulting in both noisy chroma (direction) and brightness (magnitude) of the RGB vector. The directions are processed with the diffusion flow, while the magnitude is processed with the scalar anisotropic flow in [7]. Isotropic and anisotropic direction diffusion flows on the chroma produce similar results as long as the magnitude is pro-

---

† See [50, 51] for a discrete median filtering approach to chroma denoising.

cessed with an edge preserving denoising algorithm. See [49] for additional examples, comparisons with the literature, and details on the applications of direction diffusion to color images.

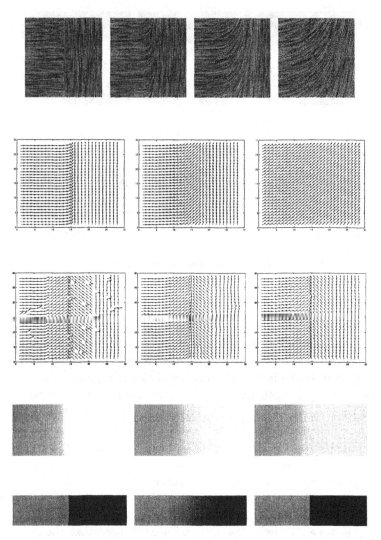

Fig. 4.1. *Toy examples illustrating the ideas in this Chapter. See text for details.*

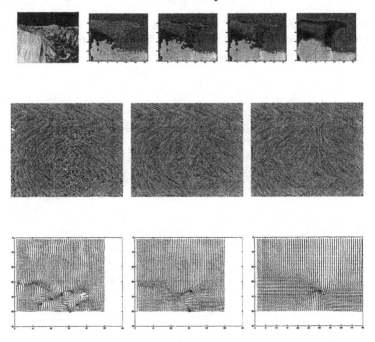

Fig. 4.2. *Optical flow example. See text for details.*

5 Vector probability diffusion

In [47], we introduced the use of anisotropic diffusion for image classification. The basic idea was to compute posterior probabilities of each class present in the image (the number of classes is given apriori), then anisotropically smooth the posteriors, and then apply a simple MAP classification rule. In [47], each posterior (one per class) was independently anisotropically smoothed, and a normalization step was added to guarantee that the posteriors add to one. Two main difficulties are encountered in this. First, each posterior probability (one per class) is independently diffused, thereby ignoring the intrinsic correlation between them. Second, due to the independent processing, the posterior probabilities are not guaranteed to add to one even after a very short diffusion time. To overcome this, we need to normalize the posterior probabilities after each discrete iteration, normalization that has a nontrivial effect on the diffusion process itself.

To solve these problems we need then to find a set of coupled PDE's that guarantees that the vector of posterior probabilities remains a prob-

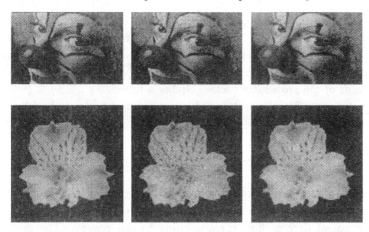

Fig. 4.3. *Denoising of a color image using 3D directional diffusion. See text for details.*

ability vector, that is, with all its components positive and adding to one. This is a particular and simple case of harmonic maps, where instead of having the data defined on a hypersphere as in the direction diffusion case, the data is defined on an hyperplane. We develop this now. Moreover, as we shall see, the numerical implementation of this system of coupled PDE's also preserves these properties, thereby removing the necessity to project back into the semi-hyperplane. This approach then overcomes both difficulties mentioned above and can be directly incorporated in the segmentation technique, replacing the component-by-component diffusion. Although the work is here discussed in the framework of posterior diffusion for image classification, it is clear that the technique can be applied to the diffusion of other probabilities in other applications as well.

5.1 Problem formulation and basic equations

Assume that a vector of a-posteriori probabilities p, mapping the image domain Ω in $\mathbb{R}^2$ to the manifold $\mathcal{P} = \{p \in \mathbb{R}^m : \|p\|_1 = 1, p_i \geq 0\}$, is given. Each component p_i of p equals the posterior probability of a class $c_i \in \mathcal{C} = \{c_i : i = 1, ..., m\}$. These posterior probabilities can be obtained for example via Bayes rule.

If we view the vectors p as a vector field, one possible way to add spatial coherence into the classification process is to diffuse the distance between points in $\mathcal{P}$, propagating the information in the probability space, before a pixelwise MAP decision is made. Inspired by the harmonic maps theory described above, the "distance" between two differential adjacent points in $\mathcal{P}$ depends on $\|\nabla p\| := \sqrt{\sum_{i=1}^{m} \|\nabla p_i\|^2}$. This is the *gradient of the probability vector.* Giving a function $\rho : \mathbb{R} \to \mathbb{R}$ (we will later discuss different selections of ρ), we proceed to solve the following minimization process:

$$\min_{p \in \mathcal{P}} J_\rho, \quad J_\rho := \int_\Omega \rho(\|\nabla p\|) d\Omega. \qquad (5.1)$$

Note that the minimization is restricted to the semi-hyperplane $\mathcal{P}$.

From this, the system of coupled diffusion equations is obtained via the gradient descent flow corresponding to this energy. The gradient descent of J_ρ restricted to $\mathcal{P}$ is given by

$$\frac{\partial p}{\partial t} = \nabla \cdot \left(\frac{\rho'(\|\nabla p\|)}{\|\nabla p\|} \nabla p \right). \qquad (5.2)$$

This equations is an abbreviated notation for a set of PDE's of the form $\frac{\partial p_i}{\partial t} = \nabla \cdot \left(\frac{\rho'(\|\nabla p\|)}{\|\nabla p\|} \nabla p_i \right), \quad i = 1, ..., m.$

Note again that the minimization is performed in the probability space $\mathcal{P}$ (a semi-hyperplane), and the system of equations (5.2) guarantees that $p(t) \in \mathcal{P}$ for all t. We have therefore obtained a system of coupled PDE's that preserves the unit L_1 norm which is characteristic of probability vectors (the components $p_i(t)$ are positive as well).

It can be shown that

$$\sum_{i=1}^{m} \|\nabla p_i\|^2 = 2\|\nabla p_1\|^2 - 2\sum_{i=2}^{m} \sum_{j=i+1}^{m} \nabla p_i . \nabla p_j.$$

Therefore, when the number of classes is $m = 2$, $\|\nabla p\|^2 = 2\|\nabla p_i\|^2$, $i = 1, 2$, and the method is equivalent to separately applying the diffusion to each posterior. For $m \geq 3$, the second term is like the correlation between different components of p. This term is not present if we diffuse each posterior probability on its own. This coupling is important to improve the classification results.

If we select ρ as the L_2 norm, $\rho(x) = x^2$, then (5.2) becomes the well known linear heat equation $\frac{\partial p}{\partial t} = \nabla^2 p$, which isotropically diffuses p. It is interesting to note that the heat equation that was previously used to denoise signals as well to generate the so called scale spaces, preserves the diffusion in the probability semi-hyperplane $\mathcal{P}$. This is expected,

since it is well known that this equation holds the maximum principle
and preserves linear combinations.

The isotropic flow has no coupling between the probability components, and does not respect boundaries. Therefore, as classically done
for scalar diffusion [41], a more robust norm is selected. For example,
we can select the L_1 norm, $\rho(x) = |x|$, obtaining

$$\frac{\partial p}{\partial t} = \nabla \cdot \left(\frac{\nabla p}{\|\nabla p\|} \right), \tag{5.3}$$

clearly anisotropic, with a conduction coefficient controlled by $\|\nabla p\|$.
Once again, from the basic invariants of this flow, the preservation of
the vector on the $\mathcal{P}$ space was expected.

5.2 Numerical implementation

The vector probability diffusion equations are to be implemented on a
square lattice. The vector field p is then $p_{j,k}$, where j, k is the position
in the lattice. (For simplicity, we do not write the subscript i indicating
the probability components.) First we rewrite the equation (5.2) in a
simpler way, $\frac{\partial p}{\partial t} = \nabla \cdot (g \nabla p)$, $g = \rho'(\|\nabla p\|)/\|\nabla p\|$, and then apply a
standard numerical scheme (assuming $\Delta x = \Delta y = 1$):

$$\frac{p_{j,k}^{t+\Delta t} - p_{j,k}^t}{\Delta t} = g_{j+\frac{1}{2},k}(p_{j+1,k}^t - p_{j,k}^t) - g_{j-\frac{1}{2},k}(p_{j,k}^t - p_{j-1,k}^t) +$$
$$g_{j,k+\frac{1}{2}}(p_{j,k+1}^t - p_{j,k}^t) - g_{j,k-\frac{1}{2}}(p_{j,k}^t - p_{j,k-1}^t). \tag{5.4}$$

The condition on Δt for stability is easily computed to be

$$\Delta t \leq \frac{1}{4 \max_{j,k}\{g_{j+\frac{1}{2},k}, g_{j-\frac{1}{2},k}, g_{j,k+\frac{1}{2}}, g_{j,k-\frac{1}{2}}\}}. \tag{5.5}$$

To allow the existence of discontinuities in the solution we use the
approximation of the gradient developed in [33].

If Δt fulfills the stability condition (5.5) then $p_{j,k}^{t+\Delta t} \geq 0$ and $\|p_{j,k}^{t+\Delta t}\|_1 =$
1, so the evolution given by (5.4) lives always in the manifold $\mathcal{P}$. More-
over, if we define $p_m^t := \min\{p_{j+1,k}^t, p_{j-1,k}^t, p_{j,k+1}^t, p_{j,k-1}^t\}$ and also
$p_M^t := \max\{p_{j+1,k}^t, p_{j-1,k}^t, p_{j,k+1}^t, p_{j,k-1}^t\}$, the solution satisfies a maximum (minimum) principle: $p_m^t \leq p_{j,k}^{t+\Delta t} \leq p_M^t$.

We then conclude that also the discrete equation (5.4) lives in the
manifold $\mathcal{P}$, and there is no need for a projection back into the semi-
hyperplane, in contrast with the scalar approach by Teo *et al.*, as discussed before. To complete the implementation details, we need to ad-

dress possible problems when $\|\nabla p\|$ vanishes or becomes very small. As it is standard in the literature, we define $\|\nabla p\|_\beta = \sqrt{\beta^2 + \|\nabla p\|^2}$, and use this instead of the traditional gradient. (The stability condition (5.5) becomes $\Delta t \leq \beta/4$.) To select the value of β we propose to look at β as a lower diffusion scale, since in the discrete case probability differences lower than β will be diffused with a conduction coefficient approximately inverse to β. We set the value of the lower scale β in the range $[0.001, 0.01]$.

5.3 Examples

We now present examples of the vector probability diffusion approach here presented applied to image segmentation. As for the scalar case, first, posterior probabilities are computed for each class using Bayes rule. These are diffused with the anisotropic vector probability diffusion flow (instead of the scalar one used in [47]), and then the MAP decision rule is applied.

The first example is based on synthetic data. The original image, containing four classes, is shown on the top-left of Figure 5.1. We then add Gaussian noise, and the objective is to segment the image back into four classes. The figure then shows the classification results without posterior diffusion, followed by the classification results corresponding to two time steps for the vector posterior diffusion approach and two time steps for the scalar approach. Using the correlation between the posterior probabilities, as done by the vector approach here introduced and not by the scalar one in [47], is of particular importance when one of the classes has less weight than the others. As shown by this example, this class (white dots) will be mostly missed by the scalar approach. Comparing the classification errors for class 4 (white dots) in the table, we see that the lowest classification error is with vector probability diffusion, 15.82% against 18.46% for the scalar approach. Furthermore, the classification error is more stable with the vectorial approach, and for the same average error (approx. 10%), the lowest classification error for class 4 is obtained with the vectorial method, 26.95% against 61.04% (see table below).

Figure 5.2 shows the results of the vectorial approach applied to the segmentation of video SAR data. Three classes are considered, {shadow, object, background}, each one modeled as a Gaussian distributions. The first image in the sequence is segmented by hand to obtain an estimation of the parameters of the distributions. For the rest of the images

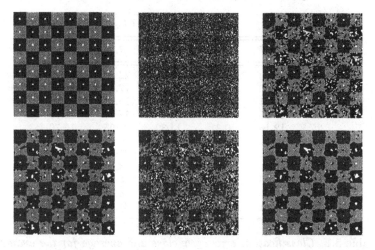

Fig. 5.1. *Left to right and top to down: Original image, classification without diffusion, results of vector probability diffusion for 18 and 25 iterations respectively, and results of the scalar approach for 10 and 15 iterations.*

we perform the following process: (a) The diffused posteriors of the previous image are used as priors and after 30 iterations the new means and variances are estimated; (b) To obtain a better segmentation which does not depend on the previous priors, a new diffusion is applied using uniform priors. In this way the small details are captured; (c) The steps (a) and (b) are iterated and in the last iteration the idea of (a) is used.

As we can see in the Figure 5.2 the results are very good, and the small details are captured. These results outperform those previously reported, where the scalar approach was used.

6 Concluding remarks

In this chapter we have shown how to use the well studied area of harmonic maps to address fundamental problems in image processing, computer vision, and computer graphics. To further exploit this topic for these applications, more advanced numerical techniques have to be developed. Work toward this direction has recently been performed for example in [1, 4, 32], while more needs to be done. For example, if we are interested in the harmonic map, and not in the evolution, additional work needs to be done to achieve the minima directly. Further studies are needed for non-quadratic norms, and further physical interpretations

Iter.	Class 1	Class 2	Class 3	Class 4	Average
0	53.77	26.67	53.68	27.34	46.65
10	47.63	17.47	47.40	22.46	39.70
15	33.26	4.76	33.30	17.77	26.02
18	24.01	2.33	24.83	15.82	18.95
20	19.75	1.50	20.88	17.77	15.78
25	12.16	0.81	14.39	26.95	10.70
5	53.70	26.51	53.61	27.34	46.55
7	50.02	18.74	49.85	24.90	41.85
10	35.06	6.24	35.03	18.46	27.69
15	17.48	1.51	18.99	24.32	14.40
20	18.11	0.76	13.14	61.04	10.34

Table 5.1. *Classification errors per class and average for the example in the figure above. The first row corresponds to the classical MAP approach, where no diffusion has been applied to the posterior probabilities. The next fife rows correspond to four different time steps of the vector probability diffusion approach here presented, followed by fife time steps for the scalar approach of Teo et al..*

of the singularities of the harmonic map are needed. In spite of these important open problems, the current knowledge on harmonic maps theory already leads to state-of-the-art practical results, as demonstrated in this chapter and the mentioned literature.

Acknowledgments

The work on harmonic maps and its applications to computer vision and image processing started as a collaboration with Vicent Caselles and Bei Tang. The work on vector probability diffusion is in collaboration with Alvaro Pardo. Portions of the work reported here have been first published in [35, 42, 48, 49]. This work was partially supported by a grant from the Office of Naval Research ONR-N00014-97-1-0509, the Office of Naval Research Young Investigator Award, the Presidential Early Career Awards for Scientists and Engineers (PECASE), a National Science Foundation CAREER Award, and by the National Science Foundation Learning and Intelligent Systems Program (LIS).

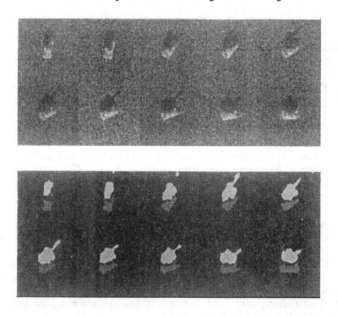

Fig. 5.2. *Original images and classification results with MAP vector probability diffusion.*

References

[1] Alouges. F. (1991). An energy decreasing algorithm for harmonic maps, in *Nematics* (J.M. Coron *et al.*, eds), Nato ASI Series, (Kluwer, Dordrecht), 1–13.

[2] Alvarez, L., Lions, P.L. and Morel, J.M. (1992). Image selective smoothing and edge detection by nonlinear diffusion, *SIAM J. Numer. Anal.* **29**, 845–866.

[3] Angenent, S., Haker, S., Tannenbaum, A. and Kikinis, R. (1998). Laplace-Beltrami operator and brain flattening, ECE Tech. Rep., University of Minnesota.

[4] Bertalmio, M., Cheng, L.T., Osher, S. and Sapiro, G. (2000). PDE's on implicit (level-sets) manifolds: The framework and applications in image processing and computer graphics, preprint.

[5] Bethuel, F. and Zheng, X. (1988). Density of smooth functions between two manifolds in Sobolev spaces, *Journal of Functional Analysis* **80**, 60–75.

[6] Black, M. and Anandan, P. (1993). A framework for the robust estimation of optical flow, in *Fourth International Conf. on Computer Vision*, (Berlin), 231–236.

[7] Black, M., Sapiro, G., Marimont, D. and Heeger, D. (1998). Robust anisotropic diffusion, *IEEE Trans. Image Processing* **7**,421–432.

[8]   Blomgren, P. and Chan, T. (1998). Color TV: Total variation methods for restoration of vector valued images, *IEEE Trans. Image Processing* **7**, 304–309.

[9]   Brezis, H., Coron, J.M. and Lieb, E.H. (1986). Harmonic maps with defects, *Comm. Math. Phys.* **107**, 649–705.

[10] Cabral, B. and Leedom, C. (1993). Imaging vector fields using line integral convolution, *ACM Computer Graphics (SIGGRAPH '93)* **27**, 263-272.

[11] Chan, T. and Shen, J. (1999). Variational restoration of non-flat image features: Models and algorithms, Tech. Rep. UCLA CAM-TR 99-20.

[12] Chang, K.C., Ding, W.Y. and Ye, R. (1992). Finite-time blow-up of the heat flow of harmonic maps from surfaces, *J. Diff. Geometry* **36**, 507–515.

[13] Chen, Y. (1989). The weak solutions of the evolution problems of harmonic maps, *Math. Z.* **201**, 69–74.

[14] Chen, Y., Hong, M.C. and Hungerbuhler, N. (1994). Heat flow of p-harmonic maps with values into spheres, *Math. Z.* **205**, 25–35.

[15] Chen, Y., Li, J. and Lin, F.H. (1995). Partial regularity for weak heat flows into spheres, *Comm. Pure & Appl. Maths* **48**, 429–448.

[16] Cohen, R., Hardt, R.M., Kinderlehrer, D., Lin, S.Y. and Luskin, M. (1987). Minimum energy configurations for liquid crystals: Computational results, in *Theory and Applications of Liquid Crystals* (J.L. Ericksen and D. Kinderlehrer, eds), IMA Volumes in Mathematics and its Applications (Springer-Verlag, New York), 99–121.

[17] Coron, J.M. (1990). Nonuniqueness for the heat flow of harmonic maps, *Ann. Inst. H. Poincaré, Analyse Non Linéaire* **7**, 335–344.

[18] Coron, J.M. and Gulliver, R. (1989). Minimizing p-harmonic maps into spheres, *J. Reine Angew. Mathem.* **401**, 82–100.

[19] Eck, M., DeRose, T., Duchamp, T., Hoppe, H., Lounsbery, M. and Stuetzle, W. (1995). Multiresolution analysis of arbitrary meshes, in *Computer Graphics (SIGGRAPH '95) Proceedings*, 173–182.

[20] Eells, J. and Lemarie, L. (1978). A report on harmonic maps, *Bull. London Math. Soc.* **10**, 1–68.

[21] Eells, J. and Lemarie, L. (1988). Another report on harmonic maps, *Bull. London Math. Soc.* **20**, 385–524.

[22] Eells, J. and Sampson, J.H. (1964). Harmonic mappings of Riemannian manifolds, *Am. J. Math.* **86**, 109–160.

[23] Feldman, M. (1994). Partial regularity for harmonic maps of evolutions into spheres, *Comm. in PDEs* **19**, 761–790.

[24] Freire, A. (1995). Uniqueness for the harmonic map flow in two dimensions, *Calc. Var.* **3**, 95–105.

[25] Giaquinta, M., Modica, G. and Soucek, J. (1993). Variational problems for maps of bounded variation with values in S^1, *Cal. Var.* **1**, 87–121.

[26] Granlund, G.H. and Knutsson, H. (1995). *Signal Processing for Computer Vision* (Kluwer, Boston).

[27] Haker, S., Angenent, S., Tannenbaum, A., Kikinis, R., Sapiro,G. and Halle, M. (1999). Conformal surface parametrization for texture mapping, Tech. Rep. 1611, IMA Preprint Series, Univ. Minnesota.

[28] Hardt, R.M. (1997). Singularities of harmonic maps, *Bull. Amer. Math. Soc.* **34**, 15–34.

[29] Hardt, R.M. and Lin, F.H. (1987). Mappings minimizing the L^p norm of the gradient, *Comm. Pure and Appl. Maths* **40**, 555–588.

[30] Hughes, T. (1987). *The Finite Element Method* (Prentice-Hall, New Jersey).

[31] Koenderink, J.J. (1984). The structure of images, *Biological Cybernetics* **50**, 363–370.

[32] Memoli, F., Sapiro, G. and Osher, S. (2000). Harmonic maps onto implicit manifolds. Preprint.

[33] Osher, S. and Rudin, L.I. (1990). Feature-oriented image enhancement using shock filters, *SIAM J. Numer. Anal.* **27**, 919–940.

[34] Osher, S. and Vese, L. (1999). Personal communication.

[35] Pardo, A. and Sapiro, G. (1999) Vector probability diffusion, Tech. Rep., Univ. Minnesota ECE TR. Also in *Proc. IEEE Int. Conf. Image Proc.*, (Vancouver, Canada, September 2000).

[36] Perona, P. (1998). Orientation diffusion, *IEEE Trans. Image Processing* **7**, 457–467.

[37] Perona, P. and Malik, J. (1990). Scale-space and edge detection using anisotropic diffusion, *IEEE Trans. Pattern. Anal. Machine Intell.* **12**, 629–639.

[38] Pismen, L.M. and Rubinstein, J. (1991). Dynamics of defects, in *Nematics* (J.M. Coron *et al.*, eds), Nato ASI Series (Kluwer, Dordrecht), 303–326.

[39] Qing, J. (1995). On singularities of the heat flow for harmonic maps from surfaces into spheres, *Comm. Analysis & Geometry* **3**, 297–315.

[40] Rudin, L.I. and Osher, S. (1994). Total variation based image restoration with free local constraints, *Proc. IEEE-ICIP* **I**, 31–35.

[41] Rudin, L.I., Osher, S. and Fatemi, E. (1992). Nonlinear total variation based noise removal algorithms, *Physica D* **60**, 259–268.

[42] G. Sapiro, G. (2000). *Geometric Partial Differential Equations and Image Analysis* (Cambridge University Press, Cambridge), in press.

[43] Sapiro, G. and Ringach, D. (1996). Anisotropic diffusion of multivalued images with applications to color filtering, *IEEE Trans. Image Processing* **5**, 1582–1586.

[44] Sochen, M., Kimmel, R. and Malladi, R. (1998). A general framework for low level vision, *IEEE Trans. Image Processing* **7**, 310–318.

[45] Struwe, M. (1985). On the evolution of harmonic mappings of Riemannian surfaces, *Comment. Math. Helvetici* **60**, 558–581.

[46] Struwe, M. (1990). *Variational Methods* (Springer Verlag, New York).

[47] Teo, P., Sapiro, G. and Wandell, B. (1997). Creating connected representations of cortical gray matter for functional MRI visualization, *IEEE Trans. Medical Imaging* **16**, 852–863.

[48] Tang, B., Sapiro, G. and Caselles, V. (2000). Diffusion of general data on non-flat manifolds via harmonic maps theory: The direction diffusion case, *Int. J. Comp. Vision* **36**, 149–161.

[49] Tang, B., Sapiro, G. and Caselles, V. (1999). Color image enhancement via chromaticity diffusion. Preprint.

[50] Trahanias, P.E. and Venetsanopoulos, A.N. (1993). Vector directional filters - A new class of multichannel image processing filters, *IEEE Trans. Image Processing* **2**, 528–534.

[51] Trahanias, P.E., Karakos, D. and Venetsanopoulos, A.N. (1996). Directional processing of color images: Theory and experimental results, *IEEE Trans. Image Processing* **5**, 868–880.

[52] Weickert, J. (1996). Foundations and applications of nonlinear

anisotropic diffusion filtering, *Zeitscgr. Angewandte Math. Mechan.* **76**, 283–286.

[53] Weickert, J. (1999). Coherence-enhancing diffusion of color images, *Image and Vision Computing* **17**, 201-212.

[54] Whitaker, R.T. and Gerig, G. (1994). Vector-valued diffusion, in *Geometry Driven Diffusion in Computer Vision* (B. ter Haar Romeny, ed.) (Kluwer, Boston).

[55] Witkin, A.P. (1983). Scale-space filtering, *Int. Joint. Conf. Artificial Intelligence* **2**, 1019–1021.

[56] Yezzi, A. (1998). Modified curvature motion for image smoothing and enhancement, *IEEE Trans. Image Processing* **7**, 345–352.

[57] You, Y.L., Xu, W., Tannenbaum, A. and Kaveh, M. (1996). Behavioral analysis of anisotropic diffusion in image processing, *IEEE Trans. Image Processing* **5**, 1539–1553.

[58] Zhang, D. and Hebert, M. (1999). Harmonic maps and their applications in surface matching, *Proc. CVPR '99* (Colorado).

Statistics From Computations

Hersir Sigurgeirsson

Scientific Computing and Computational Mathematics Program
Gates 288, Stanford University
Stanford CA94305-4040, USA
Email: hersir@stanford.edu

A. M. Stuart

Mathematical Institute
Warwick University
Coventry CV4 7AL, UK
Email: stuart@maths.warwick.ac.uk

Abstract

The study of numerical methods for initial value problems by considering their approximation properties from a dynamical systems viewpoint is now a well-established field; a substantial body of knowledge, developed over the past two decades, can be found in the literature. Nonetheless many open questions remain concerning the meaning of long-time simulations performed by approximating dynamical systems. In recent years various attempts to analyse the statistical content of these long-time simulations have emerged, and the purpose of this article is to review some of that work. The subject area is far from complete; nonetheless a certain unity can be seen in what has been achieved to date and it is therefore of value to give an overview of the field.

Some mathematical background concerning the propagation of probability measures by discrete and continuous time dynamical systems or Markov chains will be given. In particular the Frobenius-Perron and Fokker-Planck operators will be described. Using the notion of ergodicity two different approaches, *direct* and *indirect*, will be outlined. The majority of the review is concerned with indirect methods, where the initial value problem is simulated from a single initial condition and the statistical content of this trajectory studied. Three classes of problems will be studied: deterministic problems in fixed finite dimension, stochastic problems in fixed finite dimension, and deterministic problems with random data in dimension $n \to \infty$; in the latter case ideas from statistical mechanics can be exploited to analyse or interpret numerical schemes.

Throughout, the ideas are illustrated by simple numerical experiments. The emphasis is on understanding underlying concepts at a high level and mathematical detail will not be given a high priority in this review.

1 Introduction

It is well-known that, in general, individual trajectories of dynamical systems cannot be well approximated over long time-intervals, since well-posed initial-value problems admit exponentially divergent trajectories.†
However, experimental evidence strongly suggests that often statistics can be robustly computed even when trajectories cannot. The aim of the paper is to survey various avenues of investigation which attempt to justify this observation. Definitive results are few and far between and the overall picture is far from complete. However a general framework is starting to emerge and it is therefore useful to set it out here. Ultimately, study of the robustness of statistics generated through computations of dynamical systems will both add significantly to the body of existing theoretical results concerning the approximation of dynamical systems [30, 34] and lend weight to the large number of initial-value calculations routinely performed in the sciences and engineering where statistical information is extracted from the data. Our presentation is informal and is intended to be suggestive of the important ideas, rather than being precise and detailed.

We will study both the effect of rounding error on maps and of time-discretization on (ordinary and stochastic) differential equations. To illustrate the kinds of issues to be discussed consider the following

Example – The Hénon map Figure 1.1 shows some results from computer simulation of the map

$$
\begin{aligned}
x_{n+1} &= 1 - 1.4x_n^2 + 0.3y_n, & x(0) &= 0, \\
y_{n+1} &= x_n, & y(0) &= 0.
\end{aligned}
\tag{1.1}
$$

Figure 1.1 top left shows the attractor for this map, generated by a double-precision arithmetic simulation. Figure 1.1 top right shows the difference between the two trajectories, one with single precision arithmetic, the other with double. Since these trajectories are so different

† In the special case of hyperbolic systems shadowing ideas can be applied to give long-time approximation for problems with divergent trajectories.

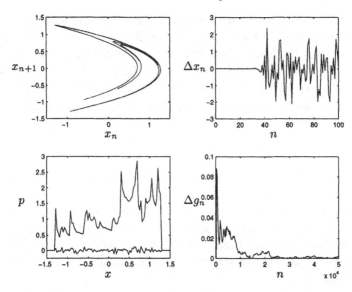

Fig. 1.1. The Hénon map. Upper left: Attractor, generated by a double-precision arithmetic simulation. Upper right: The difference between two trajectories, one computed with single precision arithmetic, the other with double. Lower left: Histogram of the empirical probability density for the x variable, computed from very long-time simulations in single precision, and the difference between the histogram computed in single and double precision. Lower right: Difference in the time averages $\frac{1}{n+1}\sum_{k=0}^{n} g(x_k)$ with $g(x) = \left(\frac{x+1.5}{3}\right)^2$.

it is natural to question whether they contain any meaningful information about (1.1). In this paper we will concentrate on the statistical properties of such simulations. To this end it is natural to compare the empirical probability densities generated by the single and double precision calculations. Histograms for the x variable, computed from very long-time simulations in both single and double precision, and their difference are shown in Figure 1.1 bottom left, and the difference in the time averages $\frac{1}{n+1}\sum_{k=0}^{n} g(x_k)$ with $g(x) = \left(\frac{x+1.5}{3}\right)^2$ in Figure 1.1 bottom right. The closeness of the single and double precision calculations in Figures 1.1 (bottom) is remarkable and suggests that, even when the trajectories themselves are unreliable, the statistics contained within them may be meaningful. ∎

In many of the examples that follow we show figures similar to Figure 1.1. In each case we perform two computer simulations of the same trajectory using single and double precision in the case of maps, and two different time steps in the case of differential equations. The upper left corner shows a single trajectory or its projection onto some two dimensional space, the upper right the difference between the x coordinates of the two trajectories, the lower left corner shows the densities of the x coordinate and their difference, and the lower right figure shows the difference between some time averages.

In section 2 we outline some mathematical background which will suggest what is required to substantiate observations about the robustness of statistics such as those exemplified by the Hénon map. We will introduce the Frobenius-Perron operator and the Fokker-Planck or Liouville equations which describe the propagation of measures by deterministic or random dynamical systems. Using this mathematical background we will distinguish between direct and indirect approaches to the problem of approximating statistics and we will highlight three important classes of problem:

A) deterministic problems in $\mathbb{R}^n$ with n fixed;

B) stochastic problems in $\mathbb{R}^n$ with n fixed;

C) deterministic problems in dimension $n \to \infty$ with random initial data.

In sections 3,4 and 5 we will discuss problems from categories A), B) and C) respectively.

2 Mathematical Background

We start by outlining how probability measures are propagated by a variety of deterministic and random dynamical systems. For simplicity we first consider deterministic maps. Imagine that

$$x_{n+1} = G(x_n), \qquad (2.1)$$

where x_0 is a random variable with density p_0. If we define the *Frobenius-Perron* operator

$$(\mathcal{L}p)(x) = \sum_{y \in G^{-1}(x)} \frac{p(y)}{|\det \, dG(y)|},$$

and provided sufficient smoothness allows us to construct a density p_n for x_n, then this density satisfies

$$p_{n+1} = \mathcal{L}p_n.$$

A smooth *invariant measure* has density p:

$$p = \mathcal{L}p.$$

In both cases suitable decay of the densities at infinity is required.

It is interesting to note that, in general, this iteration may admit quite non-smooth densities. The following example shows this.

Example – The Tent Map If

$$G(x) = \begin{cases} 2x, & x \in [0, \frac{1}{2}), \\ 2 - 2x, & x \in [\frac{1}{2}, 1], \end{cases} \qquad (2.2)$$

then

$$G^{-1}(x) = \left\{ \frac{x}{2}, \frac{2-x}{2} \right\}.$$

Thus

$$p_{n+1}(x) = \frac{1}{2} \left[p_n \left(\frac{x}{2} \right) + p_n \left(\frac{2-x}{2} \right) \right], \qquad x \in [0, 1].$$

Thus any singularity in p_n will produce a singularity in p_{n+1}. However, some limited smoothing is present due to the repeated scaling by a factor of 2 and a smooth invariant measure is

$$p(x) = 1, \qquad x \in [0, 1].$$

In fact this invariant measure is unique within a sufficiently smooth class of functions and $p_n \to p$ in this class – see [42]. ∎

We now randomise (2.1) to get the Markov chain (θ_n i.i.d. ν)

$$x_{n+1} = G_{\theta_n}(x_n) \qquad (2.3)$$

where x_0 is distributed as before. For simplicity we assume that $G_\theta(\cdot)$ is invertible for all θ. We define the Frobenius-Perron operator

$$(\mathcal{L}p)(x) = \int \frac{p(G_\theta^{-1}(x))}{|\det \, dG_\theta(G_\theta^{-1}(x)|} \nu(d\theta).$$

Then, if it exists, the probability density for x_n satisfies

$$p_{n+1}(x) = (\mathcal{L}p_n)(x).$$

Again a smooth invariant measure has density p:

$$p = \mathcal{L}p.$$

Often the randomisation smoothes through the averaging process though this need not necessarily happen. The following example illustrates such smoothing.

Example −AR(1) Process Consider the following random map, which arises in autoregressive time series analysis:

$$G_\theta(x) = \frac{1}{2}x + \theta$$

with θ picked from the standard normal distribution $\mathcal{N}(0,1)$. Thus

$$G_\theta^{-1}(x) = 2(x - \theta)$$

so that

$$p_{n+1}(x) = \int \frac{\exp(-\theta^2/2)}{\sqrt{2\pi}} 2p_n(2x - 2\theta)d\theta.$$

The invariant measure is given by the law of the random variable

$$y := \lim_{n \to \infty} \sum_{j=0}^{n} \frac{1}{2^j}\theta_j.$$

See [7] for further examples of this type.   ∎

Now we move to continuous time where the issues are similar: in the absence of randomness, non-smoothness in probability densities abounds; randomness can often act to introduce smoothing.

Consider the Itó stochastic differential equation

$$dx = f(x)dt + \sigma(x)dW, \tag{2.4}$$

with $x \in \mathbb{R}^m$, $W \in \mathbb{R}^d$ being standard d-dimensional Brownian motion and $f : \mathbb{R}^m \to \mathbb{R}^m$, $\sigma : \mathbb{R}^m \to \mathbb{R}^{m \times d}$ and $B : \mathbb{R}^m \to \mathbb{R}^{m \times m}$ defined by $B(u) = \sigma(u)\sigma(u)^T$, where $m \geq d$. Again we assume that $x(0)$ is distributed randomly with density p_0.

If we define the *Fokker-Planck* operator† by

$$\mathcal{A}p = -\nabla \cdot (fp) + \frac{1}{2}\nabla \cdot \nabla \cdot (Bp) \tag{2.5}$$

then $x(t)$ generates a probability measure with density satisfying

$$\frac{\partial p}{\partial t} = \mathcal{A}p, \quad p(x,0) = p_0(x),$$

together with suitable decay conditions as $|x| \to \infty$. A smooth invariant measure has density p:

$$0 = \mathcal{A}p,$$

---

† The divergence of a matrix A is here defined in the fashion standard in the continuum mechanics literature, that is $(\nabla \cdot A)_i = A_{ij,j}$, see [11].

again together with decay conditions at infinity.

In the absence of noise, highly non-smooth p is possible, as the following example shows.

Example – Linear Decay If

$$f(x) = -x, \quad \sigma(x) \equiv 0$$

then

$$\frac{\partial p}{\partial t} = -\frac{\partial}{\partial x}\{-xp\}$$

with solution

$$p(x,t) = e^t p_0(xe^t),$$

where $p_0(x)$ is the initial density, assumed to have compact support. Then, as $t \to \infty$, the density approaches a point mass at the origin – the invariant measure for this problem.

However, if

$$f(x) = -x, \quad \sigma(x) = \epsilon$$

then

$$\frac{\partial p}{\partial t} = -\frac{\partial}{\partial x}\{-xp\} + \frac{\epsilon^2}{2}\frac{\partial^2 p}{\partial x^2}.$$

The addition of the noise, however small, gives parabolic smoothing for p and the unique invariant measure is now a Gaussian with density

$$p(x) \propto \exp(-x^2/\epsilon^2). \quad \blacksquare$$

Invariant measures play a fundamental role, not only over many realizations of the noise and/or random data, but also for individual realizations. This is because of *ergodicity*. Roughly, the map (2.1) ergodic if for some class of real valued functions g

$$\frac{1}{n}\sum_{j=0}^{n-1} g(x_j) \to \int g(x)p(x)dx \tag{2.6}$$

or, for continuous time (2.4),

$$\frac{1}{t}\int_0^t g(x(s))ds \to \int g(x)p(x)dx, \tag{2.7}$$

together with some assumptions on the set of initial conditions for which this holds.

Birkhoff's ergodic theorem states that the limit in (2.6) exists for almost every point with respect to any invariant measure for the map, with

330

330 similar results in continuous time [42]. If for some x_0 this limit exists for every continuous bounded function g, it defines a linear functional and can hence be viewed as a Borel measure. If

$$\frac{1}{n}\sum_{j=0}^{n-1}\delta_{x_j} \Rightarrow \mu$$

(with $\Rightarrow$ denoting weak convergence) and the limit is independent of the initial point x_0 when it is taken in some set of positive Lebesgue measure, then the resulting measure is called a physical measure, or SRB (for Sinai-Ruelle-Bowen) measure, since it can then be physically observed, see [42].

To compute time averages in general we may either:

directly: calculate p and hence the right hand side in (2.6) or (2.7);

indirectly: calculate a trajectory and sum/integrate to get the left hand side of (2.6) or (2.7).

For a general discussion of the direct and indirect methodologies for the calculation of invariants of dynamical systems see the preface in [34]. Direct methods are not the subject of this review, but we give a brief bibliography before moving to the subject of indirect methods. In [40] the idea of discretising $\mathbb{R}^n$ to obtain a finite state space Markov chain approximation to (2.3) was introduced. Subsequent analysis of this approach was given in [22] and [19]. More recent theoretical work, and numerical experiments, includes [18, 17] and [16]. In the last few years these ideas have been employed as an effective computational tool, in particular by use of adaptive choice of the spatial discretization – see [4, 3, 5] for example, together with [15] and [31] where applications to molecular dynamics are described. A recent overview of the subject may be found in [9].

We will concentrate here on indirect methods and split our study into the three classes of problems described at the end of the last section. The most satisfactory theory is for problems in Category B) where, for example, the parabolic smoothing of probability densities for (2.4) is exploited. A coherent theory of ergodicity for Category A) is very hard, and hence so is any perturbation theory; much of what is currently known is limited to expansive maps and certain hyperbolic maps – see [42], [43], [23] for more details. Category C) is, somewhat surprisingly at first glance, more tractable than A) in some instances. This is because ideas from statistical mechanics allow certain problems in Category C) to be approximated by Markov processes; their stochastic stability is a strong property with implications for the robustness of statistics generated by approximation schemes.

3 Deterministic Problems

Problems in Category A) are the hardest in which to treat the effect of any approximation, including numerical. For the Hénon map SRB measures are known to exist [43] and it is reasonable to conjecture that the robust statistics observed in numerical simulations such as those shown in section 1 are due to robust calculation of an SRB measure. However, although the example of the Henon map is very suggestive that strong results might be provable in the deterministic case, the following example should strike a note of caution.

Example – The Tent Map Consider again the map (2.1), (2.2). If the initial condition x_0 is irrational the resulting trajectory is chaotic, and hence the problem is chaotic for a.e. initial condition. However, if x_0 is of the form $u/2^p$, where u is an odd integer, then $x_n = 0$ for $n > p + 1$ and thus any computer simulation of this map (using floating point arithmetic) will have trivial dynamics. To create the effect of the irrational initial conditions which lead to chaos one can add random noise slightly larger than machine precision at each step, and Figure 1.2 shows results from two such simulations using single and double precision. Consistent statistics are observed at the two levels of precision and the empirical density is close to the true (Lebesgue) density generated by chaotic solutions of the original map. No rigorous justification of this is known as yet, but work of Kifer [20] comes close. He shows that replacing (2.1) by its restriction to a uniform lattice of scale δ, and adding noise with scale ϵ, yields a Markov chain with invariant measure close to an invariant measure of the original map (2.1), provided that $\delta = \mathcal{O}(\epsilon^{1+c}), c > 0$. Further study of the effect of rounding error on iterated maps, concentrating on attractors, may be found in [6]. ∎

However, despite this cautionary example, the numerical evidence for many deterministic problems is similar to that seen for the Henon map in the introduction - namely that statistics are well-reproduced, even when trajectories are not. To illustrate this we give a simple example from ODEs:

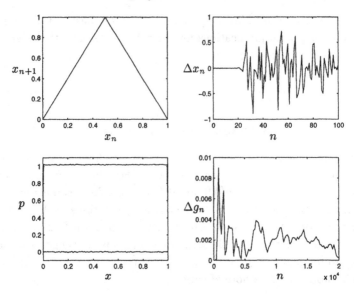

Fig. 1.2. The tent map. Upper left: x_{n+1} vs. x_n, generated by a double-precision arithmetic simulation. Upper right: The difference between two trajectories, one computed with single precision arithmetic, the other with double. Lower left: Histogram of the probability density for x, computed from very long-time simulations in single precision, and the difference between the histogram computed in single and double precision. Lower right: Difference in the time averages $\frac{1}{n+1}\sum_{k=0}^{n} g(x_k)$ with $g(x) = x^2$.

Example – Lorenz Equations Consider the equations

$$
\begin{aligned}
\dot{x} &= 10(y - x) \\
\dot{y} &= 28x - y - xz \\
\dot{z} &= xy - \tfrac{8}{3}z
\end{aligned}
$$

Figure 1.3 shows two calculations using the Euler method, one with time-step twice the other. It is clear that the method reproduces statistics well even though trajectories are completely wrong. There is no satisfactory theory to rigorously justify this observation at present but the recent work of Tucker [39] may facilitate rigorous justification. ∎

The previous problem is dissipative with a global attractor, but the observation about robustness of statistics extends to some Hamiltonian problems also†. The following example illustrates this:

† Note that, in a generic sense, such problems are not ergodic on every energy shell [24].

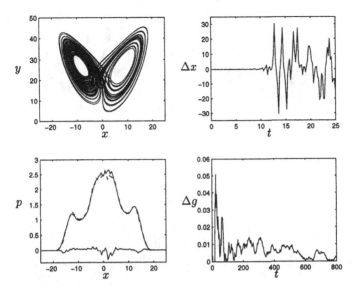

Fig. 1.3. The Lorenz equations. Upper left: The attractor computed using Euler's method. Upper right: The difference between two trajectories, one computed with time step twice the other. Lower left: Histogram of the probability density for x, computed from very long-time simulations, and the difference between the histogram computed with two different time steps, one twice the other. Lower right: Difference in the time averages $\frac{1}{n+1}\sum_{k=0}^{n} g(x_k)$ with $g(x) = \left(\frac{x+25}{50}\right)^2$.

Example – Three Interacting Particles Consider a system of three particles with pairwise interaction potential $U_i(r) = 1/r$, where $r > 0$ is the distance between the two particles, and potential energy

$$U_b(q_x, q_y) = e^{(q_x/4)^2} + e^{(q_y/4)^2} - 2e$$

added to keep the particles in a finite area [27].

Figure 1.4 shows calculations with two different time-steps (related by a factor of two) using the Verlet method [41]. Once again it is clear that the method reproduces statistics even though trajectories are completely wrong. Reich [27] has an explanation for why symplectic methods reproduce time-averages well which, in outline, goes like this: hypothesise that some Poincaré suspension of the flow satisfies the conditions developed by, for example Young and Liverani (see [43, 23, 42]) which yield decay of correlations, the law of large numbers or the central limit theorem. Note that, for times of order $\exp(-1/\Delta t)$, a symplectic method approximates a nearby Hamiltonian problem [30]; then use stochastic stability

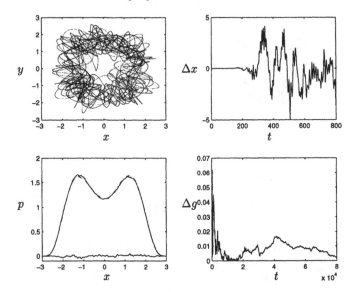

Fig. 1.4. Particle system. Upper left: Trajectories of the particles computed using the Verlet method. Upper right: The difference between two trajectories of a single particle, one computed with time step twice the other. Lower left: Histogram of the probability density for x, computed from very long-time simulations, and the difference between the histogram computed with two different time steps, one twice the other. Lower right: Difference in the time averages $\frac{1}{n+1}\sum_{k=0}^{n} g(x_k)$ with $g(x) = \left(\frac{x+3}{6}\right)^2$.

of the original Hamiltonian problem, which follows from the assumptions on the flow, to study time averages for the perturbed Hamiltonian problem. The conclusion is that time averages are well approximated for times exponentially long in Δt^{-1}. (Standard convergence theory would give times only logarithmic in Δt^{-1}.) Note however that Reich's assumptions appear very hard to verify on any concrete examples and so his ideas, whilst very suggestive, fall short of a completely satisfactory explanation. ■

4 Stochastic Problems

Because of the parabolic smoothing induced by white noise (see (2.5)) the picture here is fairly well-developed from a theoretical standpoint. The simplest examples are of the following type:

Example – Lorenz with Noise Consider the Lorenz equations with additional white noise:

$$\begin{aligned}
dx &= & 10(y - x)dt & + & dW_1 \\
dy &= & (28x - y - xz)dt & + & dW_2 \\
dz &= & \left(xy - \tfrac{8}{3}z\right) dt & + & dW_3
\end{aligned}$$

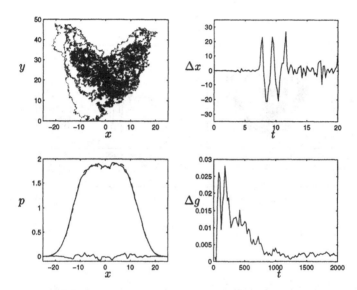

Fig. 1.5. The Lorenz equations with white noise. Upper left: A trajectory computed using Euler's method. Upper right: The difference between two trajectories, one computed with time step twice the other. Lower left: Histogram of the probability density for x, computed from very long-time simulations, and the difference between the histogram computed with two different time steps, one twice the other. Lower right: Difference in the time averages $\frac{1}{n+1} \sum_{k=0}^{n} g(x_k)$ with $g(x) = \left(\frac{x+25}{50}\right)^2$.

Figure 1.5 shows simulations, for a single realization of the noise, using the Euler-Maryuma method [21]. Once again we see that the method appears to reproduces statistics well even though trajectories are completely wrong. However, in contrast to the deterministic case, for this problem there is a very well-developed and satisfactory theory. Since $B = I$ the Fokker-Planck equation with generator $\mathcal{A}$ is uniformly parabolic; this fact, combined with dissipativity of the deterministic vector field, gives ergodicity [13]. Exploiting this structure, Talay [36, 37, 10] proves theorems which show that the densities and time averages shown in Figure 1.5 are indeed accurate approximations of their true counterparts in the SDE.†

We now modify the previous problem so that noise is only present in the x and y equations

$$\begin{aligned}
dx &= 10(y-x)dt &+& dW_1, \\
dy &= (28x - y - xz)dt &+& dW_2, \\
dz &= \left(xy - \tfrac{8}{3}z\right)dt.
\end{aligned}$$

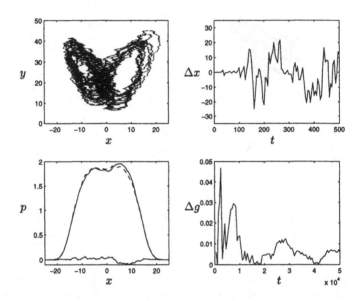

Fig. 1.6. The Lorenz equations with white noise in two components. Upper left: A trajectory computed using Euler's method. Upper right: The difference between two trajectories, one computed with time step twice the other. Lower left: Histogram of the probability density for x, computed from very long-time simulations, and the difference between the histogram computed with two different time steps, one twice the other. Lower right: Difference in the time averages $\frac{1}{n+1}\sum_{k=0}^{n} g(x_k)$ with $g(x) = \left(\frac{x+25}{50}\right)^2$.

The theory of Talay in [36, 37, 10] no longer applies since B is now singular so that the Fokker-Planck equation with generator $\mathcal{A}$ is not uniformly parabolic. Nonetheless the SDE itself is provably geometrically ergodic (see [25]) and the numerical evidence in Figure 1.6 suggests that the Euler-Maryuma method reproduces statistics well even though

---

† Note that this theory requires globally Lipschitz vector fields. Recent work by Talay [38] is aimed at removing this restriction.

trajectories are completely wrong. A theory for the approximation of such degenerate diffusions is being developed in [32] and [14], using the approach to Markov chains described in [26], and by [38] using Malliavin calculus. ∎

5 Statistical Mechanics

We have seen that for stochastic problems study of the statistics of trajectories is made simpler than for deterministic problems, essentially because of smoothing in the propagation of probability densities. This smoothing in Markov chains/processes can, in some cases, be exploited to study deterministic problems with random data. In particular such problems often simplify, as the dimension $n \to \infty$, to yield the Markov property; random noise in the initial data becomes near memoryless noise when only a subset of the variables is observed. We now exploit this basic idea from statistical mechanics to study numerical schemes.

The general picture here is to consider ODEs partitioned in the form

$$
\begin{aligned}
\dot{x} &= f(x,y), \quad x(0) = x_0 \\
\dot{y} &= g(x,y), \quad y(0) = y_0
\end{aligned}
\tag{5.1}
$$

with (x_0, y_0) a random variable distributed according to some measure ν.

Thus

$$
y(t) = \mathcal{G}(y_0, \{x(s)\}_{0 \le s \le t})
$$

and so

$$
\dot{x} = f(x, \mathcal{G}(y_0, \{x(s)\}_{0 \le s \le t})), \quad x(0) = x_0.
\tag{5.2}
$$

In previous sections we have concentrated on whether we can calculate statistics accurately over long times, where trajectories are not accurate. Here we shift emphasis slightly and ask whether we can accurately compute the statistics of the x variable without accurately computing y. There are many problems where only a subset of the variables is of interest so that this question is natural – it may be be desirable to under-resolve y for reasons of computational efficiency.

A recent approach to this problem by Chorin *et al* [2] (1998) is to develop an equation for $X(t) = \mathbb{E}x(t)$, where expectation is with respect to ν on (x_0, y_0). Assuming stationarity leads to the approximate equations

$$
\dot{X} = F(X), \quad X(0) = x_0,
$$

where

$$F(X) = \mathbb{E}\{f(x,y)|x = X\},$$

and now expectation is with respect to ν on (x,y) There is strong experimental evidence to show that this gives good approximations for some problems, but no substantial theory as yet. Some preliminary theoretical investigations may be found in [12]. Another approach is to try and identify low-dimensional stochastic models from numerical simulation of the full problem, without explicit *a priori* knowledge of which variables will comprise the low-dimensional approximation. This work may be found in [15] and [31].

We will consider a less ambitious approach where the complete system for x, y is still integrated, but methods used which are cheap and formally inaccurate trajectory-wise in the y component. The reason to expect that we might still accurately compute x, or at least its statistics, is that, often, for dimension $n \to \infty$, one obtains from (5.2) a Markov process or chain such as

$$dx = f_0(x)dt + \sigma_0(x)dW.$$

This can be accurately approximated without approximating y accurately. We look at three examples which illustrate this idea.

Example – Construction of an SDE [1] Consider the equations, $j = 0, \ldots, N$

$$\ddot{u}_j + j^2 u_j = 0,$$

$$\dot{z}_N = f(z_N) + \sum_{j=0}^{N} u_j(t),$$

$$u_j(0) = \sqrt{\frac{2}{\pi}}\eta_j, \quad \dot{u}_j(0) = 0, \quad z_N(0) = z_0$$

where the η_j are i.i.d. $\mathcal{N}(0,1)$. Thus $x = z_N$ and $y = \{(u_j, \dot{u}_j)\}_{j=0}^{N}$.
For large N, z_N approximates the solution of the SDE

$$dz = f(z)dt + dW, \quad z(0) = z_0,$$

where W is standard Brownian motion. Precisely we have that, for $T \in [0, \pi]$,

$$\mathbb{E}\|z(T) - z_N(T)\|^2 \leq \frac{C(T)}{N}.$$

Let $z^n = z(n\Delta t)$ and let $Z^n \approx z_N(n\Delta t)$ be our numerical approximation. In order to make precise the notion of computing inaccurately with respect to the y variable we take the limit

$$N \to \infty, \quad \Delta t \to 0, \quad N\Delta t = \zeta. \tag{5.3}$$

Under this limit process the oscillators u_j are not well resolved for large j. We apply Leap-Frog to the oscillators with $\zeta \leq 2$† and the theta-method to the z equation. Then, for $n\Delta t \in [0, \pi]$,

$$\mathbb{E}\|z^n - Z^n\|^2 \leq C(n\Delta t)\Delta t^{2/3}.$$

For sufficiently smooth g and for linear f

$$|\mathbb{E}g(z^n) - \mathbb{E}g(Z^n)| \leq C(n\Delta t)\Delta t^{2/3}.$$

Precise statements of these theorems, together with numerical results illustrating them, can be found in [1]. ∎

A more physically realistic model – based on a mechanical description of a *heat bath* due to Ford and Kac [8] – is the following:

Example – Heat Bath [35] Consider the equations, for $j = 1, \ldots, N$

$$\ddot{u}_j + j^2(u_j - q) = 0,$$

$$\ddot{q} + V'(q) = \sum_{j=1}^{N} \gamma^2(u_j - q)$$

$$u_j(0) = \sqrt{\frac{2}{\pi}}\eta_j, \; \dot{u}_j(0) = 0, \; q(0) = q_0, \; \dot{q}(0) = p_0$$

where $x = \{p, q\}$ and $y = \{\{v_j, u_j\}_{j=1}^N, p, q\}$, and the η_j are as before. When N is large the motion of q is governed by an SDE:‡

$$\ddot{Q} + V'(Q) - \frac{\gamma^2}{2}Q + \frac{\gamma^2\pi}{2}\dot{Q} = \dot{B}$$

$$Q(0) = q_0, \quad \dot{Q}(0) = p_0 - \frac{\gamma^2\pi}{2}q_0,$$

$$B(t) = \gamma\sqrt{\frac{\pi}{2\beta}}[W(t) - \frac{t}{\pi}W(\pi)].$$

Here $W(t)$ is as before. In fact, for any $T \in [0, \pi]$,

$$\mathbb{E}\{\|q - Q\|^2_{L^\infty(0,T)} + \|\dot{q} - \dot{Q}\|^2_{L^2(0,T)}\} \leq C(T)\frac{\ln(N)^2}{N}.$$

† The method then gives exact solutions to a nearby linear oscillator problem; see [28]
‡ This formal equation can be put in first order form and cast as an Itô SDE.

Again we compute q numerically by solving the large system in the limit (5.3). Precise statement of results, together with numerical experiments showing that under-resolved simulations can produce accurate (i.e. close to Q) simulations of q, can be found in [35]. ■

Example – Billiards [33] Small particles suspended in a fluid with velocity field v can be modelled by having the particles obey Stokes' law between elastic collisions. Briefly we have

$$m\ddot{z}_i = \alpha[v(z_i, t) - \dot{z}_i], \quad i = 1, \ldots, N,$$

$$+\text{ELASTIC COLLISIONS},$$

$$z_i(0) \sim U([0,1] \times [0,1]), \quad \dot{z}_i(0) \sim \mathcal{N}(0, I).$$

We may think of $x = (z_1, \dot{z}_1)$; y comprises the positions and velocities of the remaining particles.

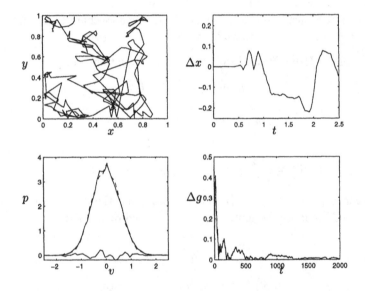

Fig. 1.7. The Billiards Problem. Upper left: A trajectory of a single particle, computed in double precision. Upper right: The difference between two trajectories, one computed with single precision arithmetic, the other with double. Lower left: Histogram of the probability density for the velocity of a single particle, computed from very long-time simulations in single precision, and the difference between the histogram computed in single and double precision. Lower right: Difference in the time averages $\frac{1}{n+1} \sum_{k=0}^{n} g(x_k)$ with $g(x) = \left(\frac{x+3}{6}\right)^2$.

Figure 1.7 shows data generated from a simulation of this problem with 210 particles and $\alpha = 0$. All data is generated by considering one particle. Figure 1.7 top left shows the particle trajectory, generated by a double-precision arithmetic simulation. Figure 1.7 top right shows the difference between two trajectories, in double and single precision; as is to be expected the trajectories rapidly diverge. However, once again, the statistics are well-reproduced: Figure 1.7 bottom left shows for that the empirical density is close in both precisions and that hence time-averages are insensitive to precision – see Figure 1.7 bottom right.

There is no rigourous explanation for the observation. However, it is believed that, for $N \gg 1$ $z_1, \dot{z}_1$ can in some situations be modelled by a Markov stochastic process of the form:

$$m\ddot{Z}_1 = \alpha[v(Z_1, t) - \dot{Z}_1] + \sum_{j=1}^{\infty} \delta(t - t_j)\mathcal{F}_j,$$

Here $\{t_j, \mathcal{F}_j)\}_{j \geq 1}$ form a family of i.i.d. random variables. Their joint distribution is calculated by arguments similar to those Boltzmann used to derive his equation. Assuming that this equation has a unique exponentially attracting invariant measure then its statistics are likely to be insensitive to perturbation; arguments showing the effect of rounding error on invariant measures of Markov chains may be found in [29] and the effect of time-discretization on time-continuous Markov chains may be found in [32]. ■

6 Conclusions

The numerical experiments and accompanying mathematical discussion in this paper have illustrated the following main points concerning the statistical content in numerical simulations of initial value problems:

- Experimental evidence strongly suggests that, often, statistics are robust under time-discretization and round-off error even when trajectories are not.
- Substantiating this is typically far harder for deterministic problems than for stochastic ones.
- Large deterministic problems with random data fall somewhere in between deterministic and stochastic problems; some progress has been made using ideas from statistical mechanics, but this work is in its infancy.
- Much remains to be done.

342   *Sigurgeirsson and Stuart*

Acknowledgement

We are grateful to Michael Dellnitz, Gary Froyland, Oliver Jünge and Sebastian Reich for helpful readings of drafts of this manuscript.

References

[1]   Cano, B., Süli, E., Stuart, A.M. and Warren, J. (2000). Stiff oscillatory systems, delta jumps and white noise, *J. FoCM,* to appear.
[2]   Chorin, A.J., Kast, A. and Kupferman, R. (1998). Optimal prediction of underresolved dynamics, *Proc. Nat. Acad. Sci. USA* **95**, 4094–4098.
[3]   Dellnitz, M. and Junge, O. (1998). An adaptive subdivision technique for the approximation of attractors and invariant measures, *Comput. Visual. Sci.* **1**, 63–68.
[4]   Dellnitz, M. and Junge, O. (1999). On the approximation of complicated dynamical behavior, *SIAM J. Num. Anal.* **36**, 491–515.
[5]   Deuflhard, P., Dellnitz, M., Junge, O. and Schütte, Sh. (1998). Computation of essential molecular dynamics by subdivision techniques I: basic concept, in *Computational Molecular Dynamics: Challenges, Methods, Ideas* (P. Deuflhard, J. Hermans, B. Leimkuhler, A.E. Mark, S. Reich and R.D. Skeel, eds), LNiCSE 4 (Springer-Verlag, Berlin), 98–115.
[6]   Diamond, P. and Kloeden, P. (1993). Spatial discretization of mappings. *Comp. Math. Applic.* **25**, 85–94.
[7]   Duflo, M. (1997). *Random Iterative Models* (Springer-Verlag, Berlin).
[8]   Ford, G.W. and Kac, M. (1987). On the quantum Langevin equation, *J. Stat. Phys.* **46**, 803–810.
[9]   Froyland, G. (2000). Extracting dynamical behaviour via Markov models, in *Nonlinear Dynamics and Statistics: Proceedings* (A.I. Mees, ed.) (Birkhäuser, Basle), 283–324.
[10]   Grorud, A. and Talay, D. (1996). Approximation of Lyapunov exponents of nonlinear stochastic differential equations, *SIAM J. Appl. Math.* **56**, 627–650.
[11]   Gurtin, M.E. (1981). *An Introduction to Continuum Mechanics* (Academic Press, New York).
[12]   Hald, O. (1999). Optimal prediction and the Klein–Gordon equation, *Proc. Natl. Acad. Sci.* **96**, 4774–4779.
[13]   Hasminskii, R.Z. (1980). *Stochastic Stability of Differential Equations* (Sijthoff and Noordhoff, Rockville).
[14]   Higham, D.J., Mattingly, J. and Stuart, A.M. (2000). Ergodicity for discretizations of SDEs: Locally Lipschitz vector fields and degenerate noise. To appear.
[15]   Huisinga, W., Best, C. Roitzsch, R., Schütte, C. and Corde, F. (1999). From simulation data to conformational ensembles: structure and dynamics based methods. *J. Comp. Chem.* **20**, 1760–1774.
[16]   Hunt, B.R. (1996). Estimating invariant measures and Lyapunov exponents, *Ergod. Th. & Dynam. Sys.* **16**, 735–749.
[17]   Hunt, F.Y. (1994). A Monte Carlo approach to the approximation of invariant measures, *Random and Computational Dynamics* **2**, 111–133.

[18] Hunt, F.Y. (1998). Unique ergodicity and the approximation of attractors and their invariant measures using Ulam's method. *Nonlinearity* **11**, 307–317.

[19] Keller, G. (1982). Stochastic stability in some chaotic dynamical systems, *Monatshefte für Mathematik* **94**, 313–333.

[20] Kifer, Y. (1997). Computations in dynamical systems via random perturbations, *Discrete & Continuous Dyn. Sys. 3*, 457–476.

[21] Kloeden, P.E. and Platen, E. (1991), *Numerical Solution of Stochastic Differential Equations* (Springer-Verlag, New York).

[22] Li, T.Y. (1976). Finite approximation of the Frobenius–Perron operator. A solution to Ulam's conjecture, *J. Approx. Th.* **17**, 177–186.

[23] Liverani, C. (1995). Decay of correlations, *Ann. Maths 142*, 239–301.

[24] Markus, L. and Meyer, K.R. (1974). Generic Hamiltonian dynamical systems are neither integrable nor ergodic, *Memoirs Amer. Math. Soc.* **144.**

[25] Mattingly, J. and Stuart, A.M. (2000). Ergodicity for SDEs: Locally Lipschitz vector fields and degenerate noise. To appear.

[26] Meyn, S.P. and Tweedie, R.L. (1996). *Markov Chains and Stochastic Stability*, 2nd ed. (Springer-Verlag, London).

[27] Reich, S. (1999). Backward error analysis for numerical integrators, *SIAM J. Num. Anal.* **36**, 1549–1570.

[28] Reich, S. (1999). Preservation of adiabatic invariants under symplectic discretization, *Appl. Numer. Math.* **29**, 45–55.

[29] Roberts, G., Rosenthal, J.S. and Schwarz, P.O. (1998). Convergence properties of perturbed Markov chains, *J. Appl. Prob.* **35**, 1–11.

[30] Sanz-Serna,J.-M. and Calvo, M.P. (1994). *Numerical Hamiltonian Problems* (Chapman & Hall, London).

[31] Schütte, C., Fischer, A., Huisinga, W. and Deuflhard, P. (1999). A direct approach to conformational dynamics based on hybrid Monte Carlo, *J. Comp. Phys.* **151**, 146–168.

[32] Shardlow, T. and Stuart, A.M. (2000). A perturbation theory for ergodic Markov chains and applications to numerical approximations, *SIAM J. Num. Anal.*, to appear.

[33] Sigurgeirsson, H. and Stuart, A.M. (2000). Collision detection for particles in a field. To appear.

[34] Stuart, A.M. and Humphries, A.R. (1996). *Dynamical Systems and Numerical Analysis* (Cambridge University Press, Cambridge).

[35] Stuart, A.M. and Warren, J. (2000). Analysis and experiments for a computational model of a heat bath, *J. Stat. Phys.*, to appear.

[36] Talay, D. (1990). Second-order discretization schemes for stochastic differential systems for the computation of the invariant law, *Stochastics and Stochastics Reports* **29**, 13–36.

[37] Talay, D. (1991). Approximation of upper Lyapunov exponents of bilinear stochastic differential systems, *SIAM J. Num. Anal.* **28**, 1141–1164.

[38] Talay, D. (2000). Approximation of the invariant probability measure of stochastic Hamiltonian dissipative systems with non globally Lipschitz co-efficients, Tech. Rep., INRIA.

[39] Tucker, W. (1999). The Lorenz attractor exists. Preprint.

[40] Ulam, S.M. (1960). *A Collection of Mathematical Problems*, Interscience Tracts in Pure and Applied Maths **8** (Interscience, New York).

[41] Verlet, L. (1967), Computer "experiments" on classical fluids. I. Thermodynamical properties of Lennard–Jones molecules, *Phys. Rev.* **159**, 98–103.

[42] Viana, M. (1996). *Stochastic Dynamics of Deterministic Systems* (IMPA, Rio de Janeiro).

[43] Young, L.-S. (1998). Developments in chaotic dynamics, *Notices AMS* *45*, 1318–1328.

Simulation of stochastic processes and applications

Denis Talay

INRIA
2004 Route des Lucioles
B.P. 93
06902 Sophia-Antipolis
France
Email Denis.Talay@sophia.inria.fr

Abstract

In this paper we give some examples of application of the simulation of stochastic processes, and we present a few recent results on the convergence rates of stochastic numerical methods.

1 The Euler scheme

In this paper we present stochastic numerical methods. The objective is the simulation of diffusion processes in order to estimate statistics of their probability laws, or the numerical resolution of deterministic partial differential equations by Monte Carlo or stochastic particle methods. In all these cases, the core of the algorithm is the time discretization of stochastic differential systems. In almost all the applications met by the author, it appears that the simplest scheme, that is, the Euler scheme is the most efficient discretization method. The reason is that it involves the minimal number of random variables to simulate, namely the independent Gaussian increments of the Brownian motion, and does not involve the derivatives of the coefficients of the discretized system. In addition, theoretical results show that very simple extrapolation procedures permit to dramatically accelerate the convergence rate of the Euler scheme.

Let $b : \mathbb{R}^d \to \mathbb{R}^d$ and $\sigma_j : \mathbb{R}^d \to \mathbb{R}^d$ $(1 \leq j \leq r)$ be given functions. Let (G_p^j) be a family of real valued independent Gaussian random variables with zero mean and unit variance. Fix $T > 0$. The Euler scheme is defined as follows,

$$
\begin{cases}
X_0^n(x) = x, \\[2mm]
X_{(p+1)T/n}^n(x) = X_{pT/n}^n(x) + b\left(X_{pT/n}^n(x)\right)\dfrac{T}{n} \\[4mm]
\qquad\qquad + \displaystyle\sum_{j=1}^{r} \sigma_j\left(X_{pT/n}^n(x)\right)\sqrt{\dfrac{T}{n}}G_{p+1}^j.
\end{cases}
\tag{1.1}
$$

The Euler scheme is deeply related to parabolic partial differential equations. Indeed, suppose that there exists a smooth solution with bounded derivatives to the evolution problem

$$
\begin{cases}
\dfrac{\partial u}{\partial t}(t,x) = Lu(t,x), \ t > 0, \ x \in \mathbb{R}^d, \\[4mm]
u(0,x) = f(x), \ x \in \mathbb{R}^d,
\end{cases}
\tag{1.2}
$$

where the differential operator L is defined by

$$
Lg(x) := \sum_i b_i(x)\partial_i g(x) + \frac{1}{2}\sum_{i,j}(\sigma(x)\sigma(x)^*)_j^i \partial_{ij} g(x),
$$

for all smooth real valued function $g(x)$. Then

$$
\begin{aligned}
& \mathbb{E}f(X_T^n(x)) - u(T,x) \\
& = \sum_{p=0}^{n-1} \mathbb{E}\left[u\left(T-(p+1)T/n, X_{(p+1)T/n}^n(x)\right)\right. \\
& \qquad\qquad \left. - u\left(T - pT/n, X_{pT/n}^n(x)\right)\right].
\end{aligned}
$$

Using the independence of (G_p^j), Taylor expansions of order 4 show that

$$
\begin{aligned}
& \mathbb{E}f(X_T^n(x)) - u(T,x) \\
& = \sum_{p=0}^{n-1} \mathbb{E}\left[u\left(T-(p+1)T/n, X_{pT/n}^n(x)\right) - u\left(T - pT/n, X_{pT/n}^n(x)\right)\right] \\
& \quad + \frac{T}{n}\sum_{p=0}^{n-1}\mathbb{E}\left[Lu\left(T-(p+1)T/n, X_{pT/n}^n(x)\right)\right] + \sum_{p=0}^{n-1}\mathcal{O}\left(\frac{1}{n^2}\right) \\
& = \frac{T}{n}\sum_{p=0}^{n-1}\mathbb{E}\left[Lu\left(T - pT/n, X_{pT/n}^n(x)\right) - \frac{\partial u}{\partial t}\left(T - pT/n, X_{pT/n}^n(x)\right)\right] \\
& \quad + \sum_{p=0}^{n-1}\mathcal{O}\left(\frac{1}{n^2}\right).
\end{aligned}
$$

In view of equation (1.2) we get

$$\mathbb{E}f(X_T^n(x)) - u(T,x) = \sum_{p=0}^{n-1} \mathcal{O}\left(\frac{1}{n^2}\right)$$

$$= \mathcal{O}\left(\frac{1}{n}\right). \tag{1.3}$$

Let $(X_{pT/n}^{n,k}, 0 \le p \le n)$ be independent realizations of the Markov chain $(X_{pT/n}^n, 0 \le p \le n)$. Then, in view of the Strong Law of Large Numbers, one has

$$u(pT/n, x) \simeq \frac{1}{M} \sum_{k=1}^{M} f\left(X_{pT/n}^{n,k}(x)\right).$$

The global error is

$$u(pT/n, x) - \frac{1}{N} \sum_{k=1}^{M} \mathbb{E}f\left(X_{pT/n}^{n,k}(x)\right)$$

$$= \underbrace{u(pT/n, x) - \mathbb{E}f(X_{pT/n}^n(x))}_{=:\epsilon_d(n)}$$

$$+ \underbrace{\mathbb{E}f(X_{pT/n}^n(x)) - \frac{1}{M} \sum_{k=1}^{M} f\left(X_{pT/n}^{n,k}(x)\right)}_{=:\epsilon_s(n,M)}.$$

The statistical error $\epsilon_s(n)$ depends on the number of simulations M. In addition, the variance which appears in the Central Limit Theorem, that is,

$$\mathbb{E}\left|f(X_T^n(x))\right|^2 - \left|\mathbb{E}f(X_T^n(x))\right|^2$$

can usually be bounded from above uniformly w.r.t. n.

Estimate (1.3) shows that the discretization error $\epsilon_d(n)$ is of order $1/n$. We even have a much better result. We do not precise the highly technical, but not too stringent, hypotheses. In substance, it is required that equation (1.2) has a smooth fundamental solution.

Theorem 1.1 *Let f be a measurable and bounded function. Suppose that the functions b^i and σ_j^i are of class $C^\infty(\mathbb{R}^d, \mathbb{R})$ with bounded derivatives of all order. Under conditions of 'uniform hypo-ellipticity' on the* *differential operator L, the Euler scheme error satisfies*

$$u(T,x) - \mathbb{E}\,f(X_T^n(x)) = -\frac{C_f(T,x)}{n} + \frac{Q_n(f,T,x)}{n^2}, \qquad (1.4)$$

and

$$|C_f(T,x)| + \sup_n |Q_n(f,T,x)| \le K(T)\|f\|_\infty \frac{1+\|x\|^Q}{T^q},$$

where K is an increasing function, Q and q are integers.

See Bally & Talay [1].

Estimate (1.4) justifies the use of the Richardson–Romberg extrapolation procedure: simulate paths of the Euler scheme with step sizes T/n and $T/(2n)$. Then

$$u(T,x) - \left\{2\mathbb{E}\,f(X_T^{2n}(x)) - \mathbb{E}\,f(X_T^n(x))\right\} = \mathcal{O}\left(\frac{1}{n^2}\right).$$

This suggests to use the new approximation

$$u(pT/n,x) \simeq \frac{1}{M}\sum_{k=1}^M \left\{2f\left(X_{pT/n}^{2n,k}(x)\right) - f\left(X_{pT/n}^{n,k}(x)\right)\right\}.$$

For numerical illustrations of the efficiency and comments, see Talay and Tubaro [25].

Similar results hold for the stopped or reflected Euler scheme for partial differential equations with Dirichlet or Neumann boundary conditions: see, e.g., Gobet [9], Costantini, Pacchiarotti & Sartoretto [8].

It is well known that, under suitable hypotheses on the functions f, b^i and σ_j^i, one has

$$u(\theta,x) = \mathbb{E}\,f(X_\theta(x)),$$

where $(X_t(x))$ is the unique solution to the stochastic differential equation

$$X_t(x) = x + \int_0^t b(X_s(x))\,ds + \int_0^t \sigma(X_s(x))\,dW_s.$$

Here, (W_t) denotes a standard Brownian motion. Thus the Euler scheme permits to compute approximations of quantities of the type $\mathbb{E}\,f(X_T(x))$, which appear in random mechanics problems (moments of the responses of dynamical systems), finance (option prices), etc. It also provides good estimates on the quantiles of the distribution of the random variable $X_T(x)$. Given $0 < \delta < 1$, define the quantiles $\rho(\delta)$ and $\rho^n(\delta)$ by

$$\mathbb{P}\left[X_T(x) \le \rho(\delta)\right] = \delta$$

and

$$\mathbb{P}\left[X_T^n(x) \le \rho^n\right] = \delta.$$

Under the hypotheses of Theorem 1.1, one has

$$|\rho(\delta) - \rho^n(\delta)| \le \frac{C}{q_T(\delta)n}, \tag{1.5}$$

where, if $p_T(y)$ denotes the density of $X_T(x)$,

$$q_T(\delta) := \inf_{y \in (\rho(\delta)-1, \rho(\delta)+1)} p_T(y).$$

See Talay & Zheng & [27] for the proof and the application to a model risk study in financial mathematics.

2 Approximation of invariant measures of dissipative stochastic Hamiltonian systems

Suppose that there exists a unique solution μ in the sense of the distributions to the stationary Fokker–Planck equation

$$L^*\mu = 0, \tag{2.6}$$

where L^* is defined by

$$L^*g(x) := \frac{1}{2}\sum_{i,j}\partial_{ij}\left((\sigma(x)\sigma(x)^*)_j^i g(x)\right) - \sum_i \partial_i(b_i(x)g(x)),$$

for all smooth real valued function $g(x)$.

Let f be a given μ-integrable function, and set

$$I := \int_{\mathbb{R}^d} f(y)\mu(dy). \tag{2.7}$$

Suppose that the Euler scheme is ergodic, that is, there exists a unique probability measure μ^n such that

$$\lim_{N \to \infty} \frac{1}{N}\sum_{p=1}^{N} f(X_{pT/n}^n(x)) = \int_{\mathbb{R}^d} f(y)\mu^n(dy)$$

for all μ^n integrable function f and all x. We claim that

$$I \cong I^{n,N} := \frac{1}{N}\sum_{p=1}^{N} f(X_{pT/n}^n(x)). \tag{2.8}$$

The global error $I - I^{h,N}$ depends on the two numerical parameters n and N and satisfies

$$I - I^{h,N} = \underbrace{I - \int_{\mathbb{R}^d} f(y)\mu^h(dy)}_{e_d(n)} + \underbrace{\int_{\mathbb{R}^d} f(y)\mu^n(dy) - I^{h,N}}_{e_s(n,N)}. \qquad (2.9)$$

Central Limit Theorems for the normalized error $\sqrt{N}e_s(n,N)$ (see, e.g., Revuz [18]) provide estimates on the statistical error $e_s(n,N)$.

We turn our attention to the discretization error $e_d(n)$. Suppose that the coefficients b^i and σ_j^i are bounded and have bounded derivatives of all order, the operator L is uniformly strictly elliptic and there exists a strictly positive constant β and a compact set K such that

$$\forall x \in \mathbb{R}^d - K, \qquad x \cdot b(x) \le -\beta|x|^2.$$

Talay & Tubaro [25] have shown that there exists a constant C such that

$$e_d(n) = \frac{C}{n} + \mathcal{O}\left(\frac{1}{n^2}\right) \qquad (2.10)$$

for all integer n. One thus can use the Richardson–Romberg extrapolation method to accelerate the convergence rate,

$$I \cong 2I^{2n,N} - I^{n,N} = \frac{1}{N}\sum_{p=1}^{2N} f\left(X_{pT/(2n)}^{2n}(x)\right) - \frac{1}{N}\sum_{p=1}^{N} f\left(X_{pT/n}^{n}(x)\right). \qquad (2.11)$$

The corresponding discretization error is of order $1/n^2$.

Talay [23, 24] extends the methodology and the results to the nonlinear dissipative stochastic Hamiltonian systems excited by Gaussian white noises studied by Soize [21]. For such systems, the difficulty of the analysis comes from the fact that the operator L is not uniformly elliptic. In addition, the drift coefficient is not globally Lipschitz, so that one has to use the implicit Euler scheme

$$\begin{cases} X_0^n(x) = x, \\ X_{(p+1)T/n}^n(x) = X_{pT/n}^n(x) + b\left(X_{(p+1)T/n}^n(x)\right)\dfrac{T}{n} \\ \qquad + \sum_{j=1}^r \sigma_j\left(X_{pT/n}^n(x)\right)\sqrt{\dfrac{T}{n}}G_{p+1}^j. \end{cases} \qquad (2.12)$$

Ergodic properties of discretization schemes of dissipative stochastic

Hamiltonian systems are also studied by Higham, Mattingly & Stuart [11], Shardlow & Stuart [20].

3 Stochastic particle methods for statistical solutions of nonlinear partial differential equations

Partial differential equations with random initial conditions are used for statistical descriptions of physical complex systems like turbulent systems. For example, random initial conditions for the Navier–Stokes equation represent the uncertainty on the present state of the fluid because of, e.g., measurement errors (see, e.g., Chorin *et al.* [7]).

Given the law of the initial condition, one needs to compute mean quantities of the solution. To this end, one can use closure models or classical Monte Carlo methods. We now discuss these two procedures, and justify the introduction of our original stochastic particle method.

Concerning the closure models: for nonlinear equations, the first moment of the solution solves a new nonlinear problem where correlations are involved; these correlations solve new evolution problems where third order moments are involved, etc. A closure relation permits to get a closed system of partial differential equations whose first moments are the solutions. It is often difficult to estimate the accuracy of this closure procedure (see, e.g., Mohammadi & Pironneau [15], Vishik & Fursikov [29]).

Concerning the Monte Carlo method: a naive procedure consists in sampling the law of the initial condition and, for each sample, discretizing the corresponding Navier–Stokes equation; one finally averages the approximate solutions to estimate the first moment (other statistics can be obtained by averaging appropriate functionals of the approximate solutions). The complexity of such a numerical method is much too large.

We propose a new procedure. The construction is based on the following observation. The 2D incompressible viscous Navier–Stokes equation for the vorticity with a deterministic initial condition and several other conservation laws like the Burgers equation can be considered as McKean–Vlasov equations. Therefore they have probabilistic interpretations in terms of mean field interacting stochastic particle systems: see, e.g., Marchioro & Pulvirenti [14], Osada [16], Sznitman [22]. These stochastic particle systems can be simulated on computers, and their empirical distributions approximate the desired exact solution in the sense of the distributions. For example, let $b(x,y)$ and $s(x,y)$ be two Lipschitz kernels from $\mathbb{R}^2$ to $\mathbb{R}$. Consider the the McKean–Vlasov equation

$$\begin{cases} \dfrac{\partial}{\partial t} <\mu_t, f> \, = \, <\mu_t, L(\mu_t)f> & \forall f \in C_K^\infty(\mathbb{R}), \\[2mm] \mu_{t=0} = \mu_0, \end{cases} \qquad (3.13)$$

where the differential operator $L(\mu)$ is defined by

$$L(\mu)f(x) = \frac{1}{2}\left(\int_{\mathbb{R}} \sigma(x,y)d\mu(y)\right)^2 f''(x) + \left(\int_{\mathbb{R}} b(x,y)d\mu(y)\right) f'(x).$$

The appropriate corresponding system of weakly interacting particles reads

$$\begin{cases} dX_t^{i,N} = \displaystyle\int_{\mathbb{R}} b(X_t^{i,N}, y)\mu_t^N(dy) \, dt + \int_{\mathbb{R}} \sigma(X_t^{i,N}, y)\mu_t^N(dy) \, dW_t^i, \\[2mm] X_0^{i,N} = X_0^i, \; i = 1,\dots,N, \end{cases}$$

$$(3.14)$$

where $(W_t^1),\dots,(W_t^N)$ are independent one-dimensional Brownian motions and μ_t^N is the random empirical measure

$$\mu_t^N := \frac{1}{N}\sum_{i=1}^{N} \delta_{X_t^{i,N}}.$$

When the initial distribution of the particles is symmetric and when the kernels are Lipschitz, one has the propagation of chaos property: the sequence of random probability measures (μ^N) on the space of trajectories defined by

$$\mu^N := \frac{1}{N}\sum_{i=1}^{N} \delta_{X^{i,N}}$$

converges in law as N goes to infinity to a deterministic probability measure μ; in addition, the flow (μ_t) of the one-dimensional distributions of μ solves (3.13) in the sense of the distributions: see, e.g., Sznitman [22], Graham, Kurtz et al. [10], Jourdain [12, 13]. Consequently one obtains a numerical resolution of equation (3.13) by applying the Euler scheme to the system (3.14): for error estimates see, e.g., Bossy & Talay [3, 4], and [10]. Optimal convergences rates have been recently obtained by Bossy [2] in the case of scalar conservation laws.

Following this idea, we propose a stochastic particle method with random weights to compute

$$\langle M_1(t) \, , \, f\rangle_{L^2(\mathbb{R})} := \mathbb{E}\int_{\mathbb{R}} f(x) \, \mu_t(\omega, dx), \qquad (3.15)$$

where f is a given test function and $\mu_t(\omega, dx)$ is the solution of a McKean–Vlasov equation with random initial condition $\mu_0(\omega, dx)$. Following Vishik & Fursikov [29], such a quantity is the first moment of the statistical solution of equation (3.13) with initial condition equal to the law of $\mu_0(\omega, dx)$. The algorithm also applies for moments of higher order.

We now fix some notation. For the sake of simplicity we limit ourselves to one dimensional McKean–Vlasov equations whose solution (μ_t) satisfies: for each t, the measure μ_t has a density $p(t, x)$ with respect to the Lebesgue measure. In the case of a random initial condition, we thus consider

$$
\begin{cases}
\dfrac{\partial p}{\partial t}(t, x, \omega) = -\dfrac{\partial}{\partial x}\left(u_b(t, x, \omega)p(t, x, \omega)\right) \\[2mm]
\qquad\qquad + \dfrac{1}{2}\dfrac{\partial^2}{\partial x^2}\left(\left(u_\sigma(t, x, \omega)\right)^2 p(t, x, \omega)\right), \\[3mm]
p(0, x, \omega) = p_0(x, \omega), \\[2mm]
u_b(t, x, \omega) := \int_{\mathbb{R}} b(x, y)p(t, y, \omega)dy, \\[2mm]
u_\sigma(t, x, \omega) := \int_{\mathbb{R}} \sigma(x, y)p(t, y, \omega)dy,
\end{cases}
\tag{3.16}
$$

where b and σ are smooth and bounded functions from $\mathbb{R}^2$ to $\mathbb{R}$. These hypotheses are too strong to cover the Navier–Stokes equation for which the function b is the Biot–Savart kernel which is singular at 0, but concern the equation obtained by smoothing the Biot–Savart kernel by means of a cut-off function.

For technical reasons we thereafter suppose that the possible initial conditions of (3.16) are parameterised by realizations $\theta(\omega)$ of a real valued random variable θ with law ν concentrated on a closed interval of $\mathbb{R}$, say $[-1, 1]$. One has

Theorem 3.1 *Suppose that*

(H_1) All the possible initial conditions are probability densities of class $C_b^2(\mathbb{R})$; they are parameterised by a real number a in $[-1, 1]$, and the mapping $\Phi : a \in [-1, 1] \mapsto p_0(\cdot, a) \in L^1(\mathbb{R})$ is Lipschitz.

(H_2) The functions b, σ are of class $C_b^{4+\varepsilon}(\mathbb{R}^2)$ with $0 < \varepsilon < 1$, and the function σ is bounded from below by a strictly positive constant.

Then there exists a unique solution to the nonlinear stochastic differen-

tial equation

$$\begin{cases} dX_t = \mathbb{E}\,[b\,(x, X_t)\,|\,\theta]\,|_{x=X_t}\,dt + \mathbb{E}\,[\sigma\,(x, X_t)\,|\,\theta]\,|_{x=X_t}\,dW_t, \\ \forall A \in \mathcal{B}(\mathbb{R}), \quad \mathbb{P}\,(X_0 \in A|\theta = a) = \int_A p_0(y, a)dy, \\ \theta : \; \textit{random variable of law } \nu. \end{cases}$$

(3.17)

In addition, it holds that

$$(M_1(t), f)_{L^2(\mathbb{R})} = \mathbb{E}\,f(X_t),$$

(3.18)

for all t and all measurable and bounded function $f : \mathbb{R} \to \mathbb{R}$.

We now describe the algorithm. Consider N independent random vectors (X_0^i, θ^i), $1 \le i \le N$ with common probability law $p_0(x, a)dx\ \nu(da)$, and the N processes $(X_t^{i,N})$ defined by

$$\begin{cases} dX_t^{i,N} = \sum_{j=1}^{N} \alpha_{ij} b(X_t^{i,N}, X_t^{j,n,N})dt \\ \qquad\quad + \sum_{j=1}^{N} \alpha_{ij}\sigma(X_t^{i,N}, X_t^{j,n,N})dW_t^i, \\ X_t^{i,N}\,|_{t=0} = X_0^i, \end{cases}$$

(3.19)

where the weights α_{ij} are positive functions of $\theta^1, \ldots, \theta^N$. They are defined by non parametric estimators of the regression functions

$$\mathbb{E}\,[b\,(x, X_t)\,|\,\theta = \cdot] \qquad \text{and} \qquad \mathbb{E}\,[\sigma\,(x, X_t)\,|\,\theta = \cdot].$$

In order to simulate these processes, we discretize in time and use the Euler scheme

$$\begin{cases} \overline{X}_{(p+1)T/n}^{i,n,N} = \overline{X}_{pT/n}^{i,n,N} + \sum_{j=1}^{N}\alpha_{ij}b(\overline{X}_{pT/n}^{i,n,N}, \overline{X}_{pT/n}^{j,n,N})\dfrac{T}{n} \\ \qquad\qquad + \sum_{j=1}^{N}\alpha_{ij}\sigma(\overline{X}_{pT/n}^{i,n,N}, \overline{X}_{pT/n}^{j,n,N})\left(W_{(p+1)T/n}^i - W_{pT/n}^i\right), \\ \overline{X}_0^{i,n,N} = X_0^i. \end{cases}$$

(3.20)

An example of a convergence rate estimate is as follows.

Theorem 3.2 *Let h_N be a strictly positive real number and let G be the zero mean and unit variance Gaussian kernel. Define the random* *weights by*

$$\alpha_{ij} := \frac{G\left((\theta^i - \theta^j)/h_N\right)}{\sum_{k=1}^{N} G\left((\theta^i - \theta^k)/h_N\right)}. \tag{3.21}$$

Suppose that $(H_1 - H_2)$ hold. In addition assume that

(H_3) $\lim_{N \to +\infty} \log(N)/(Nh_N^2) = 0.$

(H_4) *The measure ν has a strictly positive Lipschitz continuous density on $[-1, 1]$.*

(H_5) *For all integer p, one has*

$$\sup_{a \in [-1,1]} \int_{\mathbb{R}} |x|^p \, p_0(x, a) dx < +\infty.$$

Then, for all function ϕ of class $C_b^{4+\varepsilon}(\mathbb{R})$ with $0 < \varepsilon < 1$, there exist an integer N_0 and a constant $C(T)$ such that

$$\mathbb{E} \left| (M_1(T), \phi)_{L^2(\mathbb{R})} - \frac{1}{N} \sum_{i=1}^{N} \phi\left(\overline{X}_T^{i,n,N}\right) \right|^2$$

$$\leq C(T) \left(\frac{1}{\sqrt{N h_N}} + \sqrt{h_N} + \frac{1}{n^2} \right) \tag{3.22}$$

for all $N \geq N_0$.

See Vaillant [28], Talay & Vaillant [26].

4 Branching stochastic particle systems

Sherman & Peskin [19] have proposed a stochastic particle method for diffusion-reaction-convection equations of the type

$$\begin{cases} \dfrac{\partial V}{\partial t}(t, x) = \dfrac{1}{2} \dfrac{\partial^2 V}{\partial x^2}(t, x) + f \circ V(t, x), \\ V(0, x) = V_0(x), \end{cases} \tag{4.23}$$

where V_0 is the distribution function of a probability law on $\mathbb{R}$, and f is a differentiable function such that, for all $t > 0$, $V(t, \cdot)$ is a distribution function. The Sherman–Peskin method without time discretization is as follows.

At time 0, N particles with weight $1/N$ are located at points

$$V_0^{-1}(\frac{i}{N}), \quad i = 1, \dots, N-1,$$

and $V_0^{-1}(1 - \frac{1}{2N})$. For all time $t > 0$ let $\mathcal{N}_t^N$ denote the number of living particles at time t. Set

$$Z_t^N := \sum_{j=1}^{\mathcal{N}_t^N} \delta_{z_t^j} \quad \text{and} \quad \mu_t^N := \frac{1}{N} Z_t^N,$$

where z_t^j denotes the location of the particle j living at time t, and

$$V^N(t, x) := \frac{1}{N} \sum_{j=1}^{\mathcal{N}_t^N} H(x - z_t^j).$$

The function $V^N(t, x)$ is the stochastic approximation of the solution of (4.23). We now describe the branching law. The probability of one branching between t and $t + dt$ is

$$dt \sum_{j=1}^{\mathcal{N}_t^N} |f' \circ V^N(t, z_t^j)|,$$

and the branching particle is the particle J with probability

$$|f' \circ V^N(t, z_t^J)| \left[\sum_{j=1}^{\mathcal{N}_t^N} |f' \circ V^N(t, z_t^j)| \right]^{-1}.$$

It is replaced by 0 particle if $f' \circ V^N(t, z_t^j) \le 0$, and by 2 particles if $f' \circ V^N(t, z_t^j) > 0$. In addition, the living particles move as independent Brownian motions.

Under natural hypotheses on the function f, Chauvin & Rouault [6], Chauvin, Olivares-Rieumont & Rouault [5] have shown the convergence of the Sherman–Peskin method, and proven a fluctuation result on the normalised process $\sqrt{N}(\mu_\cdot^N - \mu_\cdot)$ in the appropriate Sobolev space, where (μ_t) is the flow of the probability measures whose the $V(t, x)$'s are the distribution functions.

We now address the question of the time discretization, and we extend the Sherman–Peskin method to equations of the type

$$\begin{cases} \dfrac{\partial V}{\partial t}(t, x) = \dfrac{1}{2}\sigma^2(x)\dfrac{\partial^2 V}{\partial x^2}(t, x) + b(x)\dfrac{\partial V}{\partial x}(t, x) + f \circ V(x), \\ u(0, x) = V_0'(x). \end{cases} \quad (4.24)$$

Between pT/n and $(p + 1)T/n$, the living particles at time pT/n move independently. The positions in $(p+1)T/n$ are described by

$$
\overline{Y}_{(p+1)T/n} = \overline{Y}_{pT/n} + \left\{ \sigma\left(\overline{Y}_{pT/n}\right) \sigma'\left(\overline{Y}_{pT/n}\right) - b\left(\overline{Y}_{pT/n}\right) \right\} \frac{T}{n}
$$
$$
+ \sigma\left(\overline{Y}_{pT/n}\right)\left(W_{(p+1)T/n} - W_{pT/n}\right)
$$
$$
+ \frac{1}{2}\sigma\left(\overline{Y}_{pT/n}\right)\sigma'\left(\overline{Y}_{pT/n}\right)\left\{ \left(W_{(p+1)T/n} - W_{pT/n}\right)^2 - \frac{T}{n} \right\}.
$$
$$(4.25)$$

This scheme is the Milstein scheme for the stochastic differential equation

$$
dY_t = (\sigma(Y_t)\sigma'(Y_t) - b(Y_t))dt + \sigma(Y_t)dW_t.
$$

For technical reasons the convergence rate estimate below has been proven for this scheme, but a reasonable conjecture is that the result also holds for the Euler scheme.

Let $\overline{N}^N_{(p+1)T/n}$ denote the number of particles living at time $(p+1)T/n$. Set

$$
\overline{V}^N((p+1)T/n, x) := \frac{1}{N} \sum_{j=1}^{\overline{N}^N_{(p+1)T/n}} H(x - \overline{y}^j_{(p+1)T/n}),
$$

where $\{\overline{y}^j_{(p+1)T/n}\}$ is the set of the positions of the simulated particles which are living at time $(p+1)T/n$. The particle number J branches with probability

$$
\frac{T}{n}|f' \circ \overline{V}^N((p+1)T/n, \overline{y}^J_{(p+1)T/n})|.
$$

It dies if $f' \circ \overline{V}^N((p+1)T/n, \overline{y}^J_{(p+1)T/n}) < 0$, otherwise it gives birth to two new particles.

Theorem 4.1 *Suppose that*

(1) The function f belongs to $C_b^1(\mathbb{R})$ and satisfies

$$
f(0) = f(1) = 0; \ f'(0) > 0;
$$
$$
f'(x) \le f'(0) \text{ for all } x \in [0,1] \ ; \ f'(x) < 0 \text{ for all } x \in \mathbb{R}_+ - [0,1].
$$

(2) The functions b and σ belong to $C_b^\infty(\mathbb{R})$. In addition, the function σ is bounded from below by a strictly positive constant.

(3) The function V_0' is a probability density such that

$$
\exists M > 0, \ \exists \eta > 0, \ \exists \alpha > 0, \ \forall |x| > M, \ |V_0'(x)| \le \eta \exp(-\alpha x^2).
$$

D. Talay

Let Φ be a positive function belonging to $L^1(\mathbb{R}) \bigcap L^\infty(\mathbb{R})$. Set

$$\|f\|_{L^{1,\Phi}(\mathbb{R})} := \int_{\mathbb{R}} |f(x)| \Phi(x) dx$$

for all Lebesgue integrable function f. Then

$$\exists C > 0, \ \forall N \geq 1, \ \forall n > T,$$

$$\sup_{p=0,\ldots,n} \mathbb{E} \|V(pT/n, \cdot) - \overline{V}^N(pT/n, \cdot)\|_{L^{1,\Phi}(\mathbb{R})} \leq \frac{C}{\sqrt{N}} + \frac{C}{\sqrt{n}}. \quad (4.26)$$

See Régnier [17].

Bibliography

[1] Bally, V. and Talay, D. (1996). The law of the Euler scheme for stochastic differential equations (I): convergence rate of the distribution function. *Prob. Th. Related Fields* **104**, 43–60.

[2] Bossy, M. (2000). Optimal rate of convergence of a stochastic particle method for scalar conservation laws equations. Submitted for publication.

[3] Bossy, M. and Talay, D. (1996). Convergence rate for the approximation of the limit law of weakly interacting particles: application to the Burgers equation. *Ann. Appl. Probab.* **6**, 818–861.

[4] Bossy, M. and Talay, D. (1997). A stochastic particle method for the McKean-Vlasov and the Burgers equation. *Math. Comp.* **66**, 157–192.

[5] Chauvin, B., Olivares-Rieumont, P. and Rouault, A. (1991) Fluctuations for spatial branching process with mean-field interaction. *Adv. Appl. Prob.* **23**, 716–732.

[6] Chauvin, B. and Rouault, A. (1990). A stochastic simulation for solving scalar reaction-diffusion equations. *Adv. Appl. Prob.* **22**, 88–100.

[7] Chorin, A.J., Kast, A.P. and Kupferman, R. (1998). Optimal prediction. *Proc. Nat. Acad. Sci. USA* **95**, 4094–4098.

[8] Costantini, C., Pacchiarotti, B. and Sartoretto, F. (1998). Numerical approximation for functionals of reflecting diffusion processes. *SIAM J. Appl. Math.* **58**, 73–102.

[9] Gobet, E. (2000). Weak approximation of killed diffusion using Euler schemes. To appear in *Stoch. Process Appl.*

[10] Graham, C., Kurtz, T., Méléard, S., Protter, P., Pulvirenti, M. and Talay, D. (1996). *Probabilistic Models for Nonlinear Partial Differential Equations* Lecture Notes in Mathematics 1627 (Springer-Verlag, Berlin).

[11] Higham, D. Mattingly, J. and Stuart, A.M. (2000). Ergodicity for discretizations of SDEs: Locally Lipschitz vector fields and degenerate noise. In preparation.

[12] Jourdain, B. (1997). Diffusions avec un coefficient de dérive non linéaire et irrégulier et interprétation probabiliste d'équations de type Burgers. *ESAIM P& S* **1**, 339–355.

[13] Jourdain, B. (1998). Convergence of moderately interacting particle systems to a diffusion convection equation. *Stoch. Proc. Appl.* **73**, 247–270.

[14] Marchioro, C. and Pulvirenti, M. (1982). Hydrodynamics in two dimensions and vortex theory. *Comm. Math. Phys.* **84**, 483–503.

[15] Mohammadi, B. and Pironneau, O. (1994). *Analysis of the K-Epsilon Turbulence Model* (Masson, Paris).

[16] Osada, H. (1987). Propagation of chaos for the two dimensional Navier–Stokes equation, in *Probabilistic Methods in Mathematical Physics*, eds K. Itô and N. Ikeda (Academic Press, New York), 303–334.

[17] Régnier, H. (1999). *Vitesse de convergence de méthodes particulaires stochastiques avec branchements* Université de Provence Ph.D. thesis.

[18] Revuz, D. (1975). *Markov Chains* (North Holland, Amsterdam).

[19] A.Sherman, A.S. and Peskin, C.S. (1986). A Monte Carlo method for scalar reaction-diffusion equations, *SIAM J. Sci. Statist. Comput.* **7**, 1360–1372.

[20] Shardlow, T. and Stuart, A.M. (2000). A perturbation theory for ergodic Markov chains with application to numerical approximation. To appear in *Siam J. Numer. Anal.*

[21] Soize, C. (1994). *The Fokker–Planck Equation for Stochastic Dynamical Systems and Its Explicit Steady State Solutions*, Series on Advances in Mathematics for Applied Sciences **17** (World Scientific, Singapore).

[22] Sznitman, A.-S. (1989). Topics in propagation of chaos, in *École d'Été de Probabilités de Saint Flour XIX*, ed. P.L. Hennequin, *Lecture Notes in Mathematics* **1464** (Springer-Verlag, Berlin).

[23] Talay D. (1999). Approximation of the invariant probability measure of stochastic Hamiltonian dissipative systems with non globally Lipschitz coefficients, in *Progress in Stochastic Structural Dynamics*, eds R. Bouc and C. Soize, Publication du L.M.A.-C.N.R.S. **152**.

[24] Talay D. (2000). The implicit Euler scheme for ergodic stochastic Hamiltonian dissipative systems with non globally Lipschitz coefficients. In preparation.

[25] Talay, D. and Tubaro, L. (1990). Expansion of the global error for numerical schemes solving stochastic differential equations. *Stoch. Anal. Appl.* **8**, 94–120.

[26] Talay, D. and Vaillant, O. (2000). A stochastic particle method with random weights for the computation of statistical solutions of McKean–Vlasov equations. Submitted for publication.

[27] Talay, D. and Zheng, Z. (2000). Approximation of quantiles of diffusion processes and application to model risk measurements. In preparation.

[28] Vaillant, O. (2000) *Une méthode particulaire stochastique à poids aléatoires pour l'approximation de solutions statistiques d'équations de Mc Kean-Vlasov-Fokker-Planck*. Université de Provence Ph.D. thesis.

[29] Vishik, M.J. and Fursikov A.V. (1988). *Mathematical Problems of Statistical Hydromechanics* (Kluwer, Dordrecht).

Real-time numerical solution to Duncan–Mortensen–Zakai equation

Shing-Tung Yau [†]

Department of Mathematics
Harvard University
Cambridge, MA 02138, USA
Email: yau@math.harvard.edu

Stephen S.-T. Yau [‡]

Department of Mathematics, Statistics
and Computer Science, (M/C 249)
University of Illinois at Chicago
851 South Morgan Street
Chicago, IL 60607-7045, USA
Email: yau@uic.edu

1 Introduction

The nonlinear filtering problem involves the estimation of a stochastic process $x = \{x_t\}$ (called the signal or state process) that cannot be observed directly. Information containing x is obtained from observations of a related process $y = \{y_t\}$ (the observation process). The goal of nonlinear filtering is to determine the conditional expectations of the form $E[\phi(x_i) : y_s, 0 \le s \le t]$, or perhaps even the computation of the entire conditional density $\rho(t, x)$ of x_t given the observation history $\{y_s, 0 \le s \le t\}$. When the observations are received sequentially, as in many practical applications, it is preferable that this computation be performed recursively in terms of a statistic $\theta = \{\theta_t\}$, which can be updated by using only the latest observations

$$\theta_{t+\tau} = \alpha\big(t, \tau, \theta_t, \{y_s, t \le s \le t + \tau\}\big) \tag{1.1}$$

and from which estimates can be calculated in a "pointwise" or "memoryless" manner:

$$E\big[\phi(x_s) : y_s, 0 \le s \le t\big] = \beta(t, y_t, \theta_t). \tag{1.2}$$

[†] Research partially supported by U.S. Army Research Office: DAAD19-99-1-0203.
[‡] Research partially supported by NSF Grant.

In many cases, θ_t is computable with a finite set of differential equations driven by y. In these cases, the practical implication of recursiveness is the possible implementation of the filter (1.1) - (1.2) in real time. We refer the readers to the excellent expository article by Marcus [26] for details. Mathematically, the unnormalized conditional density satisfies a time-varying parabolic partial differential equation (DMZ equation) driven by observation process y (cf. (2.5)). One would like to solve this DMZ equation by means of the solution of a partial differential equation which is independent of observation y and the solution of a finite system of ordinary differential equations driven by y. Indeed, in our previous paper [41], we have found the solution of DMZ equation for linear filtering system and exact filtering system with arbitrary initial condition in terms of the solution of Kolmogorov equation (independent of observation y) and the solution of a finite system of linear ordinary differential equation driven by y. More recently, Yau and Hu [38] have done the same thing for the so-called Yau filtering in the sense of Chen [5].

Historically Kalman–Bucy [24] first established the finite dimensional filters for linear filtering system with Gaussian initial distribution in 1961. Ever since the technique of the Kalman–Bucy filters was popularized, there has been an intense interest in finding new classes of finite dimensional recursive filters. In the sixties and early seventies, the basic approach to nonlinear filtering theory was via the "innovations method" originally proposed by Kailath and subsequently rigorously developed by Fujisaki, Kallianpur, and Kunita [17] in 1972. As pointed out by Mitter [27], the difficulty with this approach is that the innovations process is not, in general, explicitly computable (except in the well known Kalman–Bucy case). In view of this weakness, Brockett and Clark [3], Brockett [1], and Mitter [27] proposed independently the idea of using estimation algebras to construct finite-dimensional nonlinear filters. The idea is to imitate the Wei-Norman [31] approach of using the Lie algebraic method to solve the DMZ equation, which the unnormalized conditioned probability density of the state x_t must satisfy. This Lie algebra approach has several merits. First, it takes into account geometrical aspects of the situation. Second, it explains convincingly why it is easy to find exact recursive filters for linear dynamical systems while it is very difficult to filter something like cubic sensor described in the work of Hazewinkel, et. al. [20]. The third, and perhaps most important, merit of the Lie algebra approach is the following. As long as the estimation algebra is finite dimensional, not only can the finite dimensional recursive filter be constructed explicitly, but also the filters so constructed is universal in the sense of [4]. Moreover, the number of sufficient statistics in the Lie algebra method, which requires computing the conditional probability density, is linear in n, where n is the dimension of the state space.

In his talk at the International Congress of Mathematics in 1983 [2], Brockett proposed the problem of classification of all finite-dimensional estimation algebras. Nevertheless, the structure and classification of finite dimensional estimation algebras were studied in detail only since 1990 (cf. [30], [6], [7], [8], [9], and [10]). In [32], the Ω matrix was introduced, which is a matrix whose (i, j) entry is $\omega_{ij} = \frac{\partial f_j}{\partial x_i} - \frac{\partial f_i}{\partial x_j}$, where f is the drift term of the state evolution equation. The program of classifying finite dimensional estimation algebras of maximal rank was begun in 1990 by S.S.-T. Yau. There are four crucial steps.

Step 1. In 1990, Yau first observed that Wong's Ω-matrix plays an important role. As the first crucial step, he classifies all finite dimensional estimation algebras of maximal rank if Wong's matrix has entries in constant coefficients. His result was announced in CDC 1990 [35] and published in 1994 [36]. In 1991 Chiou and Yau [37] formally introduced the concept of finite dimensional estimation algebra of maximal rank and gave classification when the state space dimension n is at most 2. Their results were published in 1994 [6].

Step 2. The second crucial step was due to Chen and Yau in 1996 [7]. They developed quadratic structure theory in finite dimensional estimation algebras of maximal rank and laid down all the ingredients which are needed to give classification of finite dimensional estimation algebras of maximal rank. In particular, they proved that all the entries of Wong's matrix are degree one polynomials. They also introduced the notion of quadratic rank k. In this way the Wong's matrix is divided into 3 parts: (1) (ω_{ij}), $1 \leq i, j \leq k$, (2) (ω_{ij}), $k + 1 \leq i, j \leq n$ and (3) (ω_{ij}), $(1 \leq i \leq k, k + 1 \leq j \leq n)$ or $(k + 1 \leq i \leq n, 1 \leq j \leq k)$. In [7], Chen and Yau proved among many other things that part (1) (ω_{ij}), $1 \leq i, j \leq k$ is a matrix with constant coefficients.

Step 3. In their 1997 paper [10], Chen, Yau and Leung proved the weak Hessian matrix nondecomposition theorem for $n \leq 4$. As a result, the part (2) of the Wong's matrix, (ω_{ij}), $k + 1 \leq i, j \leq n$ is a matrix with constant coefficients. In their paper [34], Wu, Yau and Hu proved the weak Hessian matrix nondecomposition theorem for general n. Thus part (2) of the Wong's matrix, (ω_{ij}), $k + 1 \leq i, j \leq n$ is a matrix with constant coefficients for arbitrary n. Recently Yau, Wu and Wong [40] established the strong Hessian matrix nondecomposition theorem which implies the weak Hessian matrix nondecomposition theorem as a special case.

Step 4. In 1997, Yau and Hu [38] using the full power of the quadratic structure theory developed by Chen and Yau [7] to prove that part (2) of the Wong's matrix (ω_{ij}), $1 \leq i \leq k, k+1 \leq j \leq n$ and the matrix (ω_{ij}), $k+1 \leq i \leq n, 1 \leq j \leq k$ are with constant coefficients.

In summary, Yau and his co-workers have shown that for any finite dimensional estimation algebra of maximal rank, the Ω-matrix must be constant, i.e. the filtering system is a Yau filtering system. In particular they have proved that at least for estimation algebra of maximal rank case, the Wei-Norman approach to constant finite dimensional filters only works for Yau filtering systems. After the works of Yau and his co-workers, the main problem of nonlinear filtering becomes: Can one handle nonlinear filtering systems in general beyond Yau filtering systems?

The current paper is to give an affirmative answer to this main problem in nonlinear filtering theory. Our solutions to the nonlinear filtering problems are even better than those classical solutions of Kalman–Bucy for linear filtering with Gaussian initial conditions. First, unlike the classical Kalman–Bucy case which requires real time solution of system of ordinary differential equations of dimension n, our solution to general nonlinear filtering problem do not need on-line computation anymore. All computations are reduced to off-line computations. This answers the challenge proposed by Naval Research Office a few years ago: how can one solve the nonlinear filtering problem if adequate amount of computational resources are provided. Second, unlike the classical finite dimensional filters which require the observations for a small time interval (cf. (1.1)), our solution only requires the knowledge of the observation at time t.

In Section 2, we shall recall some basic concepts and results. In Section 3, we shall describe our formula which reduces the nonlinear filtering problem to a off-line computation. In Sections 4 and 5, we give a rigorous proof that our solution converges to the true solution in both pointwise sense and L^2-sense, respectively.

We gratefully acknowledge the long term support from Army Research Office which plays an important role in our research.

2 Some Basic Concepts and Results

In this section, we recall some basic concepts and results from our previous papers. The filtering problem considered here is based on the signal observation model

$$\begin{cases} dx(t) = f\big(x(t)\big)dt + g\big(x(t)\big)dv(t), & x(0) = x_0, \\ dy(t) = h\big(x(t)\big)dt + dw(t), & y(0) = 0, \end{cases} \quad (2.1)$$

in which x, v, y and w are respectively $\mathbb{R}^n$, $\mathbb{R}^p$, $\mathbb{R}^m$ and $\mathbb{R}^m$ valued processes and v and w have components that are independent, standard Brownian processes. We further assume that $n = p$; f, and h are C^∞ smooth vector-valued, and that g is an orthogonal matrix. We shall refer to $x(t)$ as the state of the system at time t and $y(t)$ as the observation at time t.

Let $\rho(t, x)$ denote the conditional probability density of the state given the observation $\{y(s) : 0 \leq s \leq t\}$. It is well known (see [12] for example) that $\rho(t, x)$ is given by normalizing a function, $\sigma(t, x)$, which satisfies the following Duncan–Mortensen–Zakai equation:

$$d\sigma(t, x) = L_0\sigma(t, x)dt + \sum_{i=1}^n L_i\sigma(t, x)dy_i(t), \ \sigma(0, x) = \sigma_0, \quad (2.2)$$

where

$$L_0 = \frac{1}{2}\sum_{i=1}^n \frac{\partial^2}{\partial x_i^2} - \sum_{i=1}^n f_i\frac{\partial}{\partial x_i} - \sum_{i=1}^n \frac{\partial f_i}{\partial x_i} - \frac{1}{2}\sum_{i=1}^m h_i^2 \quad (2.3)$$

and for $i = 1, \ldots, m, L_i$ is the zero degree differential operator of multiplication by h_i. (Here we have used the notation p_i to represent the ith component of the vector p). σ_0 is the probability density of the initial point x_0.

Equation (2.2) is a stochastic partial differential equation. In real applications, we are interested in constructing robust state estimators from observed sample paths with some property of robustness. Davis in [11] studied this problem and proposed some robust algorithms. In our case, his basic idea reduces to defining a new unnormalized density

$$u(t, x) = \exp\left(-\sum_{i=1}^m h_i(x)y_i(t)\right)\sigma(t, x). \quad (2.4)$$

It is easy to show that $u(t, x)$ satisfies the following time varying partial differential equation

$$
\begin{cases}
\dfrac{\partial u}{\partial t}(t,x) = L_0 u(t,x) + \displaystyle\sum_{i=1}^{m} y_i(t)[L_0, L_i]u(t,x) \\[2mm]
\qquad\qquad + \dfrac{1}{2} \displaystyle\sum_{i,j=1}^{m} y_i(t)y_j(t)\big[[L_0, L_i], L_j\big]u(t,x) \\[2mm]
u(0,x) = \sigma_0
\end{cases}
\qquad (2.5)
$$

where $[\cdot,\cdot]$ is the Lie bracket as described by the following definition.

Definition 1 *If X and Y are differential operators, the Lie bracket of X and Y, $[X,Y]$ is defined by*

$$
[X,Y]\phi = X(Y\phi) - Y(X\phi)
$$

for any C^∞ function ϕ.

We shall call (2.5) as robust DMZ equation.

Definition 2 *The estimation algebra E of a filtering system (2.1) is defined as the Lie algebra generated by $\{L_0, L_1, \ldots, L_m\}$. E is said to be an estimation algebra of maximal rank if, for any $1 \le i \le n$, there exists a constant c_i such that $x_i + c_i$ is in E.*

Definition 3 *Wong's $\Omega = (\omega_{ij})$ matrix is a $n \times n$ matrix where*

$$
\omega_{ij} = \frac{\partial f_i}{\partial x_i} - \frac{\partial f_i}{\partial x_j}. \qquad (2.6)
$$

Define

$$
D_i = \frac{\partial}{\partial x_i} - f_i \quad \text{and} \quad \eta = \sum_{i=1}^{n} \frac{\partial f_i}{\partial x_i} + \sum_{i=1}^{n} f_i^2 + \sum_{i=1}^{m} h_i^2. \qquad (2.7)
$$

Then

$$
L_0 = \frac{1}{2}\left(\sum_{i=1}^{n} D_i^2 - \eta \right). \qquad (2.8)
$$

Basing on the quadratic structure theory on estimation algebras developed by Chen and Yau [7], Yau and Hu [38] classified all finite dimensional estimation algebras of maximal rank which answers a question raised by Brockett in his invited address at the International Congress of Mathematics in 1983.

Theorem 1 ([38]) *Suppose that the state space of the filtering system (2.1) is of dimension n. If E is the finite-dimensional estimation algebra with maximal rank, then E is a real vector space of dimension $2n + 2$ with basis given by $1, x_1, \ldots, x_n, D_1, \ldots, D_n$, and L_0. Moreover, $\omega_{ij} = \frac{\partial f_j}{\partial x_i} - \frac{\partial f_i}{\partial x_j}$ is a constant for all $1 \leq i, j \leq n$, i.e. the underlying filtering system is necessarily a Yau filtering system.*

As an immediate consequence of Theorem 1, we have the following corollaries.

Corollary 2 (Mitter Conjecture) *Suppose that E is the finite-dimensional estimation algebra with maximal rank corresponding to the filtering system (2.1). Then any function in E is a polynomial of degree one.*

The following corollary is an immediate consequence of Theorem 1 and Theorem 7 of [36].

Corollary 3 *Suppose that the state space of the filtering system (2.1) is of dimension n. If E is a finite-dimensional estimation algebra with maximal rank, then the number of statistics in order to compute the conditional density by Lie algebraic method is n.*

It is easy to see (cf. p. 236 of [41]) that the robust DMZ equation (2.5) is of the form

$$
\frac{\partial u}{\partial t}(t, x) = \frac{1}{2} \sum_{i=1}^{n} \frac{\partial^2 u}{\partial x_i^2}(t, x) - \sum_{i=1}^{n} f_i(x) \frac{\partial u}{\partial x_i}(t, x) - \sum_{i=1}^{n} \frac{\partial f_i}{\partial x_i}(x) u(t, x)
$$

$$
- \frac{1}{2} \sum_{i=1}^{m} h_i^2(x) u(t, x) + \sum_{i=1}^{m} \sum_{j=1}^{n} y_i(t) \frac{\partial h_i}{\partial x_j}(t, x) \frac{\partial u}{\partial x_j}(t, x)
$$

$$
+ \frac{1}{2} \sum_{i=1}^{m} \sum_{j=1}^{n} y_i(t) \frac{\partial^2 h_i}{\partial x_j^2}(x) u(t, x)
$$

$$
- \sum_{i=1}^{m} \sum_{j=1}^{n} y_i(t) f_j(x) \frac{\partial h_i}{\partial x_j}(x) u(t, x)
$$

$$
+ \frac{1}{2} \sum_{i=1}^{m} \sum_{j=1}^{m} \sum_{k=1}^{n} y_i(t) y_j(t) \frac{\partial h_i}{\partial x_k}(x) \frac{\partial h_j}{\partial x_k}(x) u(t, x)
$$

$$
u(0, x) = \sigma_0(x). \tag{2.9}
$$

In [41], we introduced a direct method to solve the robust DMZ equation (2.9) for the linear filtering and exact filtering (i.e. $f(x) = \nabla\phi(x)$ for some function $\phi(x)$) systems with arbitrarily initial distribution. In [21], Hu and Yau used the direct method developed in [41] to solve the robust DMZ equation (2.9) for the Yau filtering systems. First recall the following observation of Yau [36].

Proposition 4 *The filtering system (2.1) is of Yau system in the sense of [5], i.e.* $\frac{\partial f_j}{\partial x_i} - \frac{\partial f_i}{\partial x_j} = c_{ij} = constant$, *for all* $1 \leq i, j \leq n$ *if and only if* $(f_1, \ldots f_n) = (\ell_1, \ldots, \ell_n) + \left(\frac{\partial F}{\partial x_1}, \ldots, \frac{\partial F}{\partial x_n}\right)$ *where* $\ell_1, \ldots, \ell_n$ *are polynomials of degree one and* F *is a* C^{∞} *function.*

Theorem 5 ([21]) *Consider the filtering system (2.1) satisfying*

(i) $(f_1, \ldots, f_n) = (\ell_1, \ldots, \ell_n) + \left(\frac{\partial F}{\partial x_1}, \ldots, \frac{\partial F}{\partial x_n}\right)$, *where* $\ell_i(x) = \sum\limits_{j-1}^{n} d_{ij}x_j$ $+ d_i$, $1 \leq i \leq n$ *are degree one polynomials*

(ii) $h_i(x) = \sum\limits_{j=1}^{n} c_{ij}x_j + c_i$, $1 \leq i \leq m$, *are degree one polynomials*

(iii) $\eta(x) := \sum\limits_{i=1}^{n} f_i^2(x) + \sum\limits_{i=1}^{n} \frac{\partial f_i}{\partial x_i}(x) + \sum\limits_{i=1}^{m} h_i^2(x) := \sum\limits_{i,j=1}^{n} \eta_{ij}x_i x_j + \sum\limits_{i=1}^{n} \eta_i x_i$ $+ \eta_0$ *is a degree-two polynomial.*

Then the solution $u(t, x)$ *for the robust DMZ equation (2.9) is reduced to the solution* $\tilde{u}(t, x)$ *for the Kolmogorov equation.*

$$
\begin{cases}
\dfrac{\partial \tilde{u}}{\partial t}(t, x) = \dfrac{1}{2}\Delta\tilde{u}(t, x) - \sum\limits_{i=1}^{n} \ell_i(x)\dfrac{\partial \tilde{u}}{\partial x_i}(t, x) \\
\qquad\qquad - \dfrac{1}{2}\left(\eta(x) - \sum\limits_{i=1}^{n}\dfrac{\partial \ell_i}{\partial x_i}(x) - \sum\limits_{i=1}^{n}\ell_i^2(x)\right)\tilde{u}(t, x) \\
\tilde{u}(0, x) = e^{-F(x)}\sigma_0(x)
\end{cases}
\tag{2.10}
$$

where

$$
\tilde{u}(t, x) = e^{\,c(t) + \sum\limits_{i=1}^{n} a_i(t)x_i - F\left(x + b(t)\right)} \, u\left(t, x + b(t)\right)
\tag{2.11}
$$

and $a_i(t)$, $b_i(t)$, *and* $c(t)$ *satisfy the following system of ODEs,*

$$
\begin{cases}
a_i'(t) - \dfrac{1}{2}\sum\limits_{j=1}^{n}(\eta_{ij} + \eta_{ji})b_j(t) + \sum\limits_{j=1}^{n} d_{ji}b_j'(t) = 0, & 1 \leq i \leq n, \\
b_i'(t) - a_i(t) - \sum\limits_{j=1}^{n} d_{ij}b_j(t) + \sum\limits_{j=1}^{m} c_{ji}y_j(t) = 0, & 1 \leq i \leq n, \\
a_i(0) = b_i(0) = 0, & 1 \leq i \leq n,
\end{cases}
\tag{2.12}
$$

$$\begin{cases} c'(t) = -\dfrac{1}{2}\sum_{i=1}^{n}(b_i'(t))^2 + \sum_{i=1}^{n}a_i(t)b_i'(t) - \sum_{i=1}^{n}d_ib_i'(t) \\ \qquad\quad + \dfrac{1}{2}\sum_{i,j=1}^{n}\eta_{ij}b_i(t)b_j(t) + \dfrac{1}{2}\sum_{i=1}^{n}\eta_i b_i(t) \\ c(0) = 0. \end{cases} \qquad (2.13)$$

Theorem 5 is basically an optimal result of the direct method developed by [41] because of Theorem 1.

3 Algorithm for real time solution of nonlinear filtering problem without memory

The fundamental problem of nonlinear filtering theory is how to solve robust DMZ equation (2.9) in real time and in truly memoryless manner. In this section we shall describe our algorithm which achieves this goal for any filtering system with arbitrary initial distribution. Our algorithm is based on Proposition 6. We remark that in the following proposition, if we replace τ_l by τ_{l-1}, the same proposition is still true. Therefore we can easily write down left-hand approximation algorithms corresponding to (3.7)–(3.10) if necessary.

Proposition 6 $\tilde{u}(t,x)$ *satisfies the parabolic equation*

$$\frac{\partial \tilde{u}}{\partial t}(t,x) = \frac{1}{2}\Delta(\tilde{u}(t,x) - \sum_{i=1}^{n}f_i(x)\frac{\partial \tilde{u}}{\partial x_i}(t,x)$$
$$- \left(\sum_{i=1}^{n}\frac{\partial f_i}{\partial x_i}(x) + \frac{1}{2}\sum_{i=1}^{m}h_i^2(x)\right)\tilde{u}(t,x) \qquad (3.1)$$

for $\tau_{l-1} \le t \le \tau_l$ if and only if

$$u(t,x) = e^{-\sum_{i=1}^{m}y_i(\tau_l)h_i(x)}\,\tilde{u}(t,x)$$

satisfies the robust DMZ equation with observation being frozen at $y(\tau_l)$,

$$\frac{\partial u}{\partial t}(t,x) = \frac{1}{2}\Delta u(t,x) + \sum_{i=1}^{n}\left(-f_i(x) + \sum_{j=1}^{m}y_j(\tau_l)\frac{\partial h_j}{\partial x_i}(x)\right)\frac{\partial u}{\partial x_i}(t,x)$$
$$- \left(\sum_{i=1}^{n}\frac{\partial f_i}{\partial x_i}(x) + \frac{1}{2}\sum_{i=1}^{m}h_i^2(x) - \frac{1}{2}\sum_{i=1}^{m}y_i(\tau_l)\Delta h_i(x)\right)$$

$$+ \sum_{i=1}^{m} \sum_{j=1}^{n} y_i(\tau_l) f_j(x) \frac{\partial h_i}{\partial x_j}(x)$$

$$-\frac{1}{2} \sum_{k=1}^{n} \sum_{i=1}^{m} \sum_{j=1}^{m} y_i(\tau_l) y_j(\tau_l) \frac{\partial h_i}{\partial x_k}(x) \frac{\partial h_j}{\partial x_k}(x) \Bigg) u(t,x). \quad (3.2)$$

Proof

$$e^{\sum\limits_{i=1}^{m} y_i(\tau_l) h_i(x)} \left[-\frac{\partial}{\partial t} + \frac{1}{2}\Delta + \sum_{i=1}^{n} \left(-f_i(x) + \sum_{j=1}^{m} y_j(\tau_l) \frac{\partial h_j}{\partial x_i} \right) \frac{\partial}{\partial x_i} \right.$$

$$- \left(\sum_{i=1}^{n} \frac{\partial f_i}{\partial x_i}(x) + \frac{1}{2} \sum_{i=1}^{m} h_i^2(x) - \frac{1}{2} \sum_{i=1}^{m} y_i(\tau_l) \Delta h_i(x) \right.$$

$$\left. + \sum_{i=1}^{m} \sum_{j=1}^{n} y_i(\tau_l) f_j(x) \frac{\partial h_i}{\partial x_j}(x) - \frac{1}{2} \sum_{k=1}^{n} \sum_{i=1}^{m} y_i(\tau_l) y_j(\tau_l) \frac{\partial h_i}{\partial x_k} \frac{\partial h_j}{\partial x_k} \right) \Bigg] u(t,x)$$

$$= -\frac{\partial \widetilde{u}}{\partial t}(t,x) + \frac{1}{2} e^{\sum\limits_{i=1}^{m} y_i(\tau_l) h_i(x)} \Delta(e^{-\sum\limits_{i=1}^{m} y_i(\tau_l) h_i(x)} \widetilde{u}(t,x))$$

$$+ \sum_{i=1}^{n} \left(-f_i(x) + \sum_{j=1}^{m} y_j(\tau_l) \frac{\partial h_j}{\partial x_i} \right) e^{\sum\limits_{i=1}^{m} y_i(\tau_l) h_i(x)}$$

$$\times \frac{\partial}{\partial x_i} (e^{-\sum\limits_{j=1}^{m} y_j(\tau_l) h_j(x)} \widetilde{u}(t,x)) - \left(\sum_{i=1}^{n} \frac{\partial f_i}{\partial x_i}(x) + \frac{1}{2} \sum_{i=1}^{m} h_i^2(x) \right.$$

$$- \frac{1}{2} \sum_{i=1}^{m} y_i(\tau_l) \Delta h_i(x) + \sum_{i=1}^{m} \sum_{j=1}^{n} y_i(\tau_l) f_j(x) \frac{\partial h_i}{\partial x_j}(x)$$

$$\left. - \frac{1}{2} \sum_{k=1}^{n} \sum_{i,j=1}^{m} y_i(\tau_l) y_j(\tau_l) \frac{\partial h_i}{\partial x_k}(x) \frac{\partial h_j}{\partial x_k}(x) \right) \widetilde{u}(t,x)$$

$$= -\frac{\partial \widetilde{u}}{\partial t}(t,x) + \frac{1}{2} e^{\sum\limits_{i=1}^{m} y_i(\tau_l) h_i(x)} [\Delta(e^{-\sum\limits_{i=1}^{m} y_i(\tau_l) h_i(x)}) \widetilde{u}(t,x)$$

$$+ 2\nabla(e^{-\sum\limits_{i=1}^{m} y_i(\tau_l) h_i(x)}) \cdot \nabla \widetilde{u}(t,x) + e^{-\sum\limits_{i=1}^{m} y_i(\tau_l) h_i(x)} \Delta \widetilde{u}(t,x)]$$

$$+ \sum_{i=1}^{n} \left(-f_i(x) + \sum_{j=1}^{m} y_j(\tau_l) \frac{\partial h_j}{\partial x_i} \right) e^{\sum\limits_{i=1}^{m} y_i(\tau_l) h_i(x)}$$

$$\times \left[\frac{\partial}{\partial x_i}\left(e^{-\sum\limits_{j=1}^{m} y_j(\tau_l)h_j(x)}\right) \cdot \tilde{u}(t,x) + e^{-\sum\limits_{j=1}^{m} y_j(\tau_l)h_j(x)} \frac{\partial \tilde{u}}{\partial x_i}(t,x) \right]$$

$$- \left(\sum_{i=1}^{n} \frac{\partial f_i}{\partial x_i}(x) + \frac{1}{2}\sum_{i=1}^{m} h_i^2(x) - \frac{1}{2}\sum_{i=1}^{m} y_i(\tau_l)\Delta h_i(x) \right.$$

$$\left. + \sum_{i=1}^{m}\sum_{j=1}^{n} y_i(\tau_l) f_j(x)\frac{\partial h_i}{\partial x_j}(x) - \frac{1}{2}\sum_{k=1}^{n}\sum_{i,j=1}^{m} y_i(\tau_l)y_j(\tau_l)\frac{\partial h_i}{\partial x_k}(x)\frac{\partial h_j}{\partial x_k}(x) \right)$$

$$\times \tilde{u}(t,x)$$

$$= -\frac{\partial \tilde{u}}{\partial t}(t,x) + \frac{1}{2}\Delta\tilde{u}(t,x) + \sum_{i=1}^{n}\left(-f_i(x) + \sum_{j=1}^{m} y_j(\tau_l)\frac{\partial h_j}{\partial x_i}(x) \right)\frac{\partial \tilde{u}}{\partial x_i}(t,x)$$

$$- \left(\sum_{i=1}^{n} \frac{\partial f_i}{\partial x_i}(x) + \frac{1}{2}\sum_{i=1}^{m} h_i^2(x) - \frac{1}{2}\sum_{i=1}^{m} y_i(\tau_l)\Delta h_i(x) \right.$$

$$\left. + \sum_{i=1}^{m}\sum_{j=1}^{n} y_i(\tau_l) f_j(x)\frac{\partial h_i}{\partial x_j}(x) - \frac{1}{2}\sum_{k=1}^{n}\sum_{i,j=1}^{m} y_i(\tau_l)y_j(\tau_l)\frac{\partial h_i}{\partial x_k}(x)\frac{\partial h_j}{\partial x_k}(x) \right)$$

$$\times \tilde{u}(t,x) + \frac{1}{2}e^{\sum\limits_{i=1}^{m} y_i(\tau_l)h_i(x)}\Delta\left[e^{-\sum\limits_{i=1}^{m} y_i(\tau_l)h_i(x)} \right]\tilde{u}(t,x)$$

$$+ e^{\sum\limits_{i=1}^{m} y_i(\tau_l)h_i(\pi)}\nabla\left[e^{-\sum\limits_{j=1}^{m} y_j(\tau_l)h_j(\pi)} \right]\cdot\nabla\tilde{u}(t,x)$$

$$+ \sum_{i=1}^{n}\left(-f_i(x) + \sum_{j=1}^{m} y_j(\tau_l)\frac{\partial h_j}{\partial x_i}(x) \right)e^{\sum\limits_{j=1}^{m} y_j(\tau_l)h_j(x)}$$

$$\times \frac{\partial}{\partial x_i}\left[e^{-\sum\limits_{j=1}^{m} y_j(\tau_l)h_j(x)} \right]\cdot\tilde{u}(t,x). \tag{3.3}$$

Observe that

$$\frac{\partial}{\partial x_i}e^{-\sum\limits_{j=1}^{m} y_j(\tau_l)h_j(x)} = -\sum_{j=1}^{m} y_j(\tau_l)\frac{\partial h_j}{\partial x_i}(x)e^{-\sum\limits_{j=1}^{m} y_j(\tau_l)h_j(x)}$$

$$\frac{\partial^2}{\partial x_i^2}e^{-\sum\limits_{j=1}^{m} y_j(\tau_l)h_j(x)} = -\sum_{j=1}^{m} y_j(\tau_l)\frac{\partial^2 h_j}{\partial x_i^2}(x)e^{-\sum\limits_{j=1}^{m} y_j(\tau_l)h_j(x)}$$

$$+ \sum_{k=1}^{m}\sum_{j=1}^{m} y_j(\tau_l)\frac{\partial h_j}{\partial x_i}(x)y_k(\tau_l)\frac{\partial h_k}{\partial x_i}(x)e^{-\sum_{j=1}^{m} y_j(\tau_l)h_j(x)}.$$

Therefore (3.3) becomes

$$e^{\sum_{i=1}^{m} y_i(\tau_l)h_i(x)}\left[-\frac{\partial}{\partial t} + \frac{1}{2}\Delta + \sum_{i=1}^{n}\left(-f_i(x) + \sum_{j=1}^{m} y_j(\tau_l)\frac{\partial h_j}{\partial x_i}(x)\right)\frac{\partial}{\partial x_i}\right.$$

$$-\left(\sum_{i=1}^{n}\frac{\partial f_i}{\partial x_i}(x) + \frac{1}{2}\sum_{i=1}^{m} h_i^2(x) - \frac{1}{2}\sum_{i=1}^{m} y_i(\tau_l)\Delta h_i(x)\right.$$

$$+ \sum_{i=1}^{m}\sum_{j=1}^{n} y_i(\tau_l)f_j(x)\frac{\partial h_i}{\partial x_j}(x)$$

$$\left.\left.- \frac{1}{2}\sum_{k=1}^{n}\sum_{i,j=1}^{m} y_i(\tau_l)y_j(\tau_l)\frac{\partial h_i}{\partial x_k}(x)\frac{\partial h_j}{\partial x_k}(x)\right)\right]u(t,x)$$

$$= -\frac{\partial\widetilde{u}}{\partial t}(t,x) + \frac{1}{2}\Delta\widetilde{u}(t,x) + \sum_{i=1}^{n}\left(-f_i(x) + \sum_{j=1}^{m} y_j(\tau_l)\frac{\partial h_j}{\partial x_i}\right)\frac{\partial\widetilde{u}}{\partial x_i}(t,x)$$

$$-\left(\sum_{i=1}^{n}\frac{\partial f_i}{\partial x_i}(x) + \frac{1}{2}\sum_{i=1}^{m} h_i^2(x) - \frac{1}{2}\sum_{i=1}^{m} y_i(\tau_l)\Delta h_i(x)\right.$$

$$\left.+ \sum_{i=1}^{m}\sum_{j=1}^{n} y_i(\tau_l)f_j(x)\frac{\partial h_i}{\partial x_j}(x) - \frac{1}{2}\sum_{k=1}^{n}\sum_{i,j=1}^{m} y_i(\tau_l)y_j(\tau_l)\frac{\partial h_i}{\partial x_k}(x)\frac{\partial h_j}{\partial x_k}(x)\right)$$

$$\times\widetilde{u}(t,x) + \frac{1}{2}e^{\sum_{i=1}^{m} y_i(\tau_l)h_i(x)}\left[-\sum_{j=1}^{m} y_j(\tau_l)\Delta h_j(x)e^{-\sum_{j=1}^{m} y_j(\tau_l)h_j(x)}\right.$$

$$\left.+ \sum_{i=1}^{n}\sum_{j,k=1}^{m} y_j(\tau_l)y_k(\tau_l)\frac{\partial h_j}{\partial x_i}(x)\frac{\partial h_k}{\partial x_i}(x)e^{-\sum_{j=1}^{m} y_j(\tau_l)h_j(x)}\right]\widetilde{u}(t,x)$$

$$- e^{\sum_{i=1}^{m} y_i(\tau_l)h_i(x)}\sum_{i=1}^{n}\sum_{j=1}^{m} y_j(\tau_l)\frac{\partial h_j}{\partial x_i}(x)\frac{\partial\widetilde{u}}{\partial x_i}(t,x)e^{-\sum_{j=1}^{m} y_j(\tau_l)h_j(x)}$$

$$- \sum_{i=1}^{n}\left(-f_i(x) + \sum_{j=1}^{m} y_j(\tau_l)\frac{\partial h_j}{\partial x_i}(x)\right)e^{\sum_{i=1}^{m} y_i(\tau_l)h_i(x)}$$

$$\times \sum_{j=1}^{m} y_j(\tau_l)\frac{\partial h_j}{\partial x_i}(x)e^{-\sum_{j=1}^{m} y_j(\tau_l)h_j(x)} \quad \widetilde{u}(t,x)$$

$$= -\frac{\partial \widetilde{u}}{\partial t}(t,x) + \frac{1}{2}\Delta \widetilde{u}(t,x) - \sum_{i=1}^{n} f_i(x)\frac{\partial \widetilde{u}}{\partial x_i}(t,x)$$

$$- \left(\sum_{i=1}^{n} \frac{\partial f_i}{\partial x_i}(x) + \frac{1}{2}\sum_{i=1}^{m} h_i^2(x) \right) \widetilde{u}(t,x). \tag{3.4}$$

Proposition 6 follows immediately from (3.4). □

We remark that (3.2) is obtained from robust DMZ equation by freezing the observation term $y(t)$ to $y(\tau_l)$. We shall show that the solution of (3.2) approximate to the solution of robust DMZ equation very well in both pointwise sense and L^2-sense.

Suppose that $u(t,x)$ is the solution of robust DMZ equation and we want to compute $u(\tau,x)$. Let $\mathcal{P}_k = \{0 = \tau_0 < \tau_1 < \tau_2 < \cdots < \tau_k = \tau\}$ be a partition of $[0,\tau]$. Let $u_i(t,x)$ be a solution of the following partial differential equation for $\tau_{i-1} \le t \le \tau_i$

$$\begin{cases} \dfrac{\partial u_i}{\partial t}(t,x) = \dfrac{1}{2}\Delta u_i(t,x) + \displaystyle\sum_{\ell=1}^{n}\left[-f_\ell(x) + \sum_{j=1}^{m} y_j(\tau_i)\frac{\partial h_j}{\partial x_\ell}(x) \right]\frac{\partial u_i}{\partial x_\ell}(t,x) \\[2mm] \qquad - \left[\displaystyle\sum_{\ell=1}^{n}\frac{\partial f_\ell}{\partial x_\ell}(x) + \frac{1}{2}\sum_{\ell=1}^{m} h_\ell^2(x) - \frac{1}{2}\sum_{j=1}^{m} y_j(\tau_i)\Delta h_j(x) \right. \\[2mm] \qquad + \displaystyle\sum_{j=1}^{m}\sum_{\ell=1}^{n} y_j(\tau_i)f_\ell(x)\frac{\partial h_j}{\partial x_\ell}(x) \\[2mm] \qquad \left. - \frac{1}{2}\displaystyle\sum_{p=1}^{n}\sum_{j=1}^{m}\sum_{\ell=1}^{m} y_j(\tau_i)y_\ell(\tau_i)\frac{\partial h_j}{\partial x_p}(x)\frac{\partial h_\ell}{\partial x_p}(x) \right] u_i(t,x), \\[2mm] u_i(\tau_{i-1},x) = u_{i-1}(\tau_{i-1},x). \end{cases}$$

$$\tag{3.5}$$

Define the norm of the partition $\mathcal{P}_k$ by $|\mathcal{P}_k| = \sup_{1 \le i \le k}\{|\tau_i - \tau_{i-1}|\}$.

In Section 4 below we shall show that

$$u(\tau,x) = \lim_{|\mathcal{P}_k|\to 0} u_k(\tau_k,x). \tag{3.6}$$

Therefore it remains to describe an algorithm to compute $u_k(\tau_k,x)$. By Proposition 6 $u_1(\tau_1,x)$ can be computed by $\widetilde{u}_1(\tau_1,x)$ where $\widetilde{u}_1(t,x)$ for

$0 \le t \le \tau_1$ satisfies the following equation

$$
\begin{cases}
\dfrac{\partial \tilde{u}_1}{\partial t}(t,x) = \dfrac{1}{2}\Delta \tilde{u}_1(t,x) - \displaystyle\sum_{j=1}^{n} f_j(x)\dfrac{\partial \tilde{u}_i}{\partial x_j}(t,x) \\[2mm]
\qquad\qquad - \left[\displaystyle\sum_{j=1}^{n}\dfrac{\partial f_j}{\partial x_j}(x) + \dfrac{1}{2}\sum_{j=1}^{m} h_j^2(x)\right]\tilde{u}_1(t,x), \qquad (3.7)\\[4mm]
\tilde{u}_1(0,x) = \sigma_0(x) e^{\sum_{j=1}^{m} y_j(\tau_1) h_j(x)}.
\end{cases}
$$

In fact,

$$
u_1(\tau_1,x) = e^{-\sum_{j=1}^{m} y_j(\tau_1) h_j(x)} \tilde{u}_1(\tau_1,x). \qquad (3.8)
$$

In general Proposition 6 tells us that for $i \ge 2$, $u_i(\tau_i,x)$ can be computed by $\tilde{u}_i(\tau_i,x)$ where $\tilde{u}_i(t,x)$ for $\tau_{i-1} \le t \le \tau_i$ satisfies the following equation

$$
\begin{cases}
\dfrac{\partial \tilde{u}}{\partial t}(t,x) = \dfrac{1}{2}\Delta \tilde{u}_i(t,x) - \displaystyle\sum_{j=1}^{n} f_j(x)\dfrac{\partial \tilde{u}_i}{\partial x_j}(t,x) \\[2mm]
\qquad\qquad - \left[\displaystyle\sum_{j=1}^{n}\dfrac{\partial f_j}{\partial x_j}(x) + \dfrac{1}{2}\sum_{j=1}^{m} h_j^2(x)\right]\tilde{u}_i(t,x) \qquad (3.9)\\[4mm]
\tilde{u}_i(\tau_{i-1},x) = e^{\sum_{j=1}^{m}(y_j(\tau_i)-y_j(\tau_{i-1}))h_j(x)}\tilde{u}_{i-1}(\tau_{i-1},x),
\end{cases}
$$

where the last initial condition comes from

$$
\begin{aligned}
\tilde{u}_i(\tau_{i-1},x) &= u_i(\tau_{i-1},x)e^{\sum_{j=1}^{m} y_j(\tau_i) h_j(x)} = u_{i-1}(\tau_{i-1},x)e^{\sum_{j=1}^{m} y_j(\tau_i) h_j(x)}\\
&= e^{-\sum_{j=1}^{m} y_j(\tau_{i-1}) h_j(x)}\tilde{u}_{i-1}(\tau_{i-1},x)e^{\sum_{j=1}^{m} y_j(\tau_i) h_j(x)}
\end{aligned}
$$

In fact,

$$
u_i(\tau_i,x) = e^{-\sum_{j=1}^{m} y_j(\tau_i) h_j(x)} \tilde{u}_i(\tau_i,x). \qquad (3.10)
$$

Observe that in our algorithm at step i, we only need the observation at time τ_{i-1} and τ_i. We do not need any other previous observation data. Observe also that the partial differential equation (3.9) is independent of observation $y(t)$. It can be computed off-line. This equation is precisely the equation we dealt with in [41] of which the fundamental solution is written down explicitly. We shall deal with this problem in a subsequent paper of this one.

4 Pointwise Convergence

By changing variables from x_i to $\sqrt{2}x_i$ and by letting

$$\overline{u}(t,x) = u\left(t, \frac{x}{\sqrt{2}}\right), \tag{4.1}$$

we get

$$\frac{\partial \overline{u}}{\partial t}(t,x) = \frac{\partial u}{\partial t}\left(t, \frac{x}{\sqrt{2}}\right), \qquad \frac{\partial \overline{u}}{\partial x_i}(t,x) = \frac{1}{\sqrt{2}}\frac{\partial u}{\partial x_i}\left(t, \frac{x}{\sqrt{2}}\right),$$

$$\frac{\partial^2 \overline{u}}{\partial x_i^2}(t,x) = \frac{1}{2}\frac{\partial^2 u}{\partial x_i^2}\left(t, \frac{x}{\sqrt{2}}\right).$$

Hence the robust DMZ equation becomes

$$\frac{\partial \overline{u}}{\partial t}(t,x) = \Delta \overline{u}(t,x) + \sum_{i=1}^{n} \sqrt{2}\left[-f_i\left(\frac{x}{\sqrt{2}}\right) + \sum_{j=1}^{m} y_j(t)\left(\frac{\partial h_j}{\partial x_i}\right)\left(\frac{x}{\sqrt{2}}\right)\right]$$

$$\times \frac{\partial \overline{u}}{\partial x_i}(t,x) - \left[\sum_{i=1}^{n}\frac{\partial f_i}{\partial x_i}\left(\frac{x}{\sqrt{2}}\right) + \frac{1}{2}\sum_{i=1}^{m}h_i^2\left(\frac{x}{\sqrt{2}}\right)\right.$$

$$- \sum_{i=1}^{m} y_i(t)(\Delta h_i)\left(\frac{x}{\sqrt{2}}\right) + \sum_{i=1}^{m}\sum_{j=1}^{n} y_i(t)f_j\left(\frac{x}{\sqrt{2}}\right)\frac{\partial h_i}{\partial x_j}\left(\frac{x}{\sqrt{2}}\right)$$

$$\left. - \frac{1}{2}\sum_{i,j=1}^{m}\sum_{k=1}^{n} y_i(t)y_j(t)\frac{\partial h_i}{\partial x_k}\left(\frac{x}{\sqrt{2}}\right)\frac{\partial h_j}{\partial x_k}\left(\frac{x}{\sqrt{2}}\right)\right]\overline{u}(t,x)$$

$$= \Delta \overline{u}(t,x) + \sum_{i=1}^{m}\overline{f}_i(t,x)\frac{\partial \overline{u}}{\partial x_i}(t,x) - \overline{V}(t,x)\overline{u}(t,x) \tag{4.2}$$

where

$$\overline{f}_i(t,x) = \sqrt{2}\left[-f_i\left(\frac{x}{\sqrt{2}}\right) + \sum_{j=1}^{m} y_j(t)\left(\frac{\partial h_j}{\partial x_i}\right)\left(\frac{x}{\sqrt{2}}\right)\right] \tag{4.3}$$

$$\overline{V}(t,x) = \sum_{i=1}^{n}\left(\frac{\partial f_i}{\partial x_i}\right)\left(\frac{x}{\sqrt{2}}\right) + \frac{1}{2}\sum_{i=1}^{m}h_i^2\left(\frac{x}{\sqrt{2}}\right) - \sum_{i=1}^{m} y_i(t)(\Delta h_i)\left(\frac{x}{\sqrt{2}}\right)$$

$$+ \sum_{i=1}^{m}\sum_{j=1}^{n} y_i(t)f_j\left(\frac{x}{\sqrt{2}}\right)\frac{\partial h_i}{\partial x_j}\left(\frac{x}{\sqrt{2}}\right)$$

$$-\frac{1}{2}\sum_{i,j=1}^{m}\sum_{k=1}^{n}y_i(t)y_j(t)\left(\frac{\partial h_i}{\partial x_k}\right)\left(\frac{x}{\sqrt{2}}\right)\left(\frac{\partial h_j}{\partial x_k}\right)\left(\frac{x}{\sqrt{2}}\right). \quad (4.4)$$

For any $\tau > 0$, we shall consider the following parabolic equations on $[0, \tau] \times \mathbb{R}^n$

$$\begin{cases} \frac{\partial \overline{u}}{\partial t}(t, x) = \Delta \overline{u}(t, x) + \sum_{i=1}^{n} \overline{f}_i(t, x)\frac{\partial \overline{u}}{\partial x_i}(t, x) - \overline{V}(t, x)\overline{u}(t, x) \\ \overline{u}(0, x) = \overline{\psi}(x) \end{cases} \quad (4.5)$$

$$\begin{cases} \frac{\partial \widetilde{u}}{\partial t}(t, x) = \Delta \widetilde{u}(t, x) + \sum_{i=1}^{n} \widetilde{f}_i(\tau, x)\frac{\partial \widetilde{u}}{\partial x_i}(t, x) - \overline{V}(\tau, x)\widetilde{u}(t, x) \\ \widetilde{u}(0, x) = \widetilde{\psi}(x) \end{cases} \quad (4.6)$$

where $\widetilde{f}_i(\tau, x)$ and $\overline{V}(\tau, x)$ are obtained from $\overline{f}_i(t, x)$ and $\overline{V}(t, x)$ by freezing the time variable at τ i.e.

$$\widetilde{f}_i(\tau, x) = -\sqrt{2}f_i\left(\frac{x}{\sqrt{2}}\right) + \sqrt{2}\sum_{j=1}^{m}y_j(\tau)\left(\frac{\partial h_j}{\partial x_i}\right)\left(\frac{x}{\sqrt{2}}\right) \quad (4.7)$$

$$\overline{V}(\tau, x) = \sum_{i=1}^{n}\left(\frac{\partial f_i}{\partial x_i}\right)\left(\frac{x}{\sqrt{2}}\right) + \frac{1}{2}\sum_{i=1}^{m}h_i^2\left(\frac{x}{\sqrt{2}}\right) - \sum_{i=1}^{m}y_i(\tau)(\Delta h_i)\left(\frac{x}{\sqrt{2}}\right)$$
$$+ \sum_{i=1}^{m}\sum_{j=1}^{n}y_i(\tau)f_j\left(\frac{x}{\sqrt{2}}\right)\frac{\partial h_i}{\partial x_j}\left(\frac{x}{\sqrt{2}}\right)$$
$$- \frac{1}{2}\sum_{i,j=1}^{m}\sum_{k=1}^{n}y_i(\tau)y_j(\tau)\left(\frac{\partial h_i}{\partial x_k}\right)\left(\frac{x}{\sqrt{2}}\right)\left(\frac{\partial h_j}{\partial x_k}\right)\left(\frac{x}{\sqrt{2}}\right). \quad (4.8)$$

We shall assume that f and $\widetilde{f}$ grow at most linearly and V and $\widetilde{V}$ grow at most quadratically satisfying

$$\left|\overline{f}(t, x)\right| \le c(1 + |x|), \qquad \left|\widetilde{f}(t, x)\right| \le c(1 + |x|) \quad (4.9)$$

$$\left|\nabla\overline{f}(t, x)\right| = \sqrt{\sum_{i=1}^{n}|\nabla\overline{f}_i|^2} \le c(1 + |x|),$$

$$\left|\nabla\widetilde{f}(\tau, x)\right| = \sqrt{\sum_{i=1}^{n}\nabla\widetilde{f}_i} \le c(1 + |x|) \quad (4.10)$$

$$\left|\overline{V}(t, x)\right| \le c(1 + |x|^2), \qquad \left|\widetilde{V}(\tau, x)\right| \le c(1 + |x|^2) \quad (4.11)$$

$$\left|\nabla V(t, x)\right| \le c(1 + |x|), \qquad \left|\nabla\widetilde{V}(\tau, x)\right| \le c(1 + |x|). \quad (4.12)$$

It is clear that if $f(x)$ and its first derivatives, $h(x)$ and its derivatives up to order 3 have linear growth (i.e., $\leq c(1 + |x|)$), then $\bar{f}$, $\tilde{f}$, $\bar{V}$, and $\tilde{V}$ satisfy (4.9)–(4.12) respectively.

The first goal of this section is to prove that if $\tilde{\psi}(x)$ is close to $\bar{\psi}(x)$ uniformly in x, then $\tilde{u}(\tau, x)$ is close to $\bar{u}(\tau, x)$ uniformly in x. From equations (4.5) and (4.6), we deduce

$$\frac{\partial(\bar{u} - \tilde{u})}{\partial t} = \Delta(\bar{u} - \tilde{u}) + \sum_{i=1}^{n} \bar{f}_i \frac{\partial(\bar{u} - \tilde{u})}{\partial x_i} - \bar{V}(\bar{u} - \tilde{u})$$

$$+ \sum_{i=1}^{n} (\bar{f}_i - \tilde{f}_i) \frac{\partial \tilde{u}}{\partial x_i} - (\bar{V} - \tilde{V})\tilde{u}$$

$$= (\Delta - \bar{V})(\bar{u} - \tilde{u}) + \sum_{i=1}^{n} \bar{f}_i \frac{\partial(\bar{u} - \tilde{u})}{\partial x_i}$$

$$+ \sum_{i=1}^{n} (\bar{f}_i - \tilde{f}_i) \frac{\partial \tilde{u}}{\partial x_i} - (\bar{V} - \tilde{V})\tilde{u}$$

$$= (\Delta - \bar{V})(\bar{u} - \tilde{u}) + \sum_{i=1}^{n} \bar{f}_i \frac{\partial(\bar{u} - \tilde{u})}{\partial x_i} + G_\tau(t, x) \quad (4.13)$$

where

$$G_\tau(t, x) = \sum_{i=1}^{n} (\bar{f}_i(t, x) - \tilde{f}_i(\tau, x)) \frac{\partial \tilde{u}}{\partial x_i}(t, x) - (\bar{V}(t, x) - \tilde{V}(\tau, x))\tilde{u}(t, x).$$

$$(4.14)$$

Lemma 7 *There exists a nonnegative function $\alpha(t, x, y)$ such that*

$$\begin{cases} \dfrac{\partial \alpha}{\partial t}(t, x, y) = \Delta_x \alpha(t, x, y) - \sum_{i=1}^{n} \bar{f}_i(\tau - t, x) \dfrac{\partial \alpha}{\partial x_i}(t, x, y) \\[2mm] \qquad\qquad - \left[\bar{V}(\tau - t, x) + \sum_{i=1}^{n} \dfrac{\partial \bar{f}_i}{\partial x_i}(\tau - t, x) \right] \alpha(t, x, y) \\[2mm] \alpha(0, x, y) = \delta_y(x) \\[2mm] \displaystyle\int_x \alpha(0, x, y)\, dx = 1. \end{cases}$$

$$(4.15)$$

Proof Let $\beta_n(x, y)$ be a sequence of Gaussian with

$$\int_x \beta_n(x, y)\, dx = 1 \quad \text{and} \quad \lim_{n \to \infty} \beta_n(x, y) = \delta_y(x). \quad (4.16)$$

In view of our previous work [42], there exists a solution $\alpha_n(t, x, y)$

with initial condition $\alpha_n(0, x, y) = \beta_n(x, y)$. By maximal principle, $\alpha_n(t, x, y) \geq 0$ for all $t \geq 0$. We take $\alpha(t, x, y) = \lim\limits_{n \to \infty} \alpha_n(t, x, y)$.   □

Theorem 8 *Let $w(t, x) = \overline{u}(t, x) - \widetilde{u}(t, x)$ where $\overline{u}$ and $\widetilde{u}$ are the solutions of the parabolic equations (4.5) and (4.6) respectively. Let $\alpha(t, x, y)$ be the nonnegative function in Lemma 7. Then*

$$w(\tau, y) = \int_x \alpha(\tau, x, y) w(0, x) dx + \int_0^\tau \int_x \alpha(t, x, y) G_\tau(t, x) dx$$

where $G_\tau(t, x)$ is given in (4.14).

Proof

$$\int_0^\tau \frac{d}{dt} \int_x \alpha(\tau - t, x, y) w(t, x) dx \qquad\qquad (4.17)$$

$$= -\int_0^\tau \int_x \frac{\partial \alpha}{\partial t}(\tau - t, x, y) w(t, x) dx + \int_0^\tau \int_x \alpha(\tau - t, x, y) \frac{\partial w}{\partial t}(t, x) dx,$$

L.H.S. of (4.17)

$$= w(\tau, y) - \int_x \alpha(\tau, x, y) w(0, x) dx,$$

R.H.S. of (4.17)

$$= -\int_0^\tau \int_x \Delta_x \alpha(\tau - t, x, y) w(t, x) dx$$

$$+ \int_0^T \int_x \sum_{i=1}^n \overline{f}_i(t, x) \frac{\partial \alpha}{\partial x_i}(\tau - t, x, y) w(t, x) dx$$

$$+ \int_0^\tau \int_x \left[\overline{V}(t, x) + \sum_{i=1}^n \frac{\partial \overline{f}_i}{\partial x_i}(t, x) \right] \alpha(\tau - t, x, y) w(t, x) dx$$

$$+ \int_0^\tau \int_x \alpha(\tau - t, x, y) \frac{\partial w}{\partial t}(t, x) dx$$

$$= \int_0^\tau \int_x \alpha(\tau - t, x, y) \left[\frac{\partial w}{\partial t}(t, x) - \Delta w(t, x) \right.$$

$$\left. - \sum_{i=1}^n \overline{f}_i(t, x) \frac{\partial w}{\partial x_i}(t, x) + \overline{V}(t, x) w(t, x) \right] dx$$

$$= \int_0^\tau \int_x \alpha(\tau - t, x, y) G_\tau(t, x) dx \qquad \text{by (4.13).}$$

In the above computation, we have used the fact proved in [42] that

$\alpha(t, x, y)$ has Gaussian decay in x. Hence

$$w(\tau, y) = \int_x \alpha(\tau, x, y) w(0, x) dx + \int_0^\tau \int_x \alpha(\tau - t, x, y) G_\tau(t, x) dx.$$

$\square$

Proposition 9 *Let $\alpha(t, x, y)$ be the nonnegative function in Lemma 7. Suppose that $\overline{V}(t, x) \geq -c_1$ for some positive constant c_1. Then*

$$\int_x \alpha(\tau, x, y) dx \leq e^{c_1 \tau}. \tag{4.18}$$

Proof

$$e^{c_1 t} \frac{d}{dt} \left(e^{-c_1 t} \int_x \alpha(t, x, y) dx \right)$$

$$= -c_1 \int_x \alpha(t, x, y) dx + \int_x \frac{\partial \alpha}{\partial t}(t, x, y) dx$$

$$= -c_1 \int_x \alpha(t, x, y) dx + \int_x \Delta_x \alpha(t, x, y) dx$$

$$- \int_x \sum_{i=1}^n \overline{f}_i(\tau - t, x) \frac{\partial \alpha}{\partial x_i}(t, x, y) dx$$

$$- \int_x \left[\overline{V}(\tau - t, x) + \sum_{i=1}^n \frac{\partial \overline{f}_i}{\partial x_i}(\tau - t, x) \right] \alpha(t, x, y) dx$$

$$= -c_1 \int_x \alpha(t, x, y) dx - \int_x \overline{V}(\tau - t, x) \alpha(t, x, y) dx$$

$$= - \int_x [V(\tau - t, x) + c_1] \alpha(t, x, y) dx$$

$$\leq 0.$$

It follows that $e^{-c_1 t} \int_x \alpha(t, x, y) dx$ is a decreasing function of t. Hence

$$e^{-c_1 \tau} \int_x \alpha(\tau, x, y) dx \leq \int_x \alpha(0, x, y) dx = 1$$

i.e. $\int_x \alpha(\tau, x, y) dx \leq e^{c_1 \tau}.$

$\square$

Theorem 10 *With the assumption in Proposition 9, let $w(t, x) = \overline{u}(t, x) - \widetilde{u}(t, x)$ where $\overline{u}$ and $\widetilde{u}$ are the solutions of the parabolic equations (4.5)*

S.-T. Yau and S.S.-T. Yau

and (4.6) respectively. If τ is small and $w(0, x)$ is small uniformly in x, then $w(\tau, x)$ is small uniformly in x. More precisely, we have

$$\sup_{y \in \mathbb{R}^n} |w(\tau, y)| \leq e^{c_1 \tau} \sup_{x \in \mathbb{R}^n} |w(0, x)| + \tau e^{c_1 \tau} \sup_{\substack{x \in \mathbb{R}^n \\ 0 \leq t \leq \tau}} |G_\tau(t, x)| \qquad (4.19)$$

where $G_\tau(t, x)$ is given in (4.14).

Proof In view of (4.7), (4.8) and (4.14), we have

$$G_\tau(t, x)$$
$$= \sum_{i=1}^{n} (\overline{f}_i(t, x) - \widetilde{f}_i(\tau, x)) \frac{\partial \widetilde{u}}{\partial x_i}(t, x) - (\overline{V}(t, x) - \widetilde{V}(\tau, x)) \widetilde{u}(t, x)$$
$$= \sum_{i=1}^{n} \sqrt{2} \sum_{j=1}^{m} (y_j(t) - y_j(\tau)) \frac{\partial h_j}{\partial x_i}\left(\frac{x}{\sqrt{2}}\right) \frac{\partial \widetilde{u}}{\partial x_i}(t, x)$$
$$+ \left[-\sum_{i=1}^{m} (y_i(t) - y_i(\tau)) \Delta h_i \left(\frac{x}{\sqrt{2}}\right) \right.$$
$$+ \sum_{i=1}^{m} \sum_{j=1}^{n} (y_i(t) - y_i(\tau)) f_j\left(\frac{x}{\sqrt{2}}\right) \frac{\partial h_i}{\partial x_j}\left(\frac{x}{\sqrt{2}}\right)$$
$$\left. - \frac{1}{2} \sum_{i,j=1}^{m} \sum_{k=1}^{n} (y_i(t) y_j(t) - y_i(\tau) y_j(\tau)) \frac{\partial h_i}{\partial x_k}\left(\frac{x}{\sqrt{2}}\right) \frac{\partial h_j}{\partial x_k}\left(\frac{x}{\sqrt{2}}\right) \right] \widetilde{u}(t, x).$$

Therefore if τ is small, then $G_\tau(t, x)$ is uniformly small in x for $0 \leq t \leq \tau$, because both $\widetilde{u}(t, x)$ and $\frac{\partial \widetilde{u}}{\partial x_i}(t, x)$ have Gaussian decay in x by [42]. By Theorem 8 we have the following estimate

$$|w(\tau, y)| \leq \left| \int_x \alpha(\tau, x, y) w(0, x) dx \right| + \left| \int_0^\tau \int_x \alpha(t, x, y) G_\tau(t, x) dx \right|$$

$$\leq \sup_{x \in \mathbb{R}^n} |w(0, x)| \int_x \alpha(\tau, x, y) dx$$

$$\quad + \sup_{\substack{x \in \mathbb{R}^n \\ 0 \leq t \leq \tau}} |G_\tau(t, x)| \int_0^\tau \int_x \alpha(t, x, y) dx$$

$$\leq e^{c_1 \tau} \sup_{x \in \mathbb{R}^n} |w(0, x)| + \tau e^{c_1 \tau} \sup_{\substack{x \in \mathbb{R}^n \\ 0 \leq t \leq \tau}} |G(t, x)|.$$

$\square$

Now we are ready to consider the global situation. For a fixed $T > 0$, we want to find the solution $\overline{u}(t, x)$ of the following parabolic equation on $[0, T] \times \mathbb{R}^n$

$$\begin{cases} \dfrac{\partial \overline{u}}{\partial t}(t, x) = \Delta \overline{u}(t, x) + \sum_{j=1}^{n} \overline{f}_j(t, x) \dfrac{\partial \overline{u}}{\partial x_j}(t, x) - \overline{V}(t, x)\overline{u}(t, x), \\ \overline{u}(0, x) = \overline{\psi}(x). \end{cases} \tag{4.20}$$

Let $\{0 < \tau_1 < \tau_2 < \cdots < \tau_n = T\}$ be a partition of $[0, T]$. Let $\widetilde{u}_i(t, x)$ be the solution of the following parabolic equation on $[\tau_{i-1}, \tau_i] \times \mathbb{R}^n$

$$\begin{cases} \dfrac{\partial \widetilde{u}_i}{\partial t}(t, x) = \Delta \widetilde{u}_i(t, x) + \sum_{j=1}^{n} \widetilde{f}_j(\tau_i, x) \dfrac{\partial \widetilde{u}_i}{\partial x_j}(t, x) - \widetilde{V}(\tau_i, x)\widetilde{u}_i(t, x) \\ \widetilde{u}_i(\tau_{i-1}, x) = \widetilde{u}_{i-1}(\tau_{i-1}, x) \end{cases} \tag{4.21}$$

where $\widetilde{u}_1(0, x) = \overline{\psi}(x)$, $\widetilde{f}_j(\tau_i, x)$ and $\widetilde{V}(\tau_i, x)$ are functions independent of t and are equal to $\overline{f}_j(\tau_i, x)$ and $\overline{V}(\tau_i, x)$ respectively.

Lemma 11 *Fix T, let $G_{\tau_i}(t, x) = \sum_{j=1}^{n} \left(\overline{f}_j(t, x) - \widetilde{f}_j(\tau_i, x) \right) \dfrac{\partial \widetilde{u}_i}{\partial x_j}(t, x) -$*
$\left(\overline{V}(t, x) - \widetilde{V}(\tau, x) \right)\widetilde{u}_i(t, x)$. *For any given $\epsilon > 0$, we can choose n sufficiently large so that*

$$\sup_{1 \le i \le n} \; \sup_{\tau_{i-1} \le t \le \tau_i} \; \sup_{x \in \mathbb{R}^n} \left| G_{\tau_i}(t, x) \right| \le \epsilon. \tag{4.22}$$

Proof This follows from the proof of Theorem 10. $\qquad \square$

We are now ready to prove the main theorem of this section.

Theorem 12 *Let $\overline{u}(t, x)$ and $\widetilde{u}_n(t, x)$ be the solutions of (4.20) and (4.21) respectively. For any $\epsilon > 0$, let n be sufficiently large so that Lemma 11 holds. Then*

$$\left| \overline{u}(T, x) - \widetilde{u}_n(T, x) \right| \le \epsilon T e^{c_1 T} \tag{4.23}$$

where c_1 is the constant in Proposition 9.

Proof In view of $\widetilde{u}_1(0, x) = \overline{\psi}(x) = \overline{u}(0, x)$ and Theorem 10, we have

$$\left| \overline{u}(\tau_1, x) - \widetilde{u}_1(\tau_1, x) \right| \le \tau_1 e^{c_1 \tau_1} \sup_{\substack{x \in \mathbb{R}^n \\ 0 \le t \le \tau_1}} \left| G_{\tau_1}(t, x) \right|.$$

By Theorem 10, and induction we have

$$
\left|\bar{u}(\tau_2,x)-\tilde{u}_2(\tau_2,x)\right| \leq \tau_1 e^{c_1\tau_1}e^{c_1(\tau_2-\tau_1)} \sup_{\substack{x\in\mathbb{R}^n \\ 0\leq t\leq\tau_1}} \left|G_{\tau_1}(t,x)\right|
$$
$$
+ (\tau_2-\tau_1)e^{c_1(\tau_2-\tau_1)} \sup_{\substack{x\in\mathbb{R}^n \\ \tau_1\leq t\leq\tau_2}} \left|G_{\tau_2}(t,x)\right|
$$

$$
\left|\bar{u}(\tau_3,x)-\tilde{u}_3(\tau_3,x)\right| \leq \tau_1 e^{c_1\tau_1}e^{c_1(\tau_2-\tau_1)}e^{c_1(\tau_3-\tau_2)} \sup_{\substack{x\in\mathbb{R}^n \\ 0\leq t\leq\tau_1}} \left|G_{\tau_1}(t,x)\right|
$$
$$
+ (\tau_2-\tau_1)e^{c_1(\tau_2-\tau_1)}e^{c_1(\tau_3-\tau_2)} \sup_{\substack{x\in\mathbb{R}^n \\ \tau_1\leq t\leq\tau_2}} \left|G_{\tau_2}(t,x)\right|
$$
$$
+\cdots+ (\tau_3-\tau_2)e^{c_1(\tau_3-\tau_2)} \sup_{\substack{x\in\mathbb{R}^n \\ \tau_2\leq t\leq\tau_3}} \left|G_{\tau_3}(t,x)\right|
$$

$$\vdots$$

$$
\left|\bar{u}(\tau_n,x)-\tilde{u}_n(\tau_n,x)\right| = \tau_1 e^{c_1\tau_n} \sup_{\substack{x\in\mathbb{R}^n \\ 0\leq t\leq\tau_1}} \left|G_{\tau_1}(t,x)\right|
$$
$$
+ (\tau_2-\tau_1)e^{c_1(\tau_n-\tau_1)} \sup_{\substack{x\in\mathbb{R}^n \\ \tau_1\leq t\leq\tau_2}} \left|G_{\tau_2}(t,x)\right|
$$
$$
+\cdots+ (\tau_i-\tau_{i-1})e^{c_1(\tau_n-\tau_{i-1})} \sup_{\substack{x\in\mathbb{R}^n \\ \tau_{i-1}\leq t\leq\tau_i}} \left|G_{\tau_i}(t,x)\right|
$$
$$
+\cdots+ (\tau_n-\tau_{n-1})e^{c_1(\tau_n-\tau_{n-1})} \sup_{\substack{x\in\mathbb{R}^n \\ \tau_{n-1}\leq t\leq\tau_n}} \left|G_{\tau_n}(t,x)\right|
$$
$$
\leq \epsilon\big(\tau_1 e^{c_1\tau_n} + (\tau_2-\tau_1)e^{c_1(\tau_n-\tau_1)}
$$
$$
+\cdots+ (\tau_i-\tau_{i-1})e^{c_1(\tau_n-\tau_{i-1})}
$$
$$
+\cdots+ (\tau_n-\tau_{n-1})e^{c_1(\tau_n-\tau_{n-1})}\big)
$$
$$
\leq \epsilon\big[\tau_1 + (\tau_2-\tau_1)+\cdots+ (\tau_i-\tau_{i-1})
$$
$$
+\cdots+ (\tau_n-\tau_{n-1})\big]e^{c_1 T}
$$
$$
= \epsilon T e^{c_1 T}.
$$

$\square$

Theorem 13 *Fix $T>0$, let $\mathcal{P}_n = \{0 < \tau_1 < \tau_2 < \cdots < \tau_n = T\}$ be a partition of $[0,T]$. Let $\bar{u}(t,x)$ be the solution of the following parabolic* *equation on* $[0, T] \times \mathbb{R}^n$

$$\begin{cases} \dfrac{\partial \overline{u}}{\partial t}(t, x) = \Delta \overline{u}(t, x) + \displaystyle\sum_{j=1}^{n} \overline{f}_j(t, x) \dfrac{\partial \overline{u}}{\partial x_j}(t, x) - \overline{V}(t, x) \overline{u}(t, x) \\ \overline{u}(0, x) = \psi(x). \end{cases}$$

Let $\widetilde{u}_i(t, x)$ *be the solution of the following parabolic equation on* $[\tau_{i-1}, \tau_i]$ $\times \mathbb{R}^n$

$$\begin{cases} \dfrac{\partial \widetilde{u}_i}{\partial t}(t, x) = \Delta \widetilde{u}_i(t, x) + \displaystyle\sum_{j=1}^{n} \widetilde{f}_j(\tau_i, x) \dfrac{\partial \widetilde{u}_i}{\partial x_j}(t, x) - \widetilde{V}(\tau_i, x) \widetilde{u}_i(t, x) \\ \widetilde{u}_i(\tau_{i-1}, x) = \widetilde{u}_{i-1}(\tau_{i-1}, x). \end{cases}$$

where $\widetilde{u}_1(0, x) = \psi(x)$ *and* $\widetilde{f}_j(\tau_i, x) = \overline{f}_j(\tau_i, x)$, $\widetilde{V}(\tau_i, x) = \overline{V}(\tau_i, x)$ *are obtained from* $\overline{f}_j(t, x)$ *and* $\overline{V}(t, x)$ *by freezing time variable at* τ_i. *Then*

$$\overline{u}(\tau, x) = \lim_{|\mathcal{P}_n| \to 0} \widetilde{u}_n(\tau_n, x) \text{ uniformly in } x.$$

5 L^2-Convergence

In Section 4 we have shown that the solution $\widetilde{u}(t, x)$ of (4.6) is uniformly close to the solution $\overline{u}(t, x)$ of (4.5) for $0 \le t \le \tau$ if τ is sufficiently small and $\widetilde{\psi}(x) = \widetilde{u}(0, x)$ is uniformly close to $\overline{\psi}(x) = \overline{u}(0, x)$. In this section, we shall show that $\widetilde{u}(t, x)$ is also close to $\overline{u}(t, x)$ in L^2 sense. We first recall the following lemma.

Lemma 14 *If* $\frac{d\alpha}{dt}(t) \le c\alpha(t) + \beta(t)$, *where* c *is constant, then* $e^{-ct}\alpha(t) - \alpha(0) \le \int_0^t e^{-cs}\beta(s)ds$.

Let $\overline{f}_{2R}$, $\widetilde{f}_{2R}$, $\overline{V}_{2R}$ and $\widetilde{V}_{2R}$ be the functions obtained by multiplying $\overline{f}$, $\widetilde{f}$, $\overline{V}$ and $\widetilde{V}$ respectively by a cut off function σ which is equal to one in the ball of radius $R \ge 1$ and equal to zero outside a ball of radius $2R$. We can choose σ such that

$$|\nabla \sigma(x)| \le \frac{4}{1 + |x|} \text{ and } |\Delta \sigma(x)| \le \frac{4}{1 + |x|^2}. \tag{5.1}$$

Consider the following equations

$$\frac{\partial \overline{u}_{2R}}{\partial t} = \Delta \overline{u}_{2R} + \sum_{i=1}^{n} (\overline{f}_{2R})_i \frac{\partial \overline{u}_{2R}}{\partial x_i} - \overline{V}_{2R} \overline{u}_{2R} \tag{5.2}$$

$$\frac{\partial \widetilde{u}_{2R}}{\partial t} = \Delta \widetilde{u}_{2R} + \sum_{i=1}^{n} (\widetilde{f}_{2R})_i \frac{\partial \widetilde{u}_{2R}}{\partial x_i} - \widetilde{V}_{2R} \widetilde{u}_{2R} \tag{5.3}$$

in the ball B_{2R} of radius $2R$ with the Neumann condition, where $(f_{2R})_i$ and $(\widehat{f}_{2R})_i$ denote the ith components of f_{2R} and $\widehat{f}_{2R}$ respectively. Let $\overline{\psi}_{2R}(x) = \overline{\psi}(x)\sigma(x)$ and $\widetilde{\psi}_{2R}(x) = \widetilde{\psi}(x)\sigma(x)$. Then the second initial boundary problems

$$
\begin{cases}
\dfrac{\partial \overline{u}_{2R}}{\partial t}(t,x) = \Delta \overline{u}_{2R}(t,x) + \displaystyle\sum_{i=1}^{n} (\overline{f}_{2R})_i(t,x)\dfrac{\partial \overline{u}_{2R}}{\partial x_i}(t,x) \\
\qquad\qquad\quad - \overline{V}_{2R}(t,x)\overline{u}_{2R}(t,x) \qquad \text{on } B_{2R} \times (0,T] \\
\overline{u}_{2R}(0,x) = \overline{\psi}_{2R}(x) \qquad\qquad\qquad \text{on } \overline{B}_{2R} \\
\dfrac{\partial \overline{u}_{2R}}{\partial \nu}(t,x) = 0 \qquad\qquad\qquad\quad \text{on } \partial B_{2R} \times (0,T].
\end{cases}
\tag{5.4}
$$

$$
\begin{cases}
\dfrac{\partial \widetilde{u}_{2R}}{\partial t}(t,x) = \Delta \overline{u}_{2R}(t,x) + \displaystyle\sum_{i=1}^{n} (\overline{f}_{2R})_i(t,x)\dfrac{\partial \widetilde{u}_{2R}}{\partial x_i}(t,x) \\
\qquad\qquad\quad - \overline{V}_{2R}(t,x)\overline{u}_{2R}(t,x) \qquad \text{on } B_{2R} \times (0,T] \\
\overline{u}_{2R}(0,x) = \overline{\psi}_{2R}(x) \qquad\qquad\qquad \text{on } \overline{B}_{2R} \\
\dfrac{\partial \widetilde{u}_{2R}}{\partial \nu}(t,x) = 0 \qquad\qquad\qquad\quad \text{on } \partial B_{2R} \times (0,T].
\end{cases}
\tag{5.5}
$$

have unique solutions respectively for $t \in [0,\infty)$ (cf. [16], p. 144, Theorem 2).

Lemma 15 *Assume that (4.9)–(4.12) hold and $\left|\Delta\overline{f}(t,x)\right| \leq c(1+|x|)$, $\left|\Delta\widetilde{f}(t,x)\right| \leq c(1+|x|)$ where $c \geq 4$ is a constant. Let $\widetilde{c}$ and δ be positive constants such that $\widetilde{\widetilde{c}} := \widetilde{c} + \delta < \frac{5}{254}$. Choose τ and ϵ suitably small with $\tau + \epsilon < \delta$. Then the following conclusions (i), (ii) and (iii) hold for any $0 \leq t \leq \tau$ for both $\rho \in \{\overline{\rho},\widetilde{\rho}\}$, $u \in \{\overline{u},\widetilde{u}\}$, and where $\overline{\rho}(t,x) = \frac{\widetilde{c}(1+|x|^2)}{t+\epsilon}$, $\widetilde{\rho}(t,x) = \frac{\widetilde{\widetilde{c}}(1+|x|^2)}{t+\epsilon}$*

(i) $\displaystyle\int_{\{t\}\times B_{2R}} e^{\overline{\rho}}\overline{u}_{2R}^2 \leq \int_{\{0\}\times B_{2R}} e^{\overline{\rho}}\overline{u}_{2R}^2$

(ii) $\displaystyle\int_{\{t\}\times B_{2R}} e^{\overline{\rho}}|\nabla \overline{u}_{2R}|^2 \leq \int_{\{0\}\times B_{2R}} e^{\overline{\rho}}|\nabla \overline{u}_{2R}|^2$

$\qquad + \displaystyle\int_0^t \int_{B_{2R}} e^{\overline{\rho}(x,s)}\left|\overline{u}_{2R}(s,x)\right|^2$

(iii) $\displaystyle\int_{\{t\}\times B_{2R}} e^{\overline{\rho}}|\Delta \overline{u}_{2R}|^2 \leq \int_{\{0\}\times B_{2R}} e^{\overline{\rho}}|\Delta \overline{u}_{2R}|^2$

$\qquad + O\left(\displaystyle\int_{[0,t]\times B_{2R}} e^{\overline{\rho}}|\nabla \overline{\rho}|^2|\overline{f}_{2R}|^2|\nabla \overline{u}_{2R}|^2 + \int_{[0,t]\times B_{2R}} e^{\overline{\rho}}|\nabla \overline{f}_{2R}|^2|\nabla \overline{u}_{2R}|^2\right.$

$$+ \int_{[0,t]\times B_{2R}} e^{\bar{p}}|\bar{f}_{2R}|\,|\nabla \bar{u}_{2R}|^2 |\Delta \bar{f}_{2R}| + \int_{[0,t]\times B_{2R}} e^{\bar{p}}|\bar{f}_{2R}|^4 |\nabla \bar{u}_{2R}|^2$$

$$+ \int_{[0,t]\times B_{2R}} e^{\bar{p}}|\nabla(\bar{V}_{2R}\bar{u}_{2R})|^2 + \int_{[0,t]\times B_{2R}} e^{\bar{p}}|\nabla \bar{u}_{2R}|^2 \left(\sum_{i=1}^n \frac{\partial (\bar{f}_{2R})_i}{\partial x_i} \right)^2 \Bigg).$$

Moreover the following inequalities hold for both $\{\bar{p}, \bar{f}, \bar{V}\}$, *or* $\{\bar{p}, \tilde{f}, \tilde{V}\}$, *or* $\{\tilde{p}, \bar{f}, \bar{V}\}$ *or* $\{\tilde{p}, \tilde{f}, \tilde{V}\}$ *if δ is small enough.*

(iv) $\dfrac{\partial \bar{p}}{\partial t} + 2|\nabla \bar{p}|^2 - \displaystyle\sum_{i=1}^n \bar{f}_i \frac{\partial \bar{p}}{\partial x_i} - \sum_{i=1}^n \frac{\partial \bar{f}_i}{\partial x_i} - 2\bar{V} \leq 0.$

Proof (i), (ii) and (iii) follow from Lemma 1.3 of [42] by setting $\epsilon_1 = \frac{1}{5}$ in that lemma. Let us prove (iv) for $\{\tilde{p}, \bar{f}, \bar{V}\}$. Observe that

$$\frac{\partial \tilde{p}}{\partial t} = -\frac{\overset{\approx}{c}(1+|x|^2)}{(t+\epsilon)^2}, \quad \frac{\partial \tilde{p}}{\partial x_i} = \frac{2\overset{\approx}{c}x_i}{t+\epsilon_0}, \quad |\nabla p|^2 = \frac{4\overset{\approx}{c}|x|^2}{(t+\epsilon)^2}.$$

Hence

$$\frac{\partial \tilde{p}}{\partial t} + 2|\nabla \tilde{p}|^2 - \sum_{i=1}^n \bar{f}_i \frac{\partial \tilde{p}}{\partial x_i} - \sum_{i=1}^n \frac{\partial \bar{f}_i}{\partial x_i} - 2\bar{V}$$

$$\leq -\frac{\overset{\approx}{c}(1+|x|^2)}{(t+\epsilon)^2} + \frac{8\overset{\approx}{c}^2|x|^2}{(t+\epsilon)^2} + c(1+|x|)\frac{2\overset{\approx}{c}|x|}{t+\epsilon}$$

$$\quad + nc(1+|x|) + 2c(1+|x|^2)$$

$$\leq \frac{\overset{\approx}{c}(8\overset{\approx}{c}-1)(1+|x|^2)}{(t+\epsilon)^2} + \frac{2c\overset{\approx}{c}(1+|x|)^2}{t+\epsilon} + nc(1+|x|)^2 + 2c(1+|x|)^2$$

$$\leq \left[\frac{-\overset{\approx}{c}(1-8\overset{\approx}{c})}{(t+\epsilon)^2} + \frac{2c\overset{\approx}{c}}{t+\epsilon} + (n+2)c \right](1+|x|)^2.$$

Since $1-8\overset{\approx}{c} > 0$, we see that the R.H.S. of the above inequality is less than zero for $t+\epsilon$ sufficiently small. $\qquad\square$

Proposition 16 *Consider the parabolic differential equations (4.5) and (4.6). Let ϕ be any smooth function defined on $\mathbb{R}^n$ with compact support contained in a domain Ω. Let $\bar{p}$ be any smooth function on $\mathbb{R}_+ \times \mathbb{R}^n$ satisfying*

$$\frac{\partial \bar{p}}{\partial t} + 2|\nabla \bar{p}|^2 - \sum_{i=1}^n \bar{f}_i \frac{\partial \bar{p}}{\partial x_i} - \sum_{i=1}^n \frac{\partial \bar{f}_i}{\partial x_i} - 2\bar{V} \leq 0. \qquad (5.6)$$

Then

$$\frac{d}{dt} \int_{\{t\}\times\Omega} \phi^2 e^{\overline{\rho}} (\overline{u} - \widetilde{u})^2$$

$$\leq \int_{\{t\}\times\Omega} \phi^2 e^{\overline{\rho}} (\overline{u} - \widetilde{u})^2 + 10 \int_{\{t\}\times\Omega} e^{\overline{\rho}} (\overline{u} - \widetilde{u})^2 |\nabla\phi|^2$$

$$+ 4 \int_{\{t\}\times\Omega} e^{\overline{\rho}} \left| \sum_{i=1}^n \overline{f}_i \frac{\partial\phi}{\partial x_i} \right|^2 (\overline{u} - \widetilde{u})^2$$

$$+ 4 \int_{\{t\}\times\Omega} e^{\overline{\rho}} \phi^2 \left| \sum_{i=1}^n (\overline{f}_i - \widetilde{f}_i) \frac{\partial\rho}{\partial x_i} \right|^2 \widetilde{u}^2$$

$$+ 2 \int_{\{t\}\times\Omega} e^{\overline{\rho}} \phi^2 \widetilde{u} |\overline{f} - \widetilde{f}|^2 + 4 \int_{\{t\}\times\Omega} e^{\overline{\rho}} \phi^2 \widetilde{u}^2 |\overline{V} - \widetilde{V}|^2$$

$$+ 2 \int_{\{t\}\times\Omega} e^{\overline{\rho}} \phi^2 |\overline{f} - \widetilde{f}|^2 \widetilde{u}^2$$

$$+ 4 \int_{\{t\}\times\Omega} e^{\overline{\rho}} \phi^2 \widetilde{u}^2 \left| \sum_{i=1}^n \left(\frac{\partial \overline{f}_i}{\partial x_i} - \frac{\partial \widetilde{f}_i}{\partial x_i} \right) \right|^2. \tag{5.7}$$

Proof From equations (4.5) and (4.6) we deduce

$$\frac{\partial(\overline{u} - \widetilde{u})}{\partial t} = \Delta(\overline{u} - \widetilde{u}) + \sum_{i=1}^n f_i \frac{\partial(\overline{u} - \widetilde{u})}{\partial x_i} - V(u - \widetilde{u})$$

$$+ \sum_{i=1}^n (\overline{f}_i - \widetilde{f}_i) \frac{\partial\widetilde{u}}{\partial x_i} - (\overline{V} - \widetilde{V})\widetilde{u}. \tag{5.8}$$

Then

$$\frac{d}{dt} \int_{\{t\}\times\Omega} \phi^2 (\overline{u} - \widetilde{u})^2 e^{\overline{\rho}}$$

$$= 2 \int_{\{t\}\times\Omega} \phi^2 (\overline{u} - \widetilde{u}) \left[\frac{\partial(\overline{u} - \widetilde{u})}{\partial t} \right] e^{\overline{\rho}} + \int_{\{t\}\times\Omega} \phi^2 (\overline{u} - \widetilde{u})^2 e^{\overline{\rho}} \frac{\partial\overline{\rho}}{\partial t}$$

$$= 2 \int_{\{t\}\times\Omega} e^{\overline{\rho}} \phi^2 (\overline{u} - \widetilde{u}) \Delta(\overline{u} - \widetilde{u}) + \int_{\{t\}\times\Omega} e^{\overline{\rho}} \phi^2 \sum_{i=1}^n \overline{f}_i \frac{\partial[(\overline{u} - \widetilde{u})^2]}{\partial x_i}$$

$$- 2 \int_{\{t\}\times\Omega} e^{\overline{\rho}} \phi^2 \overline{V} (\overline{u} - \widetilde{u})^2 + 2 \int_{\{t\}\times\Omega} e^{\overline{\rho}} \phi^2 \sum_{i=1}^n (\overline{f}_i - \widetilde{f}_i) \frac{\partial\widetilde{u}}{\partial x_i} (\overline{u} - \widetilde{u})$$

$$- 2 \int_{\{t\}\times\Omega} e^{\overline{\rho}} \phi^2 (\overline{V} - \widetilde{V}) \widetilde{u} (\overline{u} - \widetilde{u}) + \int_{\{t\}\times\Omega} e^{\overline{\rho}} \phi^2 (\overline{u} - \widetilde{u})^2 \frac{\partial\overline{\rho}}{\partial t}$$

$$= \int_{\{t\}\times\Omega} e^{\overline{\rho}} \phi^2 (\overline{u} - \widetilde{u})^2 \frac{\partial\rho}{\partial t} - 2 \int_{\{t\}\times\Omega} e^{\overline{\rho}} \phi^2 |\nabla(\overline{u} - \widetilde{u})|^2$$

$$-4\int_{\{t\}\times\Omega}e^{\overline{\rho}}\phi(\overline{u}-\widetilde{u})\nabla\phi\cdot\nabla(\overline{u}-\widetilde{u})-2\int_{\{t\}\times\Omega}e^{\overline{\rho}}\phi^2(\overline{u}-\widetilde{u})\nabla\overline{\rho}\cdot\nabla(\overline{u}-\widetilde{u})$$

$$-2\int_{\{t\}\times\Omega}e^{\overline{\rho}}\phi\left(\sum_{i=1}^{n}\overline{f}_i\frac{\partial\phi}{\partial x_i}\right)(\overline{u}-\widetilde{u})^2-\int_{\{t\}\times\Omega}e^{\overline{\rho}}\phi^2\left(\sum_{i=1}^{n}\overline{f}_i\frac{\partial\rho}{\partial x_i}\right)(\overline{u}-\widetilde{u})^2$$

$$-\int_{\{t\}\times\Omega}e^{\overline{\rho}}\phi^2\left(\sum_{i=1}^{n}\frac{\partial\overline{f}_i}{\partial x_i}\right)(\overline{u}-\widetilde{u})^2-2\int_{\{t\}\times\Omega}e^{\overline{\rho}}\phi^2\overline{V}(\overline{u}-\widetilde{u})^2$$

$$-4\int_{\{t\}\times\Omega}e^{\overline{\rho}}\phi(\overline{u}-\widetilde{u})\left[\sum_{i=1}^{n}(\overline{f}_i-\widetilde{f}_i)\frac{\partial\phi}{\partial x_i}\right]\widetilde{u}$$

$$-2\int_{\{t\}\times\Omega}e^{\overline{\rho}}\phi^2\left[\sum_{i=1}^{n}(\overline{f}_i-\widetilde{f}_i)\frac{\partial\rho}{\partial x_i}\right]\widetilde{u}(\overline{u}-\widetilde{u})$$

$$-2\int_{\{t\}\times\Omega}e^{\overline{\rho}}\phi^2\widetilde{u}\sum_{i=1}^{n}(\overline{f}_i-\widetilde{f}_i)\frac{\partial(\overline{u}-\widetilde{u})}{\partial x_i}$$

$$-2\int_{\{t\}\times\Omega}e^{\overline{\rho}}\phi^2(\overline{u}-\widetilde{u})\widetilde{u}\sum_{i=1}^{n}\left(\frac{\partial\overline{f}_i}{\partial x_i}-\frac{\partial\widetilde{f}_i}{\partial x_i}\right)-2\int_{\{t\}\times\Omega}e^{\overline{\rho}}\phi^2(\overline{u}-\widetilde{u})\widetilde{u}(\overline{V}-\widetilde{V})$$

$$\leq\int_{\{t\}\times\Omega}e^{\overline{\rho}}\phi^2(\overline{u}-\widetilde{u})^2\frac{\partial\overline{\rho}}{\partial t}-2\int_{\{t\}\times\Omega}e^{\overline{\rho}}\phi^2|\nabla(\overline{u}-\widetilde{u})|^2$$

$$+4\left[\frac{1}{8}\int_{\{t\}\times\Omega}|\nabla(\overline{u}-\widetilde{u})|^2e^{\overline{\rho}}\phi^2+2\int_{\{t\}\times\Omega}e^{\overline{\rho}}(\overline{u}-\widetilde{u})^2|\nabla\phi|^2\right]$$

$$+2\left[\frac{1}{4}\int_{\{t\}\times\Omega}|\nabla(\overline{u}-\widetilde{u})|^2e^{\overline{\rho}}\phi^2+\int_{\{t\}\times\Omega}e^{\overline{\rho}}\phi^2(\overline{u}-\widetilde{u})^2|\nabla\overline{\rho}|^2\right]$$

$$-2\int_{\{t\}\times\Omega}\phi e^{\overline{\rho}}\left(\sum_{i=1}^{n}\overline{f}_i\frac{\partial\phi}{\partial x_i}\right)(\overline{u}-\widetilde{u})^2-\int_{\{t\}\times\Omega}e^{\overline{\rho}}\phi^2\left(\sum_{i=1}^{n}\overline{f}_i\frac{\partial\rho}{\partial x_i}\right)(\overline{u}-\widetilde{u})^2$$

$$-\int_{\{t\}\times\Omega}e^{\overline{\rho}}\phi^2\left(\sum_{i=1}^{n}\frac{\partial\overline{f}_i}{\partial x_i}\right)(\overline{u}-\widetilde{u})^2-2\int_{\{t\}\times\Omega}e^{\overline{\rho}}\phi^2\overline{V}(\overline{u}-\widetilde{u})^2$$

$$-4\int_{\{t\}\times\Omega}e^{\overline{\rho}}\phi\sum_{i=1}^{n}(\overline{f}_i-\widetilde{f}_i)\frac{\partial\phi}{\partial x_i}\widetilde{u}(\overline{u}-\widetilde{u})$$

$$-2\int_{\{t\}\times\Omega}e^{\overline{\rho}}\phi^2\sum_{i=1}^{n}(\overline{f}_i-\widetilde{f}_i)\frac{\partial\rho}{\partial x_i}\widetilde{u}(\overline{u}-\widetilde{u})$$

$$+2\left[\frac{1}{4}\int_{\{t\}\times\Omega}e^{\overline{\rho}}\phi^2|\nabla(\overline{u}-\widetilde{u})|^2+\int_{\{t\}\times\Omega}e^{\overline{\rho}}\phi^2\widetilde{u}^2|\overline{f}-\widetilde{f}|^2\right]$$

$$-2\int_{\{t\}\times\Omega}e^{\overline{\rho}}\phi^2(\overline{u}-\widetilde{u})\widetilde{u}\sum_{i=1}^{n}\left(\frac{\partial\overline{f}_i}{\partial x_i}-\frac{\partial\widetilde{f}_i}{\partial x_i}\right)-2\int_{\{t\}\times\Omega}e^{\overline{\rho}}\phi^2(\overline{u}-\widetilde{u})\widetilde{u}(\overline{V}-\widetilde{V})$$

$$= \int_{\{t\}\times\Omega} e^{\overline{\rho}}\phi^2(\overline{u}-\widetilde{u})^2 \left(\frac{\partial\overline{\rho}}{\partial t} + 2|\nabla\overline{\rho}|^2 - \sum_{i=1}^{n}\overline{f}_i\overline{\rho}_i - \sum_{i=1}^{n}\frac{\partial\overline{f}_i}{\partial x_i} - 2\overline{V} \right)$$

$$- \frac{1}{2}\int_{\{t\}\times\Omega} e^{\overline{\rho}}\phi^2|\nabla(\overline{u}-\widetilde{u})|^2 + 8\int_{\{t\}\times\Omega} e^{\overline{\rho}}(\overline{u}-\widetilde{u})^2|\nabla\phi|^2$$

$$- 2\int_{\{t\}\times\Omega} \phi e^{\overline{\rho}}\left(\sum_{i=1}^{n}\overline{f}_i\phi_i\right)(\overline{u}-\widetilde{u})^2$$

$$- 4\int_{\{t\}\times\Omega} e^{\overline{\rho}}\phi\left[\sum_{i=1}^{n}(\overline{f}_i-\widetilde{f}_i)\phi_i\right]\widetilde{u}(\overline{u}-\widetilde{u})$$

$$- 2\int_{\{t\}\times\Omega} e^{\overline{\rho}}\phi^2\sum_{i=1}^{n}(\overline{f}_i-\widetilde{f}_i)\frac{\partial\overline{\rho}}{\partial x_i}\widetilde{u}(\overline{u}-\widetilde{u}) + 2\int_{\{t\}\times\Omega} e^{\overline{\rho}}\phi^2\widetilde{u}^2|f-\widetilde{f}|^2$$

$$- 2\int_{\{t\}\times\Omega} e^{\overline{\rho}}\phi^2(\overline{u}-\widetilde{u})\widetilde{u}\sum_{i=1}^{n}\left(\frac{\partial\overline{f}_i}{\partial x_i} - \frac{\partial\widetilde{f}_i}{\partial x_i}\right)$$

$$- 2\int_{\{t\}\times\Omega} e^{\overline{\rho}}\phi^2(\overline{u}-\widetilde{u})\widetilde{u}(\overline{V}-\widetilde{V}). \tag{5.9}$$

In view of (5.6), (5.9) implies

$$\frac{d}{dt}\int_{\{t\}\times\Omega}\phi^2 e^{\overline{\rho}}(\overline{u}-\widetilde{u})^2$$

$$\leq 8\int_{\{t\}\times\Omega} e^{\overline{\rho}}(\overline{u}-\widetilde{u})^2|\nabla\phi|^2 + 2\int_{\{t\}\times\Omega} |\phi|e^{\overline{\rho}}\left|\sum_{i=1}^{n}\overline{f}_i\frac{\partial\phi}{\partial x_i}\right|(\overline{u}-\widetilde{u})^2$$

$$+ 4\int_{\{t\}\times\Omega} e^{\overline{\rho}}|\phi|\,|\overline{f}-\widetilde{f}|\,|\nabla\phi|\,|\widetilde{u}(\overline{u}-\widetilde{u})|$$

$$+ 2\int_{\{t\}\times\Omega} e^{\overline{\rho}}\phi^2\left|\sum_{i=1}^{n}(\overline{f}_i-\widetilde{f}_i)\frac{\partial\overline{\rho}}{\partial x_i}\right||\widetilde{u}(\overline{u}-\widetilde{u})| + 2\int_{\{t\}\times\Omega} e^{\overline{\rho}}\phi^2|\widetilde{u}|^2|\overline{f}-\widetilde{f}|^2$$

$$+ 2\int_{\{t\}\times\Omega} e^{\overline{\rho}}\phi^2|(\overline{u}-\widetilde{u})\widetilde{u}|\left|\sum_{i=1}^{n}\left(\frac{\partial\overline{f}_i}{\partial x_i} - \frac{\partial\widetilde{f}_i}{\partial x_i}\right)\right|$$

$$+ 2\int_{\{t\}\times\Omega} e^{\overline{\rho}}\phi^2|(\overline{u}-\widetilde{u})\widetilde{u}|\,|\overline{V}-\widetilde{V}|$$

$$\leq 8\int_{\{t\}\times\Omega} e^{\overline{\rho}}(\overline{u}-\widetilde{u})^2|\nabla\phi|^2 + 4\int_{\{t\}\times\Omega} e^{\overline{\rho}}(\overline{u}-\widetilde{u})^2\left|\sum_{i=1}^{n}\overline{f}_i\frac{\partial\phi}{\partial x_i}\right|^2$$

$$+ \frac{1}{4}\int_{\{t\}\times\Omega} e^{\overline{\rho}}\phi^2(\overline{u}-\widetilde{u})^2 + \frac{1}{4}\int_{\{t\}\times\Omega} e^{\overline{\rho}}\phi^2(\overline{u}-\widetilde{u})^2$$

$$+ 4\int_{\{t\}\times\Omega} e^{\overline{\rho}}\phi^2\left|\sum_{i=1}^{n}(\overline{f}_i-\widetilde{f}_i)\frac{\partial\overline{\rho}}{\partial x_i}\right|^2\widetilde{u}^2 + 2\int_{\{t\}\times\Omega} e^{\overline{\rho}}\phi^2\widetilde{u}^2|\overline{f}-\widetilde{f}|^2$$

$$+ \frac{1}{4} \int_{\{t\}\times\Omega} e^{\overline{\rho}} \phi^2 (\overline{u} - \widetilde{u})^2 + 4 \int_{\{t\}\times\Omega} e^{\overline{\rho}} \phi^2 \widetilde{u}^2 |\overline{V} - \widetilde{V}|^2$$

$$+ 4 \left[\frac{1}{2} \int_{\{t\}\times\Omega} e^{\overline{\rho}} (\overline{u} - \widetilde{u})^2 |\nabla\phi|^2 + \frac{1}{2} \int_{\{t\}\times\Omega} e^{\overline{\rho}} \phi^2 |\overline{f} - \widetilde{f}|^2 \widetilde{u}^2 \right]$$

$$+ 2 \left[\frac{1}{8} \int_{\{t\}\times\Omega} e^{\overline{\rho}} \phi^2 (\overline{u} - \widetilde{u})^2 + 2 \int_{\{t\}\times\Omega} e^{\overline{\rho}} \phi^2 \widetilde{u}^2 \left| \sum_{i=1}^{n} \left(\frac{\partial \overline{f}_i}{\partial x_i} - \frac{\partial \widetilde{f}_i}{\partial x_i} \right) \right|^2 \right]$$

$$= \int_{\{t\}\times\Omega} \phi^2 e^{\overline{\rho}} (\overline{u} - \widetilde{u})^2 + 10 \int_{\{t\}\times\Omega} e^{\overline{\rho}} (\overline{u} - \widetilde{u})^2 |\nabla\phi|^2$$

$$+ 4 \int_{\{t\}\times\Omega} e^{\overline{\rho}} \left| \sum_{i=1}^{n} \overline{f}_i \frac{\partial\phi}{\partial x_i} \right|^2 (\overline{u} - \widetilde{u})^2 + 4 \int_{\{t\}\times\Omega} e^{\overline{\rho}} \phi^2 \left| \sum_{i=1}^{n} (\overline{f}_i - \widetilde{f}_i) \frac{\partial\overline{\rho}}{\partial x_i} \right|^2 \widetilde{u}^2$$

$$+ 2 \int_{\{t\}\times\Omega} e^{\overline{\rho}} \phi^2 \widetilde{u}^2 |\overline{f} - \widetilde{f}|^2 + 4 \int_{\{t\}\times\Omega} e^{\overline{\rho}} \phi^2 \widetilde{u}^2 |\overline{V} - \widetilde{V}|^2$$

$$+ 2 \int_{\{t\}\times\Omega} e^{\overline{\rho}} \phi^2 |\overline{f} - \widetilde{f}|^2 \widetilde{u}^2 + 4 \int_{\{t\}\times\Omega} e^{\overline{\rho}} \phi^2 \widetilde{u}^2 \left| \sum_{i=1}^{n} \left(\frac{\partial \overline{f}_i}{\partial x_i} - \frac{\partial \widetilde{f}_i}{\partial x_i} \right) \right|^2.$$

$$\square$$

The following theorem states when τ is sufficiently small and $\overline{\psi}(x)$ is close to $\widetilde{\psi}(x)$ in L^2-sense, then the solution $\widetilde{u}(t,x)$ of (4.6) approximates the solution $\overline{u}(t,x)$ of (4.5) well in L^2-sense.

Theorem 17 *Consider the parabolic differential equations (4.5) and (4.6). Assume that*

$$|\overline{f}(t,x)| \le c(1 + |x|), \quad |\nabla\overline{f}(t,x)| \le c, \quad |\Delta\overline{f}(t,x)| \le c,$$
$$|\widetilde{f}(t,x)| \le c(1 + |x|), \quad |\nabla\widetilde{f}(t,x)| \le c, \quad |\Delta\widetilde{f}(t,x)| \le c,$$
$$|\overline{V}(t,x)| \le c(1 + |x|^2), \quad |\nabla\overline{V}(t,x)| \le c(1 + |x|),$$
$$|\widetilde{V}(t,x)| \le c(1 + |x|^2), \quad |\nabla\widetilde{V}(t,x)| \le c(1 + |x|),$$

where $c \ge 4$ is a constant. Let $\widetilde{c}$ and δ be positive constants such that $\widetilde{\widetilde{c}} := \widetilde{c} + \delta < \frac{5}{254}$. Let

$$\overline{\rho}(t,x) = \frac{\widetilde{c}(1 + |x|^2)}{t + \epsilon}, \qquad \widetilde{\rho}(t,x) = \frac{\widetilde{\widetilde{c}}\,(1 + |x|^2)}{t + \epsilon}.$$

Suppose that

$$\int_{\mathbb{R}^n} e^{\widetilde{\rho}(0,x)} \left(|\psi(x)|^2 + |\nabla\psi(x)|^2 + |\Delta\psi(x)|^2 \right) < \infty,$$

$$\int_{\mathbb{R}^n} e^{\widetilde{\rho}(0,x)} \left(|\widetilde{\psi}(x)|^2 + |\nabla\widetilde{\psi}(x)|^2 + |\Delta\psi(x)|^2 \right) < \infty.$$

Choose τ and ϵ suitably small so that $\tau + \epsilon < \delta$ and the conclusions of Lemma 15 hold. Suppose that for $0 \leq t \leq \tau$

$$|\overline{f}(t,x) - \widetilde{f}(t,x)| \leq \widetilde{\epsilon}_1 c(1 + |x|), \tag{5.10}$$

$$\left| \sum_{i=1}^{n} \left(\frac{\partial \overline{f}_i}{\partial x_i}(t,x) - \frac{\partial \widetilde{f}_i}{\partial x_i}(t,x) \right) \right| \leq \widetilde{\epsilon}_1 c, \tag{5.11}$$

$$|\overline{V}(t,x) - \widetilde{V}(t,x)| \leq \widetilde{\epsilon}_1 c(1 + |x|^2), \tag{5.12}$$

$$\int_{\mathbb{R}^n} e^{\overline{\rho}(0,x)} |\overline{\psi}(x) - \widetilde{\psi}(x)|^2 \leq \widetilde{\epsilon}_2. \tag{5.13}$$

Then

$$\int_{\{t\} \times \mathbb{R}^n} e^{\overline{\rho}} (\overline{u} - \widetilde{u})^2 \leq \widetilde{\epsilon}_2 e^t + 16 \widetilde{\epsilon}_1^2 c^2 \widetilde{c}^2 \frac{t}{\epsilon(t + \epsilon)} e^t d_1$$

$$+ 8t \widetilde{\epsilon}_1^2 c^2 e^t d_1 + 4t \widetilde{\epsilon}_1^2 c^2 e^t d_2$$

$$\leq \widetilde{\epsilon}_2 e^\tau + \widetilde{\epsilon}_1^2 \tau e^\tau c_1$$

where $d_1 = \int_{\mathbb{R}^n} e^{\widetilde{\rho}(0,x)} (\widetilde{\psi}(x))^2$, $d_2 = \int_{\mathbb{R}^n} e^{\overline{\rho}(0,x)} (\widetilde{\psi}(x))^2$ and $c_1 = \frac{16 c^2 \widetilde{c}^2 d_1}{\epsilon^2} + 8c^2 d_1 + 4c^2 d_2$.

Proof Let ϕ and Ω be defined as in Proposition 16. In view of Lemma 14 and (5.7), we have

$$e^{-t} \int_{\{t\} \times \Omega} \phi^2 e^{\overline{\rho}} (\overline{u} - \widetilde{u})^2 - \int_{\{0\} \times \Omega} \phi^2 e^{\overline{\rho}} (\overline{u} - \widetilde{u})^2$$

$$\leq 10 \int_0^t e^{-s} \int_{\{s\} \times \Omega} e^{\overline{\rho}} (\overline{u} - \widetilde{u})^2 |\nabla \phi|^2$$

$$+ 4 \int_0^t e^{-s} \int_{\{s\} \times \Omega} e^{\rho} \left| \sum_{i=1}^{n} \overline{f}_i \frac{\partial \phi}{\partial x_i} \right|^2 (\overline{u} - \widetilde{u})^2$$

$$+ 4 \int_0^t e^{-s} \int_{\{s\} \times \Omega} e^{\overline{\rho}} \phi^2 \left| \sum_{i=1}^{n} (\overline{f}_i - \widetilde{f}_i) \frac{\partial \rho}{\partial x_i} \right|^2 \widetilde{u}^2$$

$$+ 2 \int_0^t e^{-s} \int_{\{s\} \times \Omega} e^{\overline{\rho}} \phi^2 \widetilde{u}^2 |\overline{f} - \widetilde{f}|^2$$

$$- 4 \int_0^t e^{-s} \int_{\{s\} \times \Omega} e^{\overline{\rho}} \phi^2 \widetilde{u}^2 |\overline{V} - \widetilde{V}|^2 + 2 \int_0^t e^{-s} \int_{\{s\} \times \Omega} e^{\overline{\rho}} \phi^2 |\overline{f} - \widetilde{f}|^2 \widetilde{u}^2$$

$$+ 4 \int_0^t e^{-s} \int_{\{s\} \times \Omega} e^{\overline{\rho}} \phi^2 \widetilde{u}^2 \left| \sum_{i=1}^{n} \left(\frac{\partial \overline{f}_i}{\partial x_i} - \frac{\partial \widetilde{f}_i}{\partial x_i} \right) \right|^2. \tag{5.14}$$

Let $R_0 \geq 1$ and

$$\phi(x) = \begin{cases} 1 & \text{for } |x| \leq R_0 \\ \frac{\log R - \log |x|}{\log R - \log R_0} & \text{for } R_0 \leq |x| \leq R \\ 0 & \text{for } |x| \geq R. \end{cases}$$

Then it is clear that

$$|\nabla \phi(x)|^2 \leq \frac{1}{|x|^2 (\log R - \log R_0)^2} \quad \text{for } 1 \leq R_0 \leq |x| \leq R \quad (5.15)$$

$$\left| \sum_{i=1}^{n} \overline{f}_i \frac{\partial \phi}{\partial x_i} \right| \leq |\overline{f}|^2 |\nabla \phi|^2 \leq \frac{c^2 (1 + |x|)^2}{|x|^2 (\log R - \log R_0)^2}$$

$$\leq \frac{4c^2}{(\log R - \log R_0)^2} \quad \text{for } 1 \leq R_0 \leq |x| \leq R. \quad (5.16)$$

(5.14) implies

$$e^{-t} \int_{\{t\} \times \Omega} \phi^2 e^{\overline{\rho}} (\overline{u} - \widetilde{u})^2 - \int_{\{t\} \times \Omega} \phi^2 e^{\overline{\rho}} (\overline{u} - \widetilde{u})^2$$

$$\leq \frac{10}{(\log R - \log R_0)^2} \int_0^t e^{-s} \int_{\{s\} \times B_{R_0}^c} e^{\overline{\rho}} (\overline{u} - \widetilde{u})^2$$

$$+ \frac{16c^2}{(\log R - \log R_0)^2} \int_0^t e^{-s} \int_{\{s\} \times B_{R_0}^c} e^{\overline{\rho}} (\overline{u} - \widetilde{u})^2$$

$$+ 4 \int_0^t e^{-s} \int_{\{s\} \times \Omega} e^{\overline{\rho}} \phi^2 \left| \sum_{i=1}^{n} (\overline{f}_i - \widetilde{f}_i) \frac{\partial \overline{\rho}}{\partial x_i} \right|^2 \widetilde{u}^2$$

$$+ 4 \int_0^t e^{-s} \int_{\{s\} \times \Omega} e^{\overline{\rho}} \phi^2 \widetilde{u}^2 |\overline{f} - \widetilde{f}|^2$$

$$+ 4 \int_0^t e^{-s} \int_{\{s\} \times \Omega} e^{\overline{\rho}} \phi^2 \widetilde{u}^2 |\overline{V} - \widetilde{V}|^2$$

$$+ 4 \int_0^t e^{-s} \int_{\{s\} \times \Omega} e^{\overline{\rho}} \phi^2 \widetilde{u}^2 \left| \sum_{i=1}^{n} \left(\frac{\partial \overline{f}_i}{\partial x_i} - \frac{\partial \widetilde{f}_i}{\partial x_i} \right) \right|^2 \quad (5.17)$$

where $B_{R_0}^c = \{x \in \mathbb{R}^n : |x| > R_0\}$. Thus (5.17) implies

$$e^{-t} \int_{\{t\} \times B_{R_0}} e^{\overline{\rho}} (\overline{u} - \widetilde{u})^2$$

$$\leq e^{-t} \int_{\{t\} \times \Omega} \phi^2 e^{\overline{\rho}} (\overline{u} - \widetilde{u})^2$$

$$\leq \int_{\{0\} \times B_R} e^{\overline{\rho}} (u - \widetilde{u})^2 + \frac{10}{(\log R - \log R_0)^2} \int_0^t e^{-s} \int_{\{s\} \times B_{R_0}^c} e^{\overline{\rho}} (\overline{u} - \widetilde{u})^2$$

$$
+ \frac{16c^2}{(\log R - \log R_0)^2} \int_0^t e^{-s} \int_{\{s\} \times B_{R_0}^c} e^{\overline{p}} (\overline{u} - \widetilde{u})^2
$$

$$
+ 4 \int_0^t e^{-s} \int_{\{s\} \times B_R} e^{\overline{p}} \left| \sum_{i=1}^n (\overline{f}_i - \widetilde{f}_i) \frac{\partial \overline{p}}{\partial x_i} \right|^2 \widetilde{u}^2
$$

$$
+ 4 \int_0^t e^{-s} \int_{\{s\} \times B_R} e^{\overline{p}} \widetilde{u}^2 |\overline{f} - \widetilde{f}|^2 + 4 \int_0^t e^{-s} \int_{\{s\} \times B_R} e^{\overline{p}} \widetilde{u}^2 |\overline{V} - \widetilde{V}|^2
$$

$$
+ 4 \int_0^t e^{-s} \int_{\{s\} \times B_R} e^{\overline{p}} \widetilde{u}^2 \left| \sum_{i=1}^n \left(\frac{\partial \overline{f}_i}{\partial x_i} - \frac{\partial \widetilde{f}_i}{\partial x_i} \right) \right|^2
$$

$$
\leq \int_{\{s\} \times B_R} e^{\overline{p}} (\overline{u} - \widetilde{u})^2 + \left[\frac{10 + 16c^2}{(\log R - \log R_0)^2} \right] \int_0^t e^{-s} \int_{\{s\} \times B_{R_0}^c} e^{\overline{p}} (\overline{u} - \widetilde{u})^2
$$

$$
+ 4 \int_0^t e^{-s} \int_{\{s\} \times B_R} e^{\overline{p}} |\overline{f} - \widetilde{f}|^2 |\nabla \overline{p}|^2 \widetilde{u}^2 + 4 \int_0^t e^{-s} \int_{\{s\} \times B_R} e^{\overline{p}} \widetilde{u}^2 |f - \widetilde{f}|^2
$$

$$
+ 4 \int_0^t e^{-s} \int_{\{s\} \times B_R} e^{p} \widetilde{u}^2 |\overline{V} - \widetilde{V}|^2
$$

$$
+ 4 \int_0^t e^{-s} \int_{\{s\} \times B_R} e^{\overline{p}} \widetilde{u}^2 \left| \sum_{i=1}^n \left(\frac{\partial \overline{f}_i}{\partial x_i} - \frac{\partial \widetilde{f}_i}{\partial x_i} \right) \right|^2 \tag{5.18}
$$

(5.18) implies

$$
\int_{\{t\} \times B_{R_0}} e^{\overline{p}} (\overline{u} - \widetilde{u})^2
$$

$$
\leq e^t \int_{\{0\} \times B_R} e^{\overline{p}} (\overline{u} - \widetilde{u})^2 + \frac{e^t (10 + 16c^2)}{(\log R - \log R_0)^2} \int_0^t e^{-s} \int_{\{s\} \times B_{R_0}^c} e^{p} (\overline{u} - \widetilde{u})^2
$$

$$
+ 4 e^t \int_0^t e^{-s} \int_{\{s\} \times B_R} e^{\overline{p}} |\overline{f} - \widetilde{f}|^2 |\nabla \overline{p}|^2 \widetilde{u}^2
$$

$$
+ 4 e^t \int_0^t e^{-s} \int_{\{s\} \times B_R} e^{\overline{p}} \widetilde{u}^2 |\overline{f} - \widetilde{f}|^2
$$

$$
+ 4 e^t \int_0^t e^{-s} \int_{\{s\} \times B_R} e^{\overline{p}} \widetilde{u}^2 |\overline{V} - \widetilde{V}|^2
$$

$$
+ 4 e^t \int_0^t e^{-s} \int_{\{s\} \times B_R} e^{\overline{p}} \widetilde{u}^2 \left| \sum_{i=1}^n \left(\frac{\partial \overline{f}_i}{\partial x_i} - \frac{\partial \widetilde{f}_i}{\partial x_i} \right) \right|^2
$$

$$
\leq e^t \int_{\{0\} \times B_R} e^{\overline{p}} (\overline{u} - \widetilde{u})^2 + \frac{e^t (10 + 16c^2)}{(\log R - \log R_0)^2} \int_0^t \int_{\{s\} \times B_{R_0}^c} e^{\overline{p}} (\overline{u} - \widetilde{u})^2
$$

$$+ 4e^t \int_0^t \int_{\{s\} \times B_R} e^{\bar\rho} |\bar f - \tilde f|^2 |\nabla \bar\rho|^2 \tilde u^2 + 4e^t \int_0^t \int_{\{s\} \times B_R} e^{\bar\rho} \tilde u^2 |\bar f - \tilde f|^2$$

$$+ 4e^t \int_0^t \int_{\{s\} \times B_R} e^{\bar\rho} \tilde u^2 |\bar V - \tilde V|^2$$

$$+ 4e^t \int_0^t \int_{\{s\} \times B_R} e^{\bar\rho} \tilde u^2 \left| \sum_{i=1}^n \left(\frac{\partial \bar f_i}{\partial x_i} - \frac{\partial \tilde f_i}{\partial x_i} \right) \right|^2. \tag{5.19}$$

Observe that (5.13) implies

$$e^t \int_{\{0\} \times B_R} e^{\bar\rho} (\bar u - \tilde u)^2 \le \tilde\epsilon_2 e^t. \tag{5.20}$$

By Corollary 4.1 of [42], u and $\tilde u$ decay like Gaussian in x variables. So we shall assume

$$\max_{x \in \mathbb{R}^n} (|u|, |\tilde u|) \le D_1 e^{-D_2 |x|^2} \qquad \text{for t small} \tag{5.21}$$

for some $D_1, D_2 > 0$. In view of the proof of Corollary 4.1 of [42], we can take $2D_2 \ge \frac{\tilde c}{\epsilon} + 1$ for sufficiently small t.

$$\frac{e^t (1 - +16c^2)}{(\log R - \log R_0)^2} \int_0^t \int_{\{s\} \times B_{R_0}^c} e^{\bar\rho} (\bar u - \tilde u)^2$$

$$\le \frac{4e^t D_1^2 (10 + 16c^2)}{(\log R - \log R_0)^2} \int_0^t \int_{\{s\} \times B_{R_0}^c} e^{\tilde c / \epsilon (1 + |x|^2)} e^{-2D_2 |x|^2}$$

$$= \frac{4te^t D_1^2 (10 + 16c^2)}{(\log R - \log R_0)^2} \int_{B_{R_0}^c} e^{-(2D_2 - \tilde c / \epsilon)|x|^2 + \tilde c / \epsilon}$$

$$\le \frac{4te^t D_1^2 (10 + 16c^2)}{(\log R - \log R_0)^2} \int_{B_{R_0}^c} e^{-|x|^2 + \tilde c / \epsilon}$$

$$\le \frac{4te^t D_1^2 (10 + 16c^2)}{(\log R - \log R_0)^2} \int_{\mathbb{R}^n} e^{-|x|^2 + \tilde c / \epsilon}. \tag{5.22}$$

Recall that $|\nabla \bar\rho|^2 = \frac{4\tilde c^2 |x|^2}{(t + \epsilon)^2}$. Hence (5.10) implies

$$4e^t \int_0^t \int_{\{0\} \times B_R} e^{\bar\rho} |\bar f - \tilde f|^2 |\nabla \bar\rho|^2 \tilde u^2$$

$$\le 4e^t \int_0^t \int_{\{s\} \times B_R} e^{\frac{\tilde c (1 + |x|^2)}{s + \epsilon}} \tilde\epsilon_1^2 c^2 (1 + |x|)^2 \frac{4\tilde c^2 |x|^2}{(s + \epsilon)^2} \tilde u^2$$

$$\le 16 \tilde\epsilon_1^2 c^2 \tilde c^2 e^t \int_0^t \frac{1}{(s + \epsilon)^2} \int_{\{s\} \times B_R} e^{\bar\rho} |x|^2 (1 + |x|)^2 \tilde u^2. \tag{5.23}$$

Since $s + \epsilon < \delta$, we have

$$
|x|^2(1 + |x|)^2 \le 2|x|^2 + 2|x|^4
$$

$$
\le e^{\frac{\delta}{s+\epsilon}} \left(1 + \frac{\delta|x|^2}{s+\epsilon} + \frac{\delta^2|x|^4}{(s+\epsilon)^2} + \cdots \right)
$$

$$
= e^{\frac{\delta}{s+\epsilon}} e^{\frac{\delta|x|^2}{s+\epsilon}}
$$

$$
= e^{\frac{\delta(1+|x|^2)}{s+\epsilon}}.
$$

Hence (5.23) implies

$$
4e^t \int_0^t \int_{\{s\} \times B_R} e^{\bar{\rho}} |\bar{f} - \tilde{f}|^2 |\nabla \bar{\rho}|^2 \tilde{u}^2
$$

$$
\le 16\tilde{\epsilon}_1^2 c^2 \tilde{c}^2 e^t \int_0^t \frac{1}{(s+\epsilon)^2} \int_{\{s\} \times B_R} e^{\bar{\rho}} e^{\frac{\delta(1+|x|^2)}{s+\epsilon}} \tilde{u}^2
$$

$$
= 16\tilde{\epsilon}_1^2 c^2 \tilde{c}^2 e^t \int_0^t \frac{1}{(s+\epsilon)^2} \int_{\{s\} \times B_R} e^{\tilde{\rho}} \tilde{u}^2. \tag{5.24}
$$

Similarly we can prove that

$$
e^t \int_0^t \int_{\{s\} \times B_R} e^{\bar{\rho}} \tilde{u}^2 |\bar{f} - \tilde{f}|^2
$$

$$
\le 4e^t \tilde{\epsilon}_1^2 c^2 \int_0^t \int_{\{s\} \times B_R} e^{\rho} \tilde{u}^2 (1 + |x|)^2
$$

$$
\le 4e^t \tilde{\epsilon}_1^2 c^2 \int_0^t \int_{\{s\} \times B_R} e^{\bar{\rho}} \tilde{u}^2 (1 + 2|x|^2)
$$

$$
\le 4e^t \tilde{\epsilon}_1^2 c^2 \int_0^t \int_{\{s\} \times B_R} e^{\tilde{\rho}} \tilde{u}^2 \tag{5.25}
$$

(5.12) implies

$$
4e^t \int_0^t \int_{\{s\} \times B_R} e^{\bar{\rho}} \tilde{u}^2 |\bar{V} - \tilde{V}|^2
$$

$$
\le 4e^t \tilde{\epsilon}_1^2 c^2 \int_0^t \int_{\{s\} \times B_R} e^{\bar{\rho}} \tilde{u}^2 (1 + |x|^2)^2
$$

$$
\le 4e^t \tilde{\epsilon}_1^2 c^2 \int_0^t \int_{\{s\} \times B_R} e^{\tilde{\rho}} \tilde{u}^2 \tag{5.26}
$$

(5.11) implies

$$
4e^t \int_0^t \int_{\{s\} \times B_R} e^{\bar{\rho}} \tilde{u}^2 \left| \sum_{i=1}^n \left(\frac{\partial \bar{f}_i}{\partial x_i} - \frac{\partial \tilde{f}_i}{\partial x_i} \right) \right|^2
$$

$$\leq 4e^t\tilde{\epsilon}_1^2 c^2 \int_0^t \int_{\{s\}\times B_R} e^{\bar{\rho}}\tilde{u}^2. \tag{5.27}$$

Putting (5.19), (5.20), (5.22), (5.24), (5.25), (5.26), and (5.27) together, we get

$$\int_{\{t\}\times B_{R_0}} e^{\bar{\rho}}(\bar{u}-\tilde{u})^2 \leq \tilde{\epsilon}_2 e^t + \frac{4te^t D_1^2(10+16c^2)}{(\log R - \log R_0)^2}\int_{\mathbb{R}^n} e^{-|x|^2+\tilde{c}/\epsilon}$$

$$+ 16\tilde{\epsilon}_1^2 c^2 \tilde{c}^2 e^t \int_0^t \frac{1}{(s+\epsilon)^2}\int_{\{s\}\times B_R} e^{\bar{\rho}}\tilde{u}^2$$

$$+ 4e^t\tilde{\epsilon}_1^2 c^2 \int_0^t \int_{\{s\}\times B_R} e^{\tilde{\rho}}\tilde{u}^2 + 4e^t\tilde{\epsilon}_1^2 c^2 \int_0^t \int_{\{s\}\times B_R} e^{\tilde{\rho}}\tilde{u}^2$$

$$+ 4e^t\tilde{\epsilon}_1^2 c^2 \int_0^t \int_{\{s\}\times B_R} e^{\bar{\rho}}\tilde{u}^2. \tag{5.28}$$

Let R go to infinity in (5.28), we obtain

$$\int_{\{t\}\times B_{R_0}} e^{\bar{\rho}}(\bar{u}-\tilde{u})^2$$

$$\leq \tilde{\epsilon}_2 e^t + 16\tilde{\epsilon}_1^2 c^2 \tilde{c}^2 e^t \int_0^t \frac{1}{(s+\epsilon)^2}\int_{\{s\}\times\mathbb{R}^n} e^{\tilde{\rho}}\tilde{u}^2$$

$$+ 4e^t\tilde{\epsilon}_1^2 c^2 \int_0^t \int_{\{s\}\times\mathbb{R}^n} e^{\tilde{\rho}}\tilde{u}^2 + 4e^t\tilde{\epsilon}_1^2 c^2 \int_0^t \int_{\{s\}\times\mathbb{R}^n} e^{\tilde{\rho}}\tilde{u}^2$$

$$+ 4e^t\tilde{\epsilon}_1^2 c^2 \int_0^t \int_{\{s\}\times\mathbb{R}^n} e^{\bar{\rho}}\tilde{u}^2. \tag{5.29}$$

In view of Lemma 15, (5.29) implies

$$\int_{\{t\}\times B_{R_0}} e^{\bar{\rho}}(\bar{u}-\tilde{u})^2$$

$$\leq \tilde{\epsilon}_2 e^t + 16\tilde{\epsilon}_1^2 c^2 \tilde{c}^2 e^t \int_0^t \frac{1}{(s+\epsilon)^2}\int_{\{0\}\times\mathbb{R}^n} e^{\tilde{\rho}}\tilde{u}^2$$

$$+ 8e^t\tilde{\epsilon}_1^2 c^2 \int_0^t \int_{\{0\}\times\mathbb{R}^n} e^{\tilde{\rho}}\tilde{u}^2 + 4e^t\tilde{\epsilon}_1^2 c^2 \int_0^t \int_{\{0\}\times\mathbb{R}^n} e^{\bar{\rho}}\tilde{u}^2$$

$$\leq \tilde{\epsilon}_2 e^t + \frac{16\tilde{\epsilon}_1^2 c^2 \tilde{c}^2 e^t d_1 t}{\epsilon(t+\epsilon)} + 8te^t\tilde{\epsilon}_1^2 c^2 d_1 + 4te^t\tilde{\epsilon}_1^2 c^2 d_2. \tag{5.30}$$

Let R_0 go to infinity in (5.30), we obtain the estimate in the statement of Theorem 17. $\qquad\square$

Now we are ready to consider the global situation. For a fixed $T > 0$, we want to find the solution $\bar{u}(t,x)$ of the following parabolic equation on $[0, T] \times \mathbb{R}^n$

$$
\begin{cases}
\dfrac{\partial \overline{u}}{\partial t}(t, x) = \Delta \overline{u}(t, x) + \displaystyle\sum_{j=1}^{n} \overline{f}_j(t, x) \dfrac{\partial \overline{u}}{\partial x_j}(t, x) - \overline{V}(t, x)\overline{u}(t, x), \\
\overline{u}(0, x) = \psi(x).
\end{cases}
$$

$$(5.31)$$

Let $\mathcal{P}_k = \{0 < \tau_1 < \tau_2 < \cdots < \tau_k = T\}$ be a partition of $[0, T]$. Let $\widetilde{u}_i(t, x)$ be the solution of the following parabolic equation on $[\tau_{i-1}, \tau_i] \times \mathbb{R}^n$

$$
\begin{cases}
\dfrac{\partial \widetilde{u}_i}{\partial t}(t, x) = \Delta \widetilde{u}_i(t, x) + \displaystyle\sum_{j=1}^{n} \widetilde{f}_j(\tau_i, x) \dfrac{\partial \widetilde{u}_i}{\partial x_j}(t, x) - \widetilde{V}(\tau_i, x)\widetilde{u}_i(t, x), \\
\widetilde{u}_i(\tau_{i-1}, x) = \widetilde{u}_{i-1}(\tau_{i-1}, x)
\end{cases}
$$

$$(5.32)$$

where $\widetilde{u}_1(0, x) = \psi(x) = \overline{u}(0, x)$, and $\widetilde{f}_j(\tau_i, x) = \overline{f}_j(\tau_i, x)$, $\widetilde{V}(\tau_i, x) = \overline{V}(\tau_i, x)$ are obtained from $\overline{f}_j(t, x)$ and $\overline{V}(t, x)$ by freezing time variable at τ_i respectively. We shall assume that $\overline{f}$ and $\overline{V}$ grow at most linearly and quadratically respectively. More precisely, we assume that

$$|\overline{f}t, x)| \leq c(1 + |x|),$$

$$(5.33)$$

$$|\nabla \overline{f}(t, x)| = \sqrt{\sum_{i=1}^{n} |\nabla \overline{f}_i|^2} \leq c,$$

$$(5.34)$$

$$|\overline{V}(t, x)| \leq c(1 + |x|^2).$$

$$(5.35)$$

Lemma 18 *Fix $T > 0$, let $|\mathcal{P}_k| = \sup\limits_{i}\{|t_i - t_{i-1}|\}$. For any $\widetilde{\epsilon}_1 > 0$, there exists $\widetilde{\delta} > 0$ such that if $|\mathcal{P}_k| < \widetilde{\delta}$, then*

$$|\overline{f}(t, x) - \widetilde{f}(\tau_i, x)| \leq \widetilde{\epsilon}_1 c(1 + |x|), \quad \tau_{i-1} \leq t \leq \tau_i, \quad (5.36)$$

$$\left| \sum_{j=1}^{n} \left(\dfrac{\partial \overline{f}_j}{\partial x_j}(t, x) - \dfrac{\partial \widetilde{f}_j}{\partial x_j}(\tau_i, x) \right) \right| \leq \widetilde{\epsilon}_1 c, \quad (5.37)$$

$$|\overline{V}(t, x) - \widetilde{V}(\tau_i, x)| \leq \widetilde{\epsilon}_1 c(1 + |x|^2). \quad (5.38)$$

Proof (5.36), (5.37) and (5.38) follow from the hypothesis (5.33), (5.34) and (5.35) respectively. □

We are now ready to prove the main theorem in this section.

Theorem 19 *Let $\bar{u}(t,x)$ and $\tilde{u}_k(t,x)$ be the solutions of (5.31) and (5.32) respectively. For $\tilde{\epsilon}_1 > 0$, let $|\mathcal{P}_k| = \sup_i \{|t_i - t_{i-1}|\}$ be sufficiently small so that Lemma 18 holds. Then*

$$\int_{\mathbb{R}^n} e^{\bar{\rho}(T,x)} \left(\bar{u}(T,x) - \tilde{u}_k(T,x)\right)^2 \le \tilde{\epsilon}_1^2 c_1 k |\mathcal{P}_k| e^T \le \tilde{\epsilon}_1^2 c_1 c_2(T) \quad (5.39)$$

where $\bar{\rho}(t,x) = \frac{\tilde{c}(1+|x|^2)}{t+\epsilon}$ so that the conclusion of Theorem 17 holds, c_1 is the constant in Theorem 17 and $c_2(T)$ is a constant depends only on T.

Proof In view of $\tilde{u}_1(0,x) = \psi(x) = \bar{u}(0,x)$ and Theorem 17, we have

$$\int_{\{\tau_1\} \times \mathbb{R}^n} e^{\bar{\rho}} (\bar{u} - \tilde{u})^2 \le \tilde{\epsilon}_1^2 \tau_1 e^{\tau_1} c_1$$

$$\int_{\{\tau_2\} \times \mathbb{R}^n} e^{\bar{\rho}} (\bar{u} - \tilde{u})^2 \le \tilde{\epsilon}_1^2 c_1 \left[\tau_1 e^{\tau_2} + (\tau_2 - \tau_1) e^{\tau_2 - \tau_1}\right].$$

By Theorem 17 and induction, we have

$$\int_{\{\tau_k\} \times \mathbb{R}^n} e^{\bar{\rho}} (\bar{u} - \tilde{u})^2 = \tilde{\epsilon}_1^2 c_1 \left[\tau_1 e^{\tau_k} + (\tau_2 - \tau_1) e^{\tau_k - \tau_1} + (\tau_3 - \tau_2) e^{\tau_k - \tau_2}\right.$$

$$\left. + \cdots + (\tau_k - \tau_{k-1}) e^{\tau_k - \tau_{k-1}}\right]$$

$$\le \tilde{\epsilon}_1^2 c_1 k |\mathcal{P}_k| e^T$$

$$\le \tilde{\epsilon}_1^2 c_1 c_2(T).$$

$\square$

As a consequence of Theorem 19 we have the following L^2-convergent theorem.

Theorem 20 *Fix $T > 0$, let $\mathcal{P}_k = \{0 < \tau_1 < \tau_2 < \cdots < \tau_k = T\}$ be a partition of $[0,T]$. Let $\bar{u}(t,x)$ be the solution of (5.31) on $[0,T] \times \mathbb{R}^n$. Let $\tilde{u}_i(t,x)$ be the solution of (5.32) on $[\tau_{i-1}, \tau_i] \times \mathbb{R}^n$. Let $\bar{\rho}(t,x) = \frac{\tilde{c}(1+|x|^2)}{t+\epsilon}$ so that the conclusion of Theorem 19 holds. Then*

$$\lim_{|\mathcal{P}_k| \to 0} \int_{\{T\} \times \mathbb{R}^n} \bar{\rho}(u - \tilde{u}_k)^2 = 0.$$

References

[1] Brockett, R.W. (1981). Nonlinear systems and nonlinear estimation theory, in *The Mathematics of Filtering and Identification and Applications*, ed. M. Hazewinkel and J.C. Willems (Reidel, Dordrecht).

[2] Brockett, R.W. (1983). Nonlinear control theory and differential geometry, in *Proceedings of the International Congress of Mathematics*, *August 16-24, 1983*, 1351–1368.

[3] Brockett, R.W. and Clark, J.M.C. (1980). The geometry of the conditional density functions, in *Analysis and Optimization of Stochastic Systems*, ed. O.L.R. Jacobs et.al. (Academic Press, New York), 399–309.

[4] Chaleyat-Maurel, M. and Michel, D. (1984). Des resultats de non existence de filtre de dimension finie, *Stochastics* **13**, 83–102.

[5] Chen, J. (1994). On uniquity of Yau filters, in *Proceedings of the American Control Conference, (Baltimore, Maryland), June 1994*, 252–254.

[6] Chiou, W.L. and Yau, S.S.-T. (1994), Finite-dimensional filters with nonlinear drift II: Brockett's problem on classification of finite-dimensional estimation algebras, *SIAM J. Control & Optim.* **32**, 297–310.

[7] Chen, J. and Yau, S.S.-T. (1996). Finite-dimensional filters with nonlinear drift VI: Linear structure of Ω, *Maths Control, Signals & Systems* **9**, 370–385.

[8] Chen, J. and Yau, S.S.-T. (1997). Finite-dimensional filters with nonlinear drift VII: Mitter conjecture and structure of η, *SIAM J. Control & Optim.* **35**, 1116–1131.

[9] Chen, J., Yau, S.S.-T. and Leung, C.W. (1996). Finite-dimensional filters with nonlinear drift IV: Classification of finite-dimensional estimation algebras of maximal rank with state space dimension 3, *SIAM J. Control & Optim.* **34**, 179–198.

[10] Chen, J., Yau, S.S.-T. and Leung, C.W. (1997). Finite-dimensional filters with nonlinear drift VII: Classification of finite-dimensional estimation algebras of maximal rank with state-space dimension 4, *SIAM J. Control & Optim.* **35**, 1132–1141.

[11] Davis, M.H.A. (1980). On a multiplicative functional transformation arising in nonlinear filtering theory, *Z. Wahrsh Verw. Gebiete* **54**, 125–139.

[12] Davis, M.H.A. and Marcus, S.I. (1981). An introduction to nonlinear filtering, in *The Mathematics of Filtering and Identification and Applications*, ed. M. Hazewinkel and J.S. Willems (Reidel, Dordrecht).

[13] Dong, R.T., Tam, L.F., Wong, W.S. and Yau, S.S.-T. (1991). Structure and classification theorems of finite-dimensional exact estimation algebras, *SIAM J. Control & Optim.* **29**, 866–877.

[14] Duncan, T.E. (1967). Probability densities for diffusion processes with applications to nonlinear filtering theory, Ph.D. thesis, (Stanford, CA).

[15] Fleming, W.H. and Mitter, S.K. (1982). Optimal control and nonlinear filtering for nondegenerate diffusion processes, *Stochastics* **8**, 63–77.

[16] Friedman, A. (1964). *Partial Differential Equations of Parabolic Type* (Prentice-Hall, Englewood Cliffs, NJ).

[17] Fujisaki, M., Kallianpur, G. and Kunita, H. (1972). Stochastic differential equations for the nonlinear filtering problem, *Osaka J. Maths* **2**, 19–40.

[18] Haussman, U.G. and Pardoux, E. (1988). A conditionally almost linear filtering problem with non-Gaussian initial condition, *Stochastic* **23**, 241–275.

[19] Hazewinkel, M. (1988). Lecture on linear and nonlinear filtering, in *Analysis and Estimation of Stochastic Mechanical Systems, CISM Courses and Lectures 303*, ed. W. Shiehlen and W. Wedig (Springer-Verlag, Vienna).

[20] Hazewinkel, M., Marcus, S.I. and Sussman, H.J. (1983). Nonexistence of finite dimensional filters for conditional statistics of the cubic sensor problem, *Systems Control Letter* **3**, 331–340.

[21] Hu, G.W. and Yau, S.S.-T. (????). Finite dimensional filters with nonlinear drift X: Explicit solution of DMZ equation (to appear) *IEEE Trans Automatic Control.*

[22] Hu, G.Q., Yau, S.S.-T. and Chiou W.L. (2000). Finite Dimensional Filters with Nonlinear Drift XIII: Classification of Finite-Dimensional Estimation Algebras of Maximal Rank with State Space Dimension Five, *Asian Journal Maths,* to appear.

[23] Kalman, R.E. (1960). A new approach to linear filtering and predication problem, *Trans ASME., Ser. D. J. Basic Engnrng* **82**, 35–45.

[24] Kalman, R.E. and Bucy, R.S. (1961). New results in linear filtering and prediction theory, *Trans. ASME Ser. D. J. Basic Engineering* **83**, 95–108.

[25] Makowski, A. (1986). Filtering formula for partially observed linear system with non-Gaussian initial conditions, *Stochastic* **16**, 1–24.

[26] Marcus, S. (1984). Algebraic and geometric methods in nonlinear filtering, *SIAM J. Control & Optim.* **22**, 817–844.

[27] Mitter, S.K. (1979). On the analogy between mathematical problems of nonlinear filtering and quantum physics, *Ricerche Automat.* **10**, 163–216.

[28] Mortensen, R.E. (1966). Optimal control of continuous time stochastic systems, Ph.D. thesis (UC Berkeley, CA).

[29] Ocone, D. (1980). Topics in nonlinear filtering theory, Ph.D. thesis (Mass. Inst. Tech., Cambridge, MA).

[30] Tam, L.F., Wong W.S. and Yau, S.S.-T. (1990). On a necessary and sufficient condition for finite dimensionality of estimation algebras, *SIAM J. Control & Optim.* **28**, 173–185.

[31] Wei, J. and Norman, E. (1964). On the global representation of the solutions of linear differential equations as a product of exponentials, *Proc. Amer. Math. Soc.* **15**, 327–334.

[32] Wong, W.S. (1987). On a new class of finite-dimensional estimation algebras, *Systems Control Letts* **9** 79–83.

[33] Wong, W.S. and Yau, S.S.-T. (1998). The estimation algebra of nonlinear filtering system, in *Mathematical Control Theory, Special Volume Dedicated to 60th Birthday of Brockett,* ed. J. Baillieul and J.C. Willems (Springer-Verlag, Berlin), 33–65.

[34] Wu, X., Yau, S.S.-T. and Hu, G.Q. (????). Finite Dimensional Filters With Nonlinear Drift XII: Linear and constant structure of Ω, Preprint.

[35] Yau, S.S.-T. (1990). Recent results on nonlinear filtering: New class of finite dimensional filters, in *Proceedings of the 29th Conference on Decision and Control at Honolulu, Hawaii,* 231–233.

[36] Yau, S.S.-T. (1994). Finite dimensional filters with nonlinear drift I: A class of filters including both Kalman–Bucy filters and Benes filters, *J. Math. Systems, Estimation & Control* **4**, 181–203.

[37] Yau, S.S.-T. and Chiou, W.L. (1991). Recent results on classification of finite dimensional estimation algebras: Dimension of state space 2, in *Proceedings of the 30th Conf. on Decision and Control,Brighton, England* 2758–2760.

[38] Yau, S.S.-T. and Hu, G.-Q. (????). Finite dimensional filters with nonlinear drift XIV: Classification of finite dimensional estimation algebras of maximal rank with arbitrary state space dimension and Mitter conjecture, Preprint.

[39] Yau, S.S.T. and Rasoulation, A. (1999). Classification of four-dimensional estimation algebras, *IEEE Trans Automatic Control* **44**, 2312–2318.

[40] Yau, S.S.-T., Wu, X. and Wong, W.S. (1999). Hessian matrix non-decomposition theorem, *Math. Research Letter* **6**, 1–11.

[41] Yau, S.T. and Yau, S.S.-T. (1997). Finite dimensional filters with nonlinear drift III: Duncan–Mortensen–Zakai equation with arbitrary initial condition for Kalman–Bucy filtering system and Benes filtering system, *IEEE Trans Aerospace & Electronic Systems* **33**, 1277–1294.

[42] Yau, S.T. and Yau, S.S.-T. (1998). Existence and uniqueness and decay estimates for the time dependent parabolic equation with application to Duncan–Mortensen–Zakai equation, *Asian J. Maths* **2**, 1079–1149.

[43] Zakai, M. (1969). On the optimal filtering of diffusion processes, *Z. Wahrsch. Verw. Geb.* **11**, 230–243.

Printed in the United States
By Bookmasters